The Handbook
of Environmental Chemistry

Volume 2   Part E

Edited by O. Hutzinger

Reactions and Processes

With contributions by
J. Hermens, A. Opperhuizen, D. T. H. M. Sijm,
R. P. Wayne, B. L. Worobey

With 50 Figures

Springer-Verlag
Berlin Heidelberg GmbH

Professor Dr. Otto Hutzinger

University of Bayreuth
Chair of Ecological Chemistry and Geochemistry
Postfach 101251, D-8580 Bayreuth
Federal Republic of Germany

ISBN 978-3-662-15958-3

Library of Congress Cataloging-in-Publication Data
(Revised for volume 2, pt. E)
The Handbook of environmental chemistry.
Includes bibliographies and indexes.
Contents: v.1. The natural environment and the biogeochemical cycles/with contributions by
P. Craig ... [et al.]–v.2. Reactions and processes/with contributions by W.A. Bruggeman ...
[et al.]–v.3. Anthropogenic compounds/with contributions by R. Anliker ... [et al.]
    1. Environmental chemistry.   I. Hutzinger, O.
QD31.H335   574.5′222   80-16607
ISBN 978-3-662-15958-3   ISBN 978-3-540-46161-6 (eBook)
DOI 10.1007/978-3-540-46161-6

This work is subject to copyright. All rights are reserved, whether the whole or part of the material is concerned, specifically the rights of translation, reprinting, reuse of illustrations, recitation, broadcasting, reproduction on microfilms or in other ways, and storage in data banks, Dulication of this publication or parts thereof is only permitted under the provisions of the German Copyright Law of September 9, 1965, in its version of June 24, 1985, and a copyright fee must always be paid. Violations fall under the prosecution act of the German Copyright Law.

© Springer-Verlag Berlin Heidelberg 1989
Originally published by Springer-Verlag Berlin Heidelberg New York in 1989
Softcover reprint of the hardcover 1st edition 1989

The use of registered names, trademarks, etc. in this publication does not imply, even in the absence of a specific statement, that such names are exempt from the relevant protective laws and regulations and thefore free for general use.

Typesetting: Macmillan India Ltd., Bangalore-25

2152/3020-543210

Preface

Environmental Chemistry is a relatively young science. Interest in this subject, however, is growing very rapidly and, although no agreement has been reached as yet about the exact content and limits of this interdisciplinary subject, there appears to be increasing interest in seeing environmental topics which are based on chemistry embodied in this subject. One of the first objectives of Environmental Chemistry must be the study of the environment and of natural chemical processes which occur in the environment. A major purpose of this series on Environmental Chemistry, therefore, is to present a reasonably uniform view of various aspects of the chemistry of the environment and chemical reactions occurring in the environment.

The industrial activities of man have given a new dimension to Environmental Chemistry. We have now synthesized and described over five million chemical compounds and chemical industry produces about one hundred and fifty million tons of synthetic chemicals annually. We ship billions of tons of oil per year and through mining operations and other geophysical modifications, large quantities of inorganic and organic materials are released from their natural deposits. Cities and metropolitan areas of up to 15 million inhabitants produce large quantities of waste in relatively small and confined areas. Much of the chemical products and waste products of modern society are released into the environment either during production, storage, transport, use or ultimate disposal. These released materials participate in natural cycles and reactions and frequently lead to interference and disturbance of natural systems.

Environmental Chemistry is concerned with *reactions in the environment*. It is about distribution and equilibria between environmental compartments. It is about reactions, pathways, thermodynamics and kinetics. An important purpose of this Handbook is to aid understanding of the basic distribution and chemical reaction processes which occur in the environment.

Laws regulating toxic substances in various countries are designed to assess and control risk of chemicals to man and his environment. Science can contribute in two areas to this assessment; firstly in the area of toxicology and secondly in the area of chemical expsoure. The available concentration ("environmental exposure concentration") depends on the fate of chemical compounds in the environment and thus their distribution and reaction behaviour in the environment. One very important contribution of Environmental Chemistry to the above mentioned toxic substances laws is to develop laboratory test methods, or mathematical correlations and models that predict the environmental fate of new chemical compounds. The third purpose of this Handbook is to help in the basic understanding and development of such test methods and models.

The last explicit purpose of the handbook is to present, in a concise form, the most important properties relating to environmental chemistry and hazard assessment for the most important series of chemical compounds.

At the moment three volumes of the Handbook are planned. Volume 1 deals with the natural environment and the biogeochemical cycles therein, including some background information such as energetics and ecology. Volume 2 is concerned with reactions and processes in the environment and deals with physical factors such as transport and adsorption, and chemical, photochemical and biochemical reactions in the environment, as well as some aspects of pharmacokinetics and metabolism within organisms. Volume 3 deals with anthropogenic compounds, their chemical backgrounds, production methods and information about their use, their environmental behaviour, analytical methodology and some important aspects of their toxic effects. The material for volumes 1, 2, and 3 was more than could easily be fitted into a single volume, and for this reason, as well as for the purpose of rapid publication of available manuscripts, all three volumes are published as a volume series (e.g. Vol. 1; A, B, C). Publisher and editor hope to keep the material of the volumes 1 to 3 up to date and to extend coverage in the subject areas by publishing further parts in the future. Readers are encouraged to offer suggestions and advice as to future editions of "The Handbook of Experimental Chemistry".

Most chapters in the Handbook are written to a fairly advanced level and should be of interest to the graduate student and practising scientist. I also hope that the subject matter treated will be of interest to people outside chemistry and to scientists in industry as well as government and regulatory bodies. It would be very satisfying for me to see the books used as a basis for developing graduate courses on Environmental Chemistry.

Due to the breadth of the subject matter, it was not easy to edit this Handbook. Specialists had to be found in quite different areas of science who were willing to contribute a chapter within the prescribed schedule. It is with great satisfaction that I thank all authors for their understanding and for devoting their time to this effort. Special thanks are due to the Springer publishing house and finally I would like to thank my family, students and colleagues for being so patient with me during several critical phases of preparation for the Handbook, and also to some colleagues and the secretaries for their technical help.

I consider it a privilege to see my chosen subject grow. My interest in Environmental Chemistry dates back to my early college days in Vienna. I received significant impulses during my postdoctoral period at the University of California and my interest slowly developed during my time with the National Research Council of Canada, before I was able to devote my full time to Environmental Chemistry in Amsterdam. I hope this Handbook will help deepen the interest of other scientists in this subject.

Otto Hutzinger

This preface was written in 1980. Since then publisher and editor have agreed to expand the Handbook by two new open-ended volume series: Air Pollution and Water Pollution. These broad topics could not be fitted easily into the headings of the first three volumes.

All five volume series will be integrated through the choice of topics covered and by a system of cross referencing.

The outline of the Handbook is thus as follows:

1. The Natural Environment and the Biogeochemical Cycles
2. Reactions and Processes
3. Anthropogenic Compounds
4. Air Pollution
5. Water Pollution

Bayreuth, July 1989   Otto Hutzinger

Contents

List of Contributors

Dr Joop Hermens
Dept. of Veterinary Pharmacology,
Pharmacy and Toxicology,
Biltstraat 172
Gebouw A23
NL-3508 TD Utrecht
The Netherlands

Dr Antoon Opperhuizen
Laboratory of Environmental
and Toxicological Chemistry
University of Amsterdam
Nieuwe Achtergracht 166
NL-1018 WV Amsterdam
The Netherlands

Dr Dirk T. H. M. Sijm
Laboratory of Environmental
and Toxicological Chemistry
University of Amsterdam
Nieuwe Achtergracht 166
NL-1018 WV Amsterdam
The Netherlands

Dr R. P. Wayne
Physical Chemistry Laboratory
Oxford University
South Parks Road
Oxford, OX1 3QZ
England

Dr B. L. Worobey
Food Research Division
Bureau of Chemical Safety
Food Directorate
Health Protection Branch
Health and Welfare Canada
Sir Frederick Banting 3E
Tunney's Pasture
Ottawa, Ontario
Canada K1A OL2

The Photochemistry of Ozone

R. P. Wayne

Physical Chemistry Laboratory, Oxford University, South Parks Road, Oxford, OX1 3QZ, UK

Summary

Ozone plays a major role in the atmosphere, important chemical transformations in the troposphere, stratosphere, and mesosphere all being initiated by the absorption of radiation by ozone. This article surveys laboratory data concerning the photochemistry of ozone, and indicates the relevance of the laboratory studies to interpretations of atmospheric chemistry.

The material is introduced by a brief survey of the most important atmospheric processes. The structure, spectroscopy, and excited states of ozone ultimately control the photochemistry of the molecule, and they are discussed in this context. Photofragmentation itself is the subject of the main parts of the article. Emphasis is placed on the nature of the electronic states of the atomic and molecular oxygen fragments of photolysis, and the efficiencies with which the various species are formed. The primary quantum yield for O(^{1}D) formation is certainly less than unity for $\lambda < 274$ nm, and it may thus also be less than unity in the atmospherically critical region around $\lambda = 300$ nm. Similar considerations are likely to apply to the efficiency of formation of excited singlet molecular oxygen, $O_2(^1\Delta_g)$. On the other hand, $O_2(^1\Delta_g)$ seems to be formed with high efficiency at wavelengths longer than the $\lambda \simeq 310$ nm threshold for $O(^1D)$ production. Calculations of atmospheric $[O_3]$ that depend on measurement of the intensity of the $O_2(^1\Delta_g \rightarrow {}^3\Sigma_g^-)$ Infrared Atmospheric Band may therefore be in error both if they assume a quantum yield of unity for $O_2(^1\Delta_g)$ production at $\lambda < 310$ nm and if they assume that the quantum yield is zero at longer wavelengths. The wavelength dependences of the quantum efficiencies are interpreted in terms of the spectroscopy of ozone; evidence for the breakdown of simple spin conservation arguments is presented, and some explanations for the behaviour are suggested. Consideration is given to the occurrence of isotope-selective chemistry in the photolysis of ozone that might give rise to the observed enrichment of $^{50}O_3$ in the atmosphere.

Vibrationally-excited ozone is implicated in both atmospheric processes and in laboratory experiments, and the vibrational photochemistry of ozone is explored here. Optical emission may arise from recombining $O + O_2$, and the energy-rich ozone formed may show enhanced chemical and photolytic activity.

An understanding of the details of the photodissociation of ozone, and thus the development of reliable predictive models, depends on a knowledge of the dynamics of photodissociation. Photo-fragment energy analysis and coherent Raman studies have provided information about the nascent vibrational and rotational product distributions that supplements the data about electronic states obtained from more conventional techniques. Photofragmentation appears to be rotationally impulsive and vibrationally adiabatic. A propensity for even-J rotational states in the $O_2(^1\Delta_g)$ product for photolysis by ultraviolet radiation is shown to result from radiationless transition to a surface that yields $O(^3P) + O_2(^3\Sigma_g^-)$ for ca. 10% of the reaction, but only for odd-J levels in O_2. This result is entirely consistent with the measurements of the quantum yield for $O(^1D)$ formation. Fluorescence techniques have been used to probe the ozone molecule as it is falling apart on the femtosecond time scale. The vibrational intensities reveal the nature of the upper electronic surface on which dissociation occurs and of the earliest stages of the dissociation itself. Results from some of these experiments are used to show how further information needed for atmospheric studies may eventually be won.

Introduction

For a 'minor' constituent, ozone plays a peculiarly important role in the Earth's atmospheric chemistry [1]. Perhaps the outstanding feature is the relationship between the absorption spectrum of ozone and the protection of living systems from unattenuated solar ultraviolet radiation. Macromolecules, such as proteins and nucleic acids, that are characteristic of living cells, are damaged by radiation of wavelengths shorter than about 290 nm. Major components of the atmosphere, especially O_2, are able to filter out solar ultraviolet with $\lambda < 230$ nm, at which wavelength about 1 part in 10^{16} of the intensity of the overhead Sun would be transmitted through the molecular oxygen to the ground. But at wavelengths longer than ca. 230 nm, the only species in the atmosphere capable of attenuating the Sun's radiation is ozone. Ozone has a particularly strong absorption in the critical wavelength range (230–290 nm), so that it is an effective filter in spite of its relatively small concentration.

Ozone has a positive enthalpy of formation from the standard state of molecular oxygen: that is, it is an endothermic substance

$$3/2\,O_2 \rightarrow O_3 \quad \Delta H_{298}^{\ominus} = 142.5\,\text{kJ mol}^{-1} \tag{1}$$

Reactions involving ozone thus have a tendency to be exothermic and to be thermodynamically favoured. In addition, the rates of the reactions tend to be high, so that ozone is an important chemical reaction partner. If ozone itself absorbs a quantum of light ($\lambda < 310$ nm), the energy of the system is further elevated and yet other reactions become accessible. When ozone is photodissociated by ultraviolet radiation, it yields an atomic and a molecular fragment (O and O_2) both of which are electronically excited.

$$O_3 + h\nu \rightarrow O(^1D) + O_2(^1\Delta_g) \qquad (2)$$

Atomic oxygen in the ^1D state carries an excess energy of 190 kJ mol^{-1} compared with the ^3P ground state. As a result, the atom is highly reactive. For example, processes such as

$$O(^1D) + H_2O \rightarrow OH + OH; \ \Delta H^{\ominus}_{298} = -119 \text{ kJ mol}^{-1} \qquad (3)$$

$$O(^1D) + CH_4 \rightarrow OH + CH_3; \ \Delta H^{\ominus}_{298} = -178 \text{ kJ mol}^{-1} \qquad (4)$$

$$O(^1D) + N_2O \rightarrow NO + NO; \ \Delta H^{\ominus}_{298} = -340 \text{ kJ mol}^{-1} \qquad (5)$$

are all exothermic. Reactions (3) and (4) are endothermic with ground state oxygen atoms, since the enthalpies of reaction are 190 kJ mol^{-1} less.

All three reactions are fast because they have small activation energies. Water vapour, CH_4 and N_2O are important minor atmospheric constituents, while the radicals OH, CH_3 and NO are themselves highly reactive and are involved in atmospheric chemical changes of paramount importance. In each case, the driving force for radical production is the absorption of solar radiation by ozone. Particularly in the troposphere, it is difficult to postulate alternative sources of radicals.

The molecular fragment of photodissociation of ozone by ultraviolet radiation is also electronically excited. The strongest feature of the terrestrial dayglow is due to the $^1\Delta_g \rightarrow {}^3\Sigma_g^-$ transition in O_2 at $\lambda = 1270$ nm, and is referred to as the *Infrared Atmospheric Band*. Measurements of the intensity of the band can be used to derive atmospheric ozone concentrations so long as the excitation mechanism and efficiency are known. Instrumentation on the Solar Mesosphere Explorer (SME) satellite uses this method for remote monitoring of ozone concentrations [2, 3].

Since important atmospheric chemistry is initiated by the products of ozone photolysis, it is essential to have a sound knowledge of the efficiencies with which ozone is photolysed and with which excited atoms, $O(^1D)$, are produced, and to understand what factors influence the variation of those efficiencies with wavelength of the solar radiation absorbed. Further, it is imperative that the quantum yields for excitation of molecular oxygen be known accurately if airglow intensities are to be used to derive atmospheric ozone concentrations. Quantitative information of this kind is likely to be available only from laboratory experiments. Motivation enough exists, therefore, for physical chemists to study the photochemistry of ozone, particularly with respect to the nature of the photolytic fragments and to the way in which the wavelength of the photolysing radiation— or energy of the absorbed photons—influences the pathways and efficiencies of decomposition. There is, in addition, a strong fundamental physico-chemical interest in the photochemistry of ozone that has prompted investigation of the details of dissociation at increasingly sophisticated levels. Elementary chemical reactions must be interpreted in terms of the interactions as the constituent atoms of the reactants move to the positions they occupy in the products. The laws of physics must determine the motions of the particles under the influences of the forces described by the potential energy surface for the reaction. Detailed predictions can be made of the dynamics of reactions of small polyatomic molecules such as ozone. Ultimately, one might hope to understand (and predict) inter- and intra-molecular dynamics in state-to-state detail (i.e. in terms of the internal excitation and velocity parameters as the reactants approach and the fragments separate.) Considerable interest attaches to the nascent distribution of energy amongst electronic, vibrational, and rotational internal modes, and in translational motion, immediately following a photochemical event such as photodissociation. 'Nascent' in this context means that no significant redistribution of energy has taken place subsequent to the event, so that the disposition of energy amongst the various modes can be used to infer the dynamics of the event. In a chemical reaction, the molecular system passes smoothly from reactants to products through an intermediate species; the field of reaction dynamics is concerned with how the laws of physical dynamics determine the approach of reactants and the departure of the products. In one sense, reaction dynamics is at the heart of all chemical transformations, and a better understanding of the dynamics represents a better understanding of the transformation itself. State-to-state kinetics provides some of the experimental information for theoretical approaches to reaction dynamics by attempting to investigate the rates of processes involving reactants in specific internal quantum states, and with specific velocities and starting coordinates, to form products in equally well-defined quantum states, velocities, and angular distributions and momenta. Of course, most chemistry is not performed with state-selected species, but rather with something approaching thermally equilibrated statistical distributions. Nevertheless, these statistical distributions are made up of the individual states, so that the assembly of state-to-state processes makes up the whole reaction. Techniques are just now emerging that permit the 'high-resolution' study of state-to-state photochemistry.

Photodissociation dynamics is a particularly promising part of reaction dynamics to study, because the processes following interaction between a photon and a molecule constitute a 'half collision', in which the initial properties of the photon are as well defined as possible [4]. Theoretical models of the dissociation process can be tested against the results of experiments that measure the nascent energy and momentum disposal. The models frequently indicate that there is strong sensitivity to the quantum states in the absorbing molecule, so that the fullest experimental tests will require state-selected reactants as well as state identification in the products. The observable quantities that are most sensitive to the dynamics of dissociation seem to be the rotational energy disposal, and the angular distributions and orientations of the fragments, and many of the more sophisticated studies have attempted to look at these parameters.

Ozone photochemistry has been studied in the laboratory for well over a century, and it is the results of these increasingly detailed investigations that aeronomers and atmospheric chemists have been able to apply to their own endeavours. The earliest experiments, performed with "classical" (or steady-state) techniques already provided indirect evidence for the formation, under some circumstances, of excited atomic and molecular fragments. The development of time-resolved methods, such as flash photolysis, together with direct identification of the fragments, provided the first steps towards probing the detailed dynamics of dissociation. In general, subsequent effort has been directed towards examining the excitation in the fragments, and sometimes their translational energies and angular distributions of release as well, at times increasingly close to the moment of absorption of the photon, so that the measurements reveal more about the fragments as they are born and less about what happens to them in subsequent collisions. Experiments have now been performed that examine the dynamical behaviour on the femtosecond timescale: that is, they examine the intermediate species, during and immediately after absorption, as it is in the process of falling apart [5, 6].

The significance of the new experiments is two-fold. First, the more detailed is our understanding of the phenomenon of photodissociation, the more reliable will be the predictions of behaviour under conditions not accessible in the laboratory. Such predictions may have particular pertinence for atmospheric studies. Secondly, the experiments, while not primarily directed towards the elucidation of atmospheric behaviour, yield results whose relevance to atmospheric chemistry must be considered. Interpretation of ozone chemistry in the mesosphere and thermosphere requires explicit state-to-state dynamic information, because reactants and products may not be in thermodynamic equilibrium, but rather possess steady-state distributions of molecular energy levels. The purposes of this article are to describe the development of laboratory experiments on ozone photochemistry, to point out what the potential implications for atmospheric chemistry might be, and to discuss briefly the progress being made in understanding the detailed dynamics.

Ozone in the Atmosphere

Ozone's importance in the Earth's stratosphere and troposphere are particularly well documented [1, 7–10]. We summarize here the salient features that will illustrate the need for laboratory study of ozone photochemistry.

The Stratosphere

Concentrations of ozone in the atmosphere are rather variable, the mixing ratio with respect to the entire atmosphere being a few tenths of a part per million. However, most of the ozone is found in a layer about 20 km thick, and centred on an altitude of 25–30 km. That is, atmospheric ozone is mainly in the stratosphere. Peak abundances in the ozone layer can reach 10 ppm.

The existence of a stratosphere at all in the Earth's atmosphere is a consequence of ozone's presence there. Simple thermodynamic arguments suggest that temperature should decrease with increasing altitude in the atmosphere. In the Earth's atmosphere the temperature drops by about 6.5 K for every kilometer of height increase, for roughly the first 15–20 km above the surface, but above that altitude begins to increase again. The temperature inversion leads, of course, to great vertical stability, in distinction to the behaviour of the troposphere, which is characterized by convection-driven turbulence. The stratospheric heating results mainly from solar photodissociation of ozone, and the subsequent exothermic chemical reactions that we shall discuss shortly.

The basic processes that establish the ozone layer were described by Chapman as long ago as 1930. The important 'oxygen only' reactions are

$$O_2 + h\nu \rightarrow O + O \qquad \text{for } \lambda < 242.4\,\text{nm} \tag{6}$$

$$O_3 + h\nu \rightarrow O_2 + O \qquad \text{for } \lambda < 1180\,\text{nm} \tag{7}$$

$$O + O_2 + M \rightarrow O_3 + M; \; \Delta H^\circ_{298} = -106.5\,\text{kJ mol}^{-1} \tag{8}$$

$$O + O_3 \rightarrow O_2 + O_2; \; \Delta H^\circ_{298} = -391.5\,\text{kJ mol}^{-1} \tag{9}$$

Because reactions (7) and (8) can interconvert atomic oxygen and ozone, O and O_3 are identified as the family of 'odd oxygen'. Reaction (6) creates two odd oxygens, and reaction (9) destroys two, while reactions (7) and (8) themselves obviously leave the odd oxygen concentration unaltered, although they do affect the ratio of [O] to $[O_3]$. After sunset, atomic oxygen concentrations fall rapidly at altitudes below about 40 km, since the source reactions (6) and (7) are cut off, but the sink processes (8) and (9) remain. Ozone is thus neither formed nor destroyed at night, and diurnal variations in $[O_3]$ are small at these altitudes. Higher in the atmosphere, diurnal changes are pronounced as daytime photolysis of O_3 becomes faster, and conversion of O back to O_3 becomes slower because of the lower pressure (the rate of reaction (8) is proportional to the *square* of the pressure).

The simple four-reaction scheme predicts the layer structure found in the atmosphere. At high altitudes, there is much short-wavelength ultraviolet radiation capable of dissociating molecular oxygen, but relatively little of the O_2 itself. Low in the atmosphere, there is plenty of O_2, but short-wavelength radiation is absent, because it has already been filtered out by the O_2 and O_3 lying above. Solar ultraviolet energy absorbed by the O_2 and O_3 is ultimately liberated as heat, in part through the exothermic chemical reactions (8) and (9). It is this heating that gives rise to the stratospheric temperature inversion.

Proper calculation of the ozone profile, using rate parameters for the reactions determined in the laboratory, shows that the predicted profile has the same general shape as the measured one. However, the absolute concentrations are all higher, by a factor of up to four or five, than the true atmospheric ones. The problem arises because the loss process, reaction (9), has an activation energy of $18.4\,\text{kJ mol}^{-1}$, and is too slow at stratospheric temperatures (say 220–270 K) to balance the production of ozone at the correct concentration. It is now well established that reaction (9) can be catalysed by trace constituents in the atmosphere. The idea is summed up in the reaction scheme

$$X + O_3 \rightarrow XO + O_2 \tag{10}$$

$$XO + O \rightarrow X + O_2 \tag{11}$$

$$\overline{O + O_3 \rightarrow O_2 + O_2} \qquad \text{Net}$$

The reactive species X is regenerated in the second step, so that its abundance is not affected by its participation in odd-oxygen removal. Several species have been suggested for the catalytic 'X' in the atmosphere. The most important for the natural stratosphere are $X = H$ and OH, $X = NO$, and $X = Cl$. The corresponding catalytic cycles are referred to as the HO_x, NO_x, and ClO_x cycles. All the cycles have activation energies for the individual steps that are lower than the activation energy of the direct $O + O_3$ reaction. Whether or not the catalytic reactions are actually *faster* than the direct reaction at stratospheric temperatures will depend on the relative concentrations of XO and O_3. Throughout much of the stratosphere, loss of odd oxygen turns out to be dominated by the NO_x cycle for altitudes up to about 45 km. The ClO_x cycle also becomes faster than the direct reaction at altitudes greater than about 30 km, and above 50 km the HO_x cycles become the most important loss mechanisms.

All three catalytic families, HO_x, NO_x, and Cl_x appear to be present in the 'natural' atmosphere unpolluted by man's activities. Precursors of the catalytic species have sources at the Earth's surface (supplemented in the case of NO_x by direct conversion of the N_2 and O_2 of the atmosphere at high altitudes). These precursors have to be transported through the troposphere to the stratosphere. Amongst the species of importance are H_2O, CH_4, N_2O, and CH_3Cl, which are converted to the catalyst radicals in the stratosphere.

The conversion step involves, at least in the case of production of OH and of NO, the reactions of $O(^1D)$ with H_2O, CH_4, and N_2O (reactions (3), (4) and (5)) referred to in the introduction as being some of the several important atmospheric chemical transformations driven ultimately by ozone photolysis. The stratosphere is very dry, probably because water from the troposphere has to pass through the 'cold trap' at the tropopause, and CH_4 constitutes more than a third of the total $[H_2O] + [CH_4]$ in the lower stratosphere. Reaction (4) of $O(^1D)$ with CH_4 is therefore an important source of OH, especially since the oxidation of the CH_3 radical (to CO) also yields two or three more odd-hydrogen species. Both N_2O and CH_4 are the result of biological activity (mostly microbial) on the Earth's surface. (Incidentally, the main contribution to CH_3Cl is again biological, this time in the oceans, although burning of vegetation and some volcanic eruptions are additional sources). For the ClO_x cycle, the main entry involves direct photolysis of CH_3Cl rather than reaction with $O(^1D)$.

The recognition that the radicals that ultimately control atmospheric ozone concentrations have biological sources brings us to the question of how ozone photochemistry has been intimately bound up with the evolution both of life on Earth and of the Earth's atmosphere [1,11,12]. Much evidence shows that Earth was without a primordial atmosphere. The quantities of gases liberated as a result of volcanic activity, and from the slow decay of solid radioactive elements, are sufficient to account for our atmosphere. However, molecular oxygen is not released, and the primitive atmosphere must have contained N_2, CO_2, and H_2O as its most important constituents, together with traces of reducing gases such as H_2 and CO. Simple (inorganic) photolysis of water or of CO_2 cannot lead to concentrations of O_2 above about 10^{-3} of their present levels, and the likely concentrations in the absence of life were probably much less than this limit.

Photosynthesis, dependent on life, is the only known process that can have caused the rise of $[O_2]$ to its present value. We note that the process involves consumption of carbon dioxide and water, and the concomitant liberation of oxygen.

The presence of roughly 20% of oxygen in the contemporary atmosphere itself raises some further questions. Earth possesses an atmosphere that for hundreds of millions of years appears to have been disregarding the laws of physics and chemistry. Minor oxidizable constituents of our atmosphere, such as methane, ammonia, hydrogen, carbon monoxide and nitrous oxide, survive in the presence of large concentrations of oxygen. Thermodynamic considerations would suggest virtually complete oxidation of these components. Earth's peculiar behaviour is a consequence of the existence of life on it. Biological processes, acting together with physical and chemical change, determine the composition of our atmosphere. Conversely, our unique atmosphere seems essential for the support of life in many of the forms that we know it. Oxygen, the unexpected gas of our atmosphere, is almost entirely the result of biological activity. Not only does biology provide the atmospheric oxidant, but it also continually provides the oxidizable minor gas "fuels". That is, biological processes bring about the thermodynamic disequilibrium of our atmosphere, and, in effect, reduce its entropy. Energy is needed for this entropy reduction, and virtually all of it is supplied by radiation from the Sun, mediated by the biota.

Pre-biological oxygen concentrations in the atmosphere are important in two ways. Organic molecules are susceptible to thermal oxidation and photo-oxidation and are unlikely to have accumulated in large concentrations in an oxidizing atmosphere. The low pre-biological oxygen concentrations thus seem essential to the development of the organic precursors of life. Living organisms can develop mechanisms that protect against oxidative degradation, but they are still photo-chemically sensitive to short wavelength ultraviolet. As we pointed out in the Introduction, molecular oxygen and ozone together filter out the damaging radiation in the present-day atmosphere. The amount of ozone in the atmosphere, and its altitude distribution, will depend on the concentration of the precursor oxygen, and will thus have altered markedly as the atmosphere evolved. Concentrations of ozone are also controlled by the rates of loss processes for the molecule. But, as we have just seen, loss is regulated by catalytic cycles involving other trace gases of the atmosphere, such as the oxides of nitrogen, that are themselves at least partly of biological origin. We have already noted that the oxygen in the Earth's atmosphere comes largely from biological sources. Now we see that the ozone, needed as a filter to protect life, has its concentration determined not only by the biologically-generated oxygen needed for its production, but also by the bio-logically-generated trace gases that play a part in its destruction. Such observations have led Lovelock [13] to the idea of *Gaia* (Earth Mother), in which climate and the chemical composition of the Earth's surface and atmosphere are kept at an optimum by and for the biosphere.

One interpretation of the connection between the evolution of life and of the atmosphere, due first to Berkner and Marshall [14], suggests that the development of O_2, and hence of protective O_3, controlled the migration of life onto land. At low atmospheric O_2, *liquid* water, at a depth of say 10 m, will filter out much of the damaging ultraviolet radiation, while allowing photosynthetically-active visible light to reach living organisms. Life in the oceans seems improbable at this stage, since organisms would be brought too near the surface by mechanical motions, and the biota would probably be confined to the safety of stagnant pools and lakes. When O_2 and O_3 had built up yet further, the ultraviolet zone of lethality would be restricted to a thin layer at the ocean surface, and life could thus spread to entire ocean areas, thus greatly enhancing photosynthetic activity. As the oxygen content of the atmosphere moved towards its present level, enough O_3 was available for liquid water not to be needed for protection, and life could finally be supported on dry land. Whatever the details, it is clear that the evolution of life and its continued existence on the planet depend on the absorption of the Sun's ultraviolet radiation by ozone and that the ozone concentration is determined in a subtle way by life itself.

The Troposphere

About 90 per cent of the total atmospheric mass resides in the troposphere, and the bulk of the minor trace gas burden is found there also. The Earth's surface acts as the main source of the trace gases, although some NO_x and CO may be produced in thunderstorms. Hydroxyl radicals dominate the chemistry of the troposphere in the same way that oxygen atoms and ozone dominate the stratosphere. Free radical chain reactions, initiated by OH, oxidize H_2, CH_4 and other hydrocarbons, and CO to CO_2 and H_2O. The reactions thus constitute a low-temperature combustion system. The radical-chain processes are photochemically driven, although stratospheric ozone limits the solar radiation at the Earth's surface to wavelengths longer than 280 nm. At these wavelengths, the most important photochemically active species are O_3, NO_2, and HCHO. All three can yield OH (or HO_2) indirectly, and thus initiate the oxidation chains. Ozone photolysis is, however, a critical step, since the other photolytic processes owe either their origin or their importance to it. Although only 10 per cent of the total atmospheric ozone is found in the troposphere, all *primary* initiation of oxidation chains in the natural atmosphere depends on that ozone. Some ozone is transported to the troposphere from the stratospheric ozone layer, but a mechanism also exists for generation of ozone in the troposphere itself. If NO_2 is present, then NO_2 photolysis (at $\lambda \leqslant 400$ nm)

$$NO_2 + h\nu \rightarrow O + NO \tag{12}$$

is a source of atomic oxygen that can form ozone in reaction (8)

$$O + O_2 + M \rightarrow O_3 + M \tag{8}$$

The NO can itself be oxidized back to NO_2, as we shall see shortly, so that the formation of O_3 is not stoicheometrically limited by the supply of NO_2 molecules initially present.

A first understanding of tropospheric photochemistry may be gained by considering methane as the only hydrocarbon present, and taking as our starting point the artificial situation where no CH_4 has yet been oxidized. Hydroxyl radicals must then be derived from ozone photolysis (at $\lambda \leqslant 310$ nm), in the way already described for the stratosphere

$$O_3 + h\nu \rightarrow O(^1D) + O_2(^1\Delta_g) \tag{2}$$

$$O(^1D) + H_2O \rightarrow OH + OH \tag{3}$$

Attack of OH on CH_4 yields methyl radicals, and a sequence of oxidation steps ensues that we will follow, for the time being, to the formation of HCHO (formaldehyde)

$$OH + CH_4 \rightarrow CH_3 + H_2O \tag{13}$$

$$CH_3 + O_2 + M \rightarrow CH_3O_2 + M \tag{14}$$

$$CH_3O_2 + NO \rightarrow CH_3O + NO_2 \tag{15}$$

$$CH_3O + O_2 \rightarrow HCHO + HO_2 \tag{16}$$

$$HO_2 + NO \rightarrow OH + NO_2 \tag{17}$$

Two very important features are displayed by this scheme. First, reactions (15) and (17) both provide a route for the oxidation of NO back to NO_2, and thus to a replenishment of tropospheric ozone through reactions (12) and (8). Secondly, the reactions as written are cyclic, the OH radical chain carrier being regenerated.

The aldehyde product of reaction (16) can itself be photolysed in the troposphere; the major photolytic pathway at $\lambda \leqslant 338$ nm yields two radical fragments that enter into further reactions

$$HCHO + h\nu \rightarrow H + HCO \tag{18}$$

$$HCO + O_2 \rightarrow CO + HO_2 \tag{19}$$

$$H + O_2 + M \rightarrow HO_2 + M \tag{20}$$

Finally, we may follow the oxidation through yet another step. Carbon monoxide reacts with OH

$$OH + CO \rightarrow H + CO_2 \tag{21}$$

so that the ultimate product is CO_2. The H atom re-enters the oxidation chain *via* reactions (20) and (17). In the unpolluted atmosphere, roughly 70 per cent of the OH reacts with CO, and 30 per cent with methane itself.

The oxidation steps that we have written for methane obviously have their analogues for higher hydrocarbons, but in all cases the reactions depend on the switch between peroxy- (RO_2) and oxy- (RO) radicals in an interaction with NO. Oxides of nitrogen are therefore a central part of the oxidation scheme, because they both effect the switch and are the ultimate source of ozone, and thus of OH radicals. Natural sources of NO_x include microbial actions in the soil, which produce NO as well as N_2O. Oxidation of biogenic NH_3, initiated by OH radicals, would be another significant source of NO_x. Lightning discharges appear to be responsible for less than ten per cent of the total NO_x budget. The important point is that OH radicals dominate tropospheric chemistry, that their source is attack of $O(^1D)$, mainly on H_2O, and that the excited $O(^1D)$ atoms are derived from ozone photolysis. It is evident, therefore, that a quantitative understanding of tropospheric chemistry demands a detailed knowledge of the mechanism and efficiency of ozone photolysis to yield $O(^1D)$ in the same way that the interpretation of stratospheric ozone concentrations requires the same information.

Ozone and Airglow

Excited states of O and O_2 make an extremely important series of contributions to the airglow of Earth and other planets [1,10,15]. All the relatively long-wavelength transitions that can be observed at the Earth's surface are forbidden by electric dipole selection rules, and the excited states are metastable; some of the excitation mechanisms directly involve ozone photochemistry, and so form a part of our present survey.

One important atomic feature is the $O(^1D \rightarrow {}^3P_{1,2})$ red doublet at $\lambda = 630.0$ and 636.4 nm. Some $O(^1D)$ is formed by photodissociation of molecular oxygen in the Schumann–Runge region at altitudes of 80–110 km. At longer wavelengths, and thus potentially at lower altitudes, ozone photolysis is a source of $O(^1D)$ in reaction (2). The maximum in $O(^1D)$ production rate from this reaction arises at an altitude of ca. 40 km. However, quenching of $O(^1D)$ by N_2 and O_2 is efficient, and at 40 km the collisional lifetime is less than 1 μs, so that radiation from $O(^1D)$ (lifetime 110 s) is very improbable at these altitudes. On the other hand, the quenching of $O(^1D)$ itself makes an important contribution to the molecular oxygen airglow through the energy-transfer process

$$O(^1D) + O_2(^3\Sigma_g^-) \rightarrow O_2(^1\Sigma_g^+) + O(^3P) \tag{22}$$

The Atmospheric Band system $(O_2(b^1\Sigma_g^+ \rightarrow X^3\Sigma_g^-))$ is strong in the terrestrial dayglow, and is easily observed by rocket photometry. Energy transfer in reaction (22) is the dominant excitation mechanism [15] at altitudes above 100 km and below 65 km. Excited, ^1D, oxygen atoms come from O_3 photolysis, reaction (2), in the lower altitude region.

The $^1\Delta_g$ *molecular* fragment of ozone photolysis in reaction (2) is, by contrast with the excited atomic fragment, only weakly quenched by O_2 or N_2. In the upper stratosphere, the collisional lifetime is of the order of tens to hundreds of seconds. Although the radiative lifetime is also very long (44 min) for the highly forbidden $O_2(a^1\Delta_g \rightarrow X^3\Sigma_g^-)$ Infrared Atmospheric Band at $\lambda = 1270$ nm, a sensible fraction of $O_2(^1\Delta_g)$ can emit because of the inefficiency of physical deactivation. In fact, the Infrared Atmospheric Band is the most intense feature of the dayglow, the $(0,0)$ transition (observed at altitudes high enough to avoid self-absorption [16–22]) having an intensity of 20 MR (1 rayleigh (R) is the brightness of a source emitting 10^6 photons cm^{-2} s^{-1} in all directions, and can be shown to be roughly identical to the volume emission rate in photons per cubic centimetre per second for a typical emitting layer [15]). Consideration of energy balance alone [23] points to ozone photolysis as the source of $O_2(^1\Delta_g)$, since no species other than ozone absorbs enough sunlight at the altitudes where the dayglow emission originates. Peak concentrations of $O_2(^1\Delta_g)$, determined by rocket photometry, are typically [21,22] 2×10^{10} molecule cm^{-3} at 50–60 km altitude. That is, metastable excited oxygen molecules are present at the parts per million level. Concentration–altitude profiles are most conveniently derived from rocket measurements [21, 22, 24, 25]. Data on quantum yields for singlet O_2 production in reaction (2) (see section 4.5) and on quenching rate coefficients may be combined with experimentally-determined atmospheric O_3 concentrations and solar irradiances to yield the $[O_2(^1\Delta_g)]$—altitude profile. Figure 1 shows the results of one such calculation [26] together with some data from rocket measurements [24, 25] for the Earth's dayglow. The good agreement may be taken as confirming the general outline of the O_3 photolysis mechanism for $O_2(^1\Delta_g)$ production (but see also section 4.5). Intensity variations at twilight [16,19], or, more spectacularly, in eclipses [17,18] can assist in the elucidation of excitation and quenching mechanisms. The close similarity in quenching rates determined from the eclipse observations and in the laboratory [27–29] provides further confirmation that O_3 photolysis is the source of $O_2(^1\Delta_g)$. Indeed, confidence in the model is now so great that it is frequently used in reverse, and measurements of the IR Atmospheric Band intensities in the dayglow are used to deduce atmospheric O_3 concentrations and profiles. For example, a secondary peak found [22] in the dayglow emission intensity at about 90 km is taken [18] to be associated with a secondary maximum in $[O_3]$ at this altitude. Instruments on the Solar Mesosphere Explorer (SME) satellite now routinely obtain [2, 3] atmospheric $[O_3]$ distributions from the intensity of emission at $\lambda = 1270$ nm. The possibility that the efficiency of $O_2(^1\Delta_g)$ production in reaction (2) has been wrongly assessed is discussed later in the appropriate section.

The IR Atmospheric Band is observed in the airglow of both Mars [30, 31] and Venus [32]. As on Earth, O_3 photolysis appears to be the only reasonable source of

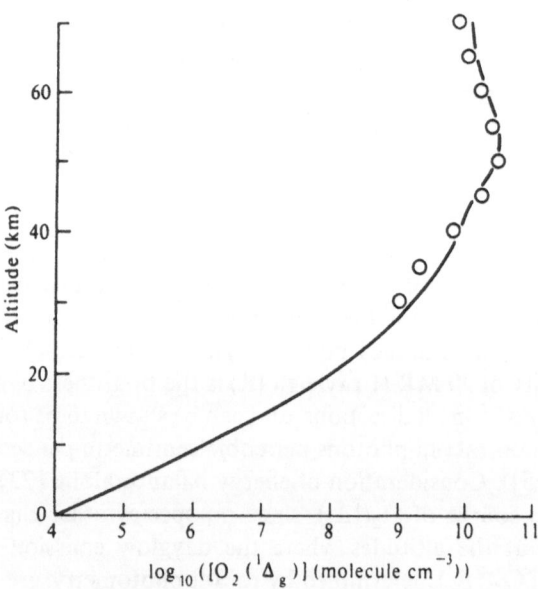

Fig. 1. Daytime atmospheric concentrations of $O_2(^1\Delta_g)$. The circles are experimental rocket measurements, while the solid line is a calculation based on laboratory spectroscopic and kinetic data [26]

the amounts of $O_2(^1\Delta_g)$ present in the Martian atmosphere. Catalytic reactions involving HO_x species play a major role [33] in O_3 loss at low latitudes, but less at high latitudes in winter. Hence O_3 concentrations at the winter pole will be higher, and the observed higher intensities of the IR Atmospheric Band at that pole can be explained. Ozone photolysis cannot be the source of $O_2(^1\Delta_g)$ on Venus because of the low O_2, and hence O_3, column densities [34], but explanations for Venusian airglow lie outside the scope of the present article.

Electronic Spectroscopy and Electronic States of Ozone

A knowledge of the spectroscopy of ozone is a pre-requisite to the quantitative understanding of the photochemistry of the molecule. Assessment of photolysis rates demands accurate absorption cross sections over the wavelength range of interest, and for modelling of atmospheric photochemistry it is also necessary to know how the absorption cross sections vary with temperature. The electronic states of ozone directly involved in the optical transitions, and those states energetically accessible by radiationless energy transfer, determine what the photochemical product channels will be (the nature of the primary products, and the efficiency of their formation is the subject of a section of this review). In this section, we examine information available on the spectroscopy and electronic states of ozone. A full and critical survey of ozone spectroscopy has been prepared by Steinfeld et al. [35].

Band Systems

Four distinct systems are recognized for the optical absorption of ozone in the near infrared to conventional ultraviolet regions of the spectrum. The longest wavelength electronic absorption system is that of the Wulf system in the near infrared, which blends into the Chappuis bands in the visible region. The Huggins bands appear in the near ultraviolet, with an inferred [36] (but not observed) origin at $\lambda = 368.7$ nm. All these three systems consist of diffuse bands. At wavelengths somewhat longer than 300 nm, the Huggins bands are replaced by the Hartley continuum. This system is the strongest of all, with a peak absorption near $\lambda = 250$ nm.

Figure 2 shows one recent absorption spectrum [37] of ozone in the ultraviolet region; the Huggins bands and the Hartley continuum are clearly visible. Table 1 summarizes information about the four absorption systems (data are from reference 35 with the exception of the predicted vibrational frequencies of Hay and Dunning [38] and the experimental vibrational frequencies of Katayama [36]).

The electronically excited states listed in Table 1 will be discussed in further detail later; we consider here some special features of the transitions themselves. The Wulf bands involve the $1^1A_2 \leftarrow 1^1A_1$ transition, which is not allowed for vibrationally unexcited species, but which becomes vibronically allowed in the asymmetric

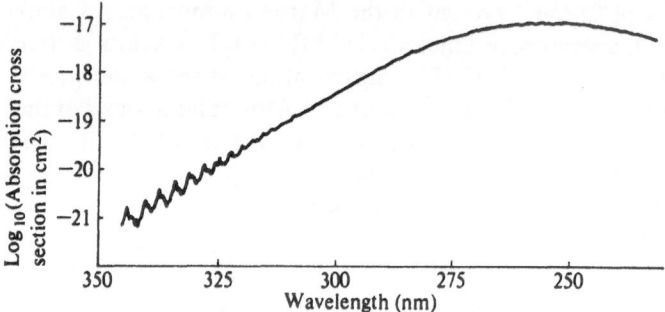

Fig. 2. Absorption spectrum of ozone in the ultraviolet region at 300 K. Data of Bass, A.M., and Paur, R.J., presented at the International Workshop on Atmospheric Spectroscopy, Rutherford Appleton Laboratory, Chilton, July 1983: see also [37]

Table 1. Absorption systems and transitions in the electronic spectroscopy of ozone[a]

System	Spectral region nm	Peak absorption σ/cm^2	λ/nm	Upper electronic state[b]	Vibrational spacing/cm^{-1} Predicted[c]	Observed
Wulf	1000–662	—	—	$1\,^1A_2(\nu_3=1)$	$537(\nu_2)$	566.7
Chappuis	739–451	5.1×10^{-21}	602	$1\,^1B_1$	$965(\nu_1)$	930[d]
					$489(\nu_2)$	460[d]
Huggins	370–300			see text	$1022(\nu_1)$	750[e]
		1.1×10^{-17}	250		$539(\nu_2)$	400[e]
					$—(\nu_3)$	830[e]
Hartley	300–200			$1\,^1B_2$	—	—

a) Data from [35] except where noted.
b) The ground electronic state is $1\,^1A_1$ with an experimental bond angle of 116.8° and R_e = 1.271–1.278 Å [see [38]; vibrational frequencies (fundamentals) in ν_1, ν_2, ν_3 are 1103.1, 700.9, 1042.1 cm^{-1} respectively.
c) Data from [38].
d) Data from [39].
e) Data from [36].

stretching mode. The lowest energy cold band of the transition lies at roughly 10 000 cm^{-1}, which implies that the electronic origin lies at (10 000–1103) = 8897 cm^{-1}, or ca.1.1 eV. At somewhat higher energies, the $1\,^1B_1 \leftarrow 1\,^1A_1$ transition gives rise to the Chappuis bands. Table 1 indicates that the observed vibrational spacing is rather larger than the prediction of Hay and Dunning [38]. Hay et al. [40] assign the Hartley band to the $1\,^1B_2 \leftarrow 1\,^1A_1$ transition, and the shape of the continuum arises largely from a symmetric stretching progression [40–43], dissociation along the asymmetric stretching mode being responsible for the vibrational broadening that produces the continuum [5, 38, 40, 44]. A locally bound region on the $1\,^1B_2$ surface may be responsible [36] for the Huggins bands, although there is also speculation [45] that the $2\,^1A_1$ state may be involved. Bands arising from odd v_3 appear in the Huggins system; such bands should be forbidden if the molecule has C_{2v} symmetry. However, recent experiments [46] on the excitation of fluorescence suggest that the upper state of the Huggins system is, indeed, effectively the 1^1B_2 state, but with C_s symmetry (as predicted by calculation [38, 40, 41]) rather than C_{2v} symmetry. The fluorescence excitation studies employ jet cooling of the ozone to ca. 3 K to remove rotational congestion, and the partial resolution of rotational structure was possible that led both to the identification [36, 46] of the upper state in the Huggins system and to an upper bound for its lifetime of 3.6 ps [46].

Absorption Cross Sections

Detailed reviews of measurements of absorption cross sections in the Chappuis, Huggins, and Hartley systems have appeared in several places [7, 35, 47].

For the Chappuis bands, the current recommendations for use in atmospheric modelling [7] derive from the measurements of Vigroux [48]. Other important measurements are those of Inn and Tanaka [49, 50] and of Griggs [51]. Although Steinfeld et al. [35] discuss arguments for the validity of the cross sections of Inn and Tanaka, which are about 10% smaller than those of Vigroux or Griggs, WMO [7] believes that the accuracy of the available data, as indicated by the agreement between the data of Vigroux and of Griggs, is sufficient for applications to atmospheric modelling. No recommendation is given [7] for the temperature dependence of the absorption cross section in the Chappuis bands, and, indeed, Vigroux [48] found virtually no variation of the cross sections over the temperature range 181–353 K.

The ultraviolet absorption regions are by far the most important for photodissociation of ozone in the stratosphere and mesosphere. The currently recommended cross sections in the Hartley and Huggins systems are those given in the WMO [7] and JPL-NASA [47] reports. The most reliable cross sections for the various wavelength intervals are based on data obtained as follows: 175–185 nm, Watanabe et al. [52] (tabulated by Ackerman [53]); 185–225 nm, Molina and Molina [54]; 225–245 nm, Inn and Tanaka [49] (tabulated by Ackerman [53]); 245–348 nm, Bass and Paur [37, 55]; 348–363 nm, Inn and Tanaka [49] (tabulated by Ackerman [53]). The high resolution studies of Bass and Paur [37, 55] were relative, and the experimental basis for normalizing the data is discussed critically by Steinfeld et al. [35]. Recent work by Mauersberger et al. [56, 57] gives a cross section at 297 K of 1.137×10^{-17} cm^2 at $\lambda = 253.7$ nm (the mercury line wavelength used for most normalization). The value is about 1% smaller than the commonly accepted value of 1.147×10^{-17} cm^2 obtained by Hearn [58], and about 2% smaller than that reported by Molina and Molina [54]. The most recent determination, by Yoshino et al. [59], yields a cross section of 1.145×10^{-17} cm^2 at 298 K. The discrepancies appear to be beyond the limits of experimental precision, and the sources remain uncertain.

At wavelengths shorter than about 260 nm, the absorption cross section in the Hartley band is *almost* independent of temperature [7, 35, 47], although the small temperature dependence expected [42] and observed [59, 60] at the normalizing wavelength ought to be taken into account [35, 54]. A significant temperature dependence is seen at longer wavelengths in the rest of the Hartley band and in the Huggins band. At around 300 nm, the ratio of cross sections for 273 K and 203 K is 1.11 [37, 55], and reaches 3.32 around $\lambda = 345$ nm. In general, there is good agreement between the various studies [37, 55, 59–63] of temperature dependence in the Hartley and Huggins bands. The dependence seems to arise from absorption by vibrationally excited, electronic ground state, ozone in the thermal population [60]. When the spectral structure associated with bending and stretching sequences is resolved in the Huggins bands, the cross section at locations between sharp peaks shows a large temperature dependence, while the changes are much smaller near absorption maxima. Again, these changes can be explained in terms of the ground state vibrational populations. An empirical quadratic temperature expression may be used to parameterize the temperature dependence [37, 54]. Adler–Golden [42] has proposed a semi-empirical function based on a theoretical analysis of the temperature dependence of the cross section. This function is stated [35, 42] to be of a Boltzmann form, with an energy term of 900 cm^{-1}, which corresponds to the average of v_1, v_2, and v_3 frequencies. However, a Boltzmann expression of the kind suggested does not allow for the variability of temperature dependence with wavelength, a feature that is emphasized by the latest experimental data [59], although it may still be useful for interpolation purposes at any fixed wavelength.

Electronic Structure and Excited States of Ozone

Each oxygen atom in ozone has two unpaired electrons, so that the central atom can form a covalent bond with the terminal atoms, each of which possesses an unpaired electron. These unpaired electrons may occupy π orbitals perpendicular to the molecular plane or σ orbitals in it, and may be coupled as singlet or triplet spin states. The ground state (1^1A_1) resembles a biradical [38] with the pπ electrons on the two terminal atoms (π_a, π_b) weakly coupled into a singlet state, and the lowest seven excited states also correspond to biradical states with the singly occupied orbitals found in the π or non-bonding σ orbitals of the terminal atoms. Thus the lowest excited state (1^3B_2) is a triplet resulting from triplet spin coupling of the π_a and π_b electrons. The next two pairs of states ($1^3A_2, 1^1A_2, 1^3B_1, 1^1B_1$) are π^5 configurations resulting from the promotion of an electron from a pσ to a pπ orbital on either terminal atom. Double excitation leads to two states with π^6 configuration (2^3B_2, 2^1A_1). After these seven biradical excited states, the next two ($1^1B_2, 3^1A_1$) are ionic in character, and involve charge transfer from the central atom to one of the terminal ones. Table 2 summarizes these conclusions by showing a simplified valence bond representation of the occupancy of the orbitals on the three oxygen atoms. Also shown are the main configurations of the corresponding molecular orbital representation. It is worth noting that the singlet coupling of the lone pair pσ orbitals in the π^6 configuration represents a weak σ bond. Thus, as the bond angle is decreased in the 2^1A_1 state, the 3-membered ring structure of ozone (oxirane) is formed.

Table 2. Electronic states of ozone

State C_{2v}	C_s	Valence bond representation[a]	Molecular orbital representation (main configuration)	Calculated vertical energy eV [64]	Calculated bond angle (°) [38]
1^1A_1	$1^1A'$	$\ell_a^2 \ell_b^2 \pi_c^2 \pi_a \pi_b$	$1b_1^2 6a_1^2 4b_2^2 1a_2^2$	0.00	116
1^3B_2	$1^3A'$	$\ell_a^2 \ell_b^2 \pi_c^2 \pi_a \pi_b$	$1b_1^2 6a_1^2 4b_2^2 1a_2^2 2b_1$	1.20	108
1^3A_2	$1^3A''$	$\ell_a \ell_b^2 \pi_c^2 \pi_a^2 \pi_b$	$1b_1^2 6a_1^2 4b_2 1a_2^2 2b_1$	1.44	100
1^1A_2	$1^1A''$	$\ell_a \ell_b^2 \pi_c^2 \pi_a^2 \pi_b$	$1b_1^2 6a_1^2 4b_2 1a_2^2 2b_1$	1.59	102
1^3B_1	$2^3A''$	$\ell_a \ell_b^2 \pi_c^2 \pi_a^2 \pi_b$	$1b_1^2 6a_1 4b_2^2 1a_2^2 2b_1$	1.59	124
1^1B_1	$2^1A''$	$\ell_a \ell_b^2 \pi_c^2 \pi_a^2 \pi_b$	$1b_1^2 6a_1 4b_2^2 1a_2^2 2b_1$	1.95	118
2^3B_2	$2^3A'$	$\ell_a \ell_b \pi_c^2 \pi_a^2 \pi_b{}^2$	$1b_1^2 6a_1 4b_2 1a_2^2 2b_1^2$	3.27	104
2^1A_1	$2^1A'$	$\ell_a \ell_b \pi_c^2 \pi_a^2 \pi_b^2$	$1b_1^2 4b_2^2 1a_2^2 2b_1^2$	3.60	60
1^1B_2	$3^1A'$	$\ell_a^2 \ell_b^2 \pi_c(+)\pi_a^2(-)\pi_b$	$1b_1 6a_1^2 4b_2^2 1a_2 2b_1^2$	4.97	108

(a) ℓ_a, ℓ_b denote the lone pair $p\sigma$ orbitals on the terminal (a, b) atoms; π_c is the lone pair $(p\pi)$ orbital on the central (c) atom.

Considerable theoretical effort [38, 64–68] has gone into calculating the energies and properties of electronically excited states and potential energy surfaces for ozone that assist in understanding the spectroscopy, photochemistry and dissociative pathways [40, 41]. Thunemann et al. [64] used energy-extrapolation methods to calculate vertical excitation energies (1.59, 2.41, 4.97 eV) for the 1^1A_2, 1^1B_1, and 1^1B_2 states that agree well with the experimental values based on the origins of the Wulf, Chappuis, and Hartley bands (1.6, 2.1, and 4.9 eV). The absolute energies thus seem more reliable than those obtained by Hay and Dunning [38] using a polarization configuration interaction (POL-CI) technique, and can be used to fix the positions of the vertical excitation energies for the states not seen in ordinary absorption spectroscopy. Figure 3 indicates the vertical energies, marked V, of the singlet and triplet states of ozone obtained in this way, and the energies are also listed in Table 2. The figure also shows the electronic states of $O + O_2$ fragments with which the electronic states correlate: this information will be needed in the discussion of the section on photodissociation of ozone.

Bond angles seem reasonably well predicted by the POL-CI method [38], the ground state prediction of 116.0 degrees matching the experimental [69] finding of 116.8 degrees. The last column of table 2 demonstrates the changes in geometry predicted [38] for the different states of ozone. It is noteworthy that these calculations give an optimum geometry for the 2^1A_1 state of roughly D_{3h} symmetry.

Adiabatic excitation energies (i.e. the differences in energy between the excited states and the ground state each at their optimum geometry) are important in investigating the existence of bound electronic states of ozone. Since Thunemann et al. [64] did not calculate adiabatic excitation energies, they must be estimated from the vertical excitation energies. One possible procedure [35] is to combine the *differences* between vertical and adiabatic energies of Hay and Dunning [38] with the absolute vertical energies of Thunemann et al. This is the procedure adopted to obtain the adiabatic energies (marked A) in Fig. 3. An alternative procedure [10] is to scale down the values of Hay and Dunning by a factor of

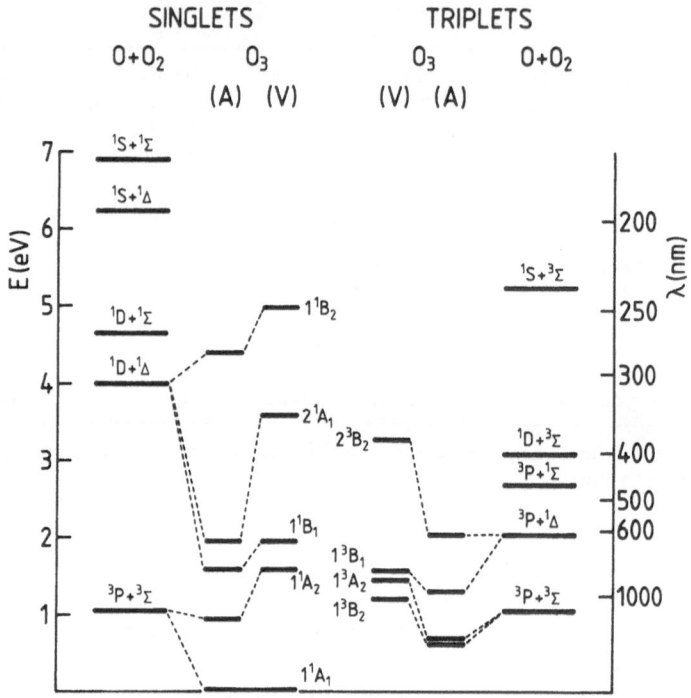

Fig. 3. Vertical (V) and adiabatic (A) excitation energies of electronic states of ozone, and the correlations with the states of product fragments

1.2–1.3 on the basis of a comparison of their vertical energy with the more sophisticated calculation of Wilson and Hopper [65] for the 1^3B_2 state. The two methods give comparable results, and provide reasonable agreement with the adiabatic excitation energy of the 1^3B_2 state calculated directly (0.74 eV) by Wilson and Hopper. Nevertheless, error limits on the adiabatic energies are clearly larger than those for vertical energies, particularly where geometry changes are appreciable. The cyclic form of the 2^1A_1 state is frequently mentioned as a candidate for a bound excited state, and if this ring state lies below the $O + O_2$ dissociation limit, it could be important in ozone recombination and dissociation. However, the adiabatic excitation energy of this state is particularly uncertain [35, 66, 67], calculations yielding the range 0.5–1.3 eV [35]. Current opinion is that, even if the energy of the ring state is low, the barrier between it and the equilibrium C_{2v} geometry lies above the $O + O_2$ dissociation limit.

Experimental evidence for bound excited states of ozone is rather sparse. The only direct observation of triplet excited states is that of Swanson and Celotta [70], who observed a broad peak near 1.65 eV in an electron impact study for a scattering angle of 90 degrees, where spin-forbidden transitions should be most prominent. The vertical excitation energies given in Table 2 show that the 1^3B_1 state is accessible, although the observed [70] features may include contributions from the lower triplets (1^3B_2, A^3A_2). McGrath and co-workers [71] have reported an absorption transient following irradiation of ozone in the visible region by a dye-pumped flash system. Enhanced absorption in the ultraviolet region was observed, at a peak wavelength of 320 nm. Arguments were presented to support the assignment of the absorption to electronically excited ozone, probably in the 1^1A_2 state, with a radiative lifetime of about 4 ms. Some circumstantial corroboration comes from earlier experiments of von Rosenberg and Trainor [72] who observed infrared emission on the recombination of O with O_2. Most of the emission results from vibrational transitions in ground electronic state ozone (see section on vibrational photochemistry), but it is suggested that part of the chemiluminescence derives from electronically excited ozone. Wraight [73] suggests that the upper state involved is the 1^1A_2, and provides explanations for why this state should be populated. Further experiments by Bair and co-workers [74] on the recombination of O with O_2 seem to be consistent with more than 60% of the ozone first formed being electronically excited, probably in the $^3B_2$ state, and the same state is invoked by Ramirez et al. [75] as one of the routes to formation of O_3 in the pulse radiolysis of oxygen. Nevertheless, it seems at present that the identification of the excited states in the enhanced absorption [71], infrared emission [72] or kinetic [74, 75] experiments is far from certain.

Photodissociation of Ozone

Electronic Excitation in the Product Fragments

The richness of ozone's photochemistry is in part a consequence of the rather weak energy of the O_2–O bond (just over 106 kJ mol^{-1} or 1 eV). Since the peak of the ozone absorption spectrum in the ultraviolet region corresponds to about 470 kJ mol^{-1}, it is apparent that much excess energy is available in the ultraviolet photolysis which might lead to the excitation of the fragment species. It has long been known that the photochemical behaviour of ozone depends on the wavelength of photolysis (see reference [76] for a survey of the early work). Photolysis in the Chappuis (visible) region seems to proceed through a simple two-step mechanism, while for photolysis in the Hartley (ultraviolet) region, large quantum yields for decomposition are observed [76], and the quantum yields are sensitive to the presence of added hydrogen-containing species [77]. The most important implication of these observations is [76–78] that ground state, ^3P, oxygen atoms are the product of photolysis by red light, and excited, ^1D, atoms are formed in the ultraviolet region, although we shall see both that excited molecular species can also be formed, and that the efficiency of excited atom production by ultraviolet photolysis may fall short of 100%.

Both atomic and molecular oxygen have several accessible excited states. Table 3 shows some of the thermochemical thresholds, expressed as wavelengths in nm, for different dissociation channels. The wavelengths indicated in the Table would be the absolute limits for the formation of any particular product pair, were it not for the possible contribution of internal energy in the molecule. Vibrational and rotational energy can, in fact, contribute towards the total energy necessary for

Table 3. Wavelength thresholds (in nm) for ozone photodissociation channels

Molecule Atom	$O_2(^3\Sigma_g^-)$	$O_2(^1\Delta_g)$	$O_2(^1\Sigma_g^+)$	$O_2(^3\Sigma_u^+)$	$O_2(^3\Sigma_u^-)$
$O(^3P)$	1180	612	463	230	173
$O(^1D)$	411	310	267	168	136
$O(^1S)$	237	199	181	129	109

dissociation, so that the onset of participation of a channel does not have to be a step-function of wavelength, and it may be a function of temperature, as we shall discuss later.

The experimental evidence shows that absorption in the visible region of the spectrum leads to the formation of *two* triplet, ground state, fragments, while in the strong part of the Hartley band, at $\lambda < 300$ nm, the products are mainly the *two* singlets shown by the table to be energetically accessible $(O_2(^1\Delta_g) + O(^1D))$. In this section some of the circumstantial evidence that first led to these conclusions is presented; other indirect and direct identification of the products of photolysis has subsequently become available, and this latter evidence is described at appropriate points in later sections.

Static determinations (i.e. performed without provision for time resolution) of the overall quantum yield for ozone decomposition provided the first evidence for the formation of ground state products in the red region of the spectrum. Limiting quantum yields of two are found [79, 80], consistent with the simple mechanism

$$O_3 + h\nu \rightarrow O_2 + O \tag{7}$$

$$O + O_3 \rightarrow O_2 + O_2 \tag{9}$$

Higher quantum yields in the Hartley region [76, 78, 81–83] are used to adduce the presence of excited products in the ultraviolet photolysis. Although there has been some discussion about whether the observed chain decomposition is an energy chain involving oxygen species alone or a radical chain resulting from the presence of adventitious impurities [78, 81], it is clear that the chains are initiated by atomic oxygen in the ^1D state. The quantum yields thus give a measure of the efficiency of $O(^1D)$ production. At low $[O_3]$, where the chains initiated by $O(^1D)$ do not contribute, the quantum yields for ozone decomposition can approach 4, a result expected if the *molecular* product of photolysis is also excited. We shall show in section 4.5 that the O_2 is, in fact, in the first singlet state, $O_2(a^1\Delta_g)$. A simplified reaction scheme can be used to interpret the results:

$$O_3 + h\nu \rightarrow O(^1D) + O_2(^1\Delta_g) \tag{2}$$

$$\phi(^1D) \quad \phi(^1\Delta_g)$$

$$\rightarrow O + O_2 \tag{7}$$

$$O(^1D) + O_3 \rightarrow \text{chain decomposition} \tag{23}$$

$$O_2(^1\Delta_g) + O_3 \rightarrow 2O_2 + O(^3P) \tag{24}$$

$$O(^3P) + O_3 \rightarrow O_2 + O_2. \tag{9}$$

This scheme predicts an overall quantum yield, Φ, of the *form*

$$\Phi = 2 + 2\phi(^1\Delta_g) + f\phi(^1D)\,[O_3]. \qquad (I)$$

where f is a constant describing the extent of chain decomposition. Thus, in the red region of the spectrum, where only $O(^3P)$ is formed (i.e. $\phi(^1D)$ is zero), the quantum yield is two, because reaction (23) does not participate. On the other hand, if both $\phi(^1D)$ and $\phi(^1\Delta_g)$ are unity, as the arguments presented so far suggest they might be in the ultraviolet region, then the quantum yield would be expected to be a linear function of $[O_3]$, with an extrapolated intercept at zero $[O_3]$ of four. Figure 4 shows some typical results obtained at three wavelengths in the ultraviolet [82, 83]. The chain nature of the decomposition for photolysis at the shortest wavelength (254 nm) is clearly indicated by the positive slope of the quantum yield—$[O_3]$ relationship, and the intercept of nearly four provides strong circumstantial evidence for the production of $O_2(^1\Delta_g)$ with a yield approaching unity. Smaller slopes are observed at longer wavelengths, suggesting a reduced contribution from $O(^1D)$ formation, as we shall discuss later.

At wavelengths shorter than 237 nm, $O(^1S)$ becomes an energetically possible product, although experiments [84] in the wavelength range 170–240 nm show that the ^1S state atom is not produced to a significant extent. Absorption of a sufficiently energetic photon does not, therefore, guarantee that an energetically accessible route will in fact be followed. Indeed, at $\lambda = 157.6$ nm, about half the ozone molecules fragment to $O(^1D) + O_2(^1\Sigma_g^+)$; the remainder of the ozone dissociates via a completely new route to $O + O + O$ [85].

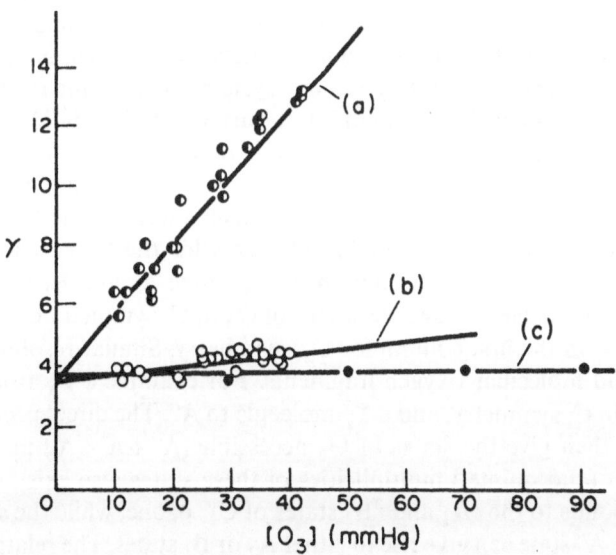

Fig. 4. Quantum yield for photolysis of ozone at three wavelengths: (a) $\lambda = 254$ nm; (b) $\lambda = 313$ nm; (c) $\lambda = 334$ nm. Data of Jones and Wayne [82, 83]

Correlation of Spin and Symmetry in Photodissociation

Thermodynamics may impose a limiting constraint on a dissociation channel, and
the entries in Table 3 are based on energy criteria. However, other conditions may
also need to be satisfied for the actual occurrence of fragmentation by any given
pathway. In particular, reactions are generally efficient only if they proceed
adiabatically: that is, if a single continuous potential energy surface connects
products with reactants. Of the correlation or conservation rules that are based on
this concept, the important ones in the decomposition of ozone are those concerned
with spin and with orbital symmetry.

Conservation of spin in a dissociation requires that the quantum vector sum of
the spins of the product fragments has at least one value in common with the spin of
the excited state from which they derive. The ground and excited states of ozone are
various singlet and triplet species. Singlet ozone $(S=0)$ correlates either with two
singlet fragments (total $S=0$) or two triplets (total $S=2,1$ or 0). Similarly, triplet
ozone $(S=1)$ correlates with one singlet and one triplet (total $S=1$), or with two
triplets through the $S=1$ component. Application of the rules to a photodissociative process usually adds the optical selection rule $(\Delta S=0)$ that determines
the population of the upper electronic state by absorption. Probable candidate
states for the absorptions of ozone have been described earlier and we note that
there is no change in electron spin on optical absorption in any of the absorption
systems described. Thus, if the O and O_2 fragments of the dissociation are to
correlate with the excited state of O_3, the argument runs that both must be singlets
or both must be triplets. With this rule imposed, some of the pairs of products
suggested by Table 3 are excluded.

As long as spin–orbit coupling is small, the correlation of symmetry and of spin
may be considered separately. The derivation of the molecular state of O_3 from the
states of O and O_2 is slightly more complicated than the determination of whether
or not spin is conserved. Both the fragments have a higher symmetry than the O_3
molecule, but the symmetry of the triatomic system is less during the fragmentation
process. Ozone itself has C_{2v} symmetry, but during the dissociation the conforma-
tion of the nuclei has the point group C_s. It is necessary, therefore, to resolve the
symmetry species of the higher symmetry point groups into those of the latter. This
resolution is achieved by finding in the character table for the higher symmetry
group the characters for the symmetry elements for the lower symmetry group.
Thus, A_1 and B_2 in C_{2v} both become A' in C_s, while A_2 and B_1 become A''. The
second column of Table 2 shows the states of O_3 in C_s symmetry corresponding to
the states given in the first column for C_{2v} symmetry. Similar resolutions apply to
the atomic and molecular oxygen fragments. For example a P_g atom resolves to
$A'+A''+A''$ in C_s symmetry, and a Σ_g^- molecule to A''. The direct products of these
latter species then give the states of O_3 accessible $(A''+A'+A')$ in C_s symmetry.
Singlet, triplet (and quintet) multiplicities of these states also exist. One set of A'
states corresponds to the $^1A_1$ and $^3B_2$ states of C_{2v} ozone, while the second set are
repulsive. The A'' state can give rise to either A_2 or B_1 states. The relative energies of
the $^{1,3}A_2$ and $^{1,3}B_1$ states depend on the bond angle in C_{2v} ozone [38] so that
ground state $O+O_2$ fragments correlate with the A_2 states for $\Theta \simeq 100°$ but with

the B_1 states for $\Theta \simeq 120°$. The lowest adiabatic excitation energies correspond to $^1A_2$ and $^3A_2$ states (it can be shown that the lowest B_1 states must correlate with excited fragments).

The correlations between the nine lowest states of O_3 and the electronic states of $O + O_2$ is indicated in Fig. 2, which also emphasizes spin adiabaticity by displaying singlet and triplet states separately. We see that the major pathway of photolysis in the Hartley bands is entirely consistent with adiabatic fragmentation. The state populated by absorption is the $^1B_2$ which correlates with the observed major products of photodissociation in this region $(O_2(a^1\Delta_g) + O(^1D))$. Note that the singlet and triplet A_2 and B_1 states are labelled according to the symmetry species for the optimum (adiabatic) geometry, and that the species of the vertical states is probably reversed. In the Chappuis bands, therefore, the $^1B_1$ state of ozone that is excited can correlate with the ground state photofragments that are formed.

A certain amount of caution is needed in applying the spin and other conservation rules. First, they are valid only so far as the quantum number, such as S, to which they pertain is a good description of the system. Spin-orbit coupling can relax the rigour with which the $\Delta S = 0$ rule must be applied in both optical and radiationless transitions. That is, the excitation process could occur without conservation of spin, and the subsequent dissociation might be non-adiabatic. The other point that should be made is that rules show *excluded* possibilities rather than permitted ones. We shall return several times later to consideration of the validity of the spin conservation rule.

With the essential theoretical background established, we now go on to discuss experimental determinations of the efficiency with which the possible product states are formed in the photolysis of ozone. In the next section relative efficiencies of $O(^1D)$ production at different wavelengths is considered.

Dependence of $O(^1D)$ Yield on Wavelength of Photolysis

As described in the sections on the atmosphere, $O(^1D)$ drives chemical transformations of great importance in the stratosphere and troposphere. Photolysis of ozone by ultraviolet radiation is the source of the excited atomic oxygen, so that a knowledge of the efficiency of $O(^1D)$ production is vital to a quantitative understanding of much of atmospheric chemistry. One particularly significant photochemical region is the long-wave tail of the Hartley-Huggins bands. In this region, solar intensity rises rapidly, but is accompanied by a decrease in absorption cross section. The variation with wavelength of quantum yield for $O(^1D)$ production thus has a profound effect on the predicted rate of excited atom formation. Static experiments, such as those described earlier, can be used to obtain the information needed. According to relation (I), slopes of the overall quantum yield $-[O_3]$ plots (e.g. Fig. 4) should be proportional to the primary quantum yield for $O(^1D)$ formation, $\phi(^1D)$. It is apparent from Fig. 4 that there is a marked decrease in $\phi(^1D)$ as λ increases from 254 nm to 334 nm, the yield being essentially zero at the longer wavelength. At $\lambda = 310$ nm, $\phi(^1D)$ is around 0.1 of its value at $\lambda = 254$ nm. Such behaviour is, at least qualitatively, that expected if photolysis is spin conserved, since, as shown in the table, $\lambda = 310$ nm is approximately the thermochemical threshold for formation of two singlet products. Subsequent experiments, both static and those using laser flash photolysis, have greatly refined the early results [86–93]. Figure 5 shows the form of the quantum yield—wavelength curve currently recommended by WMO [7] and NASA [47] for use in stratospheric modelling. The value of the absolute limiting yield will be discussed in the next section, but the figure makes it plain that there is, indeed, a sharp fall in the relative $\phi(^1D)$ at a wavelength of roughly 310 nm in accordance with the predictions.

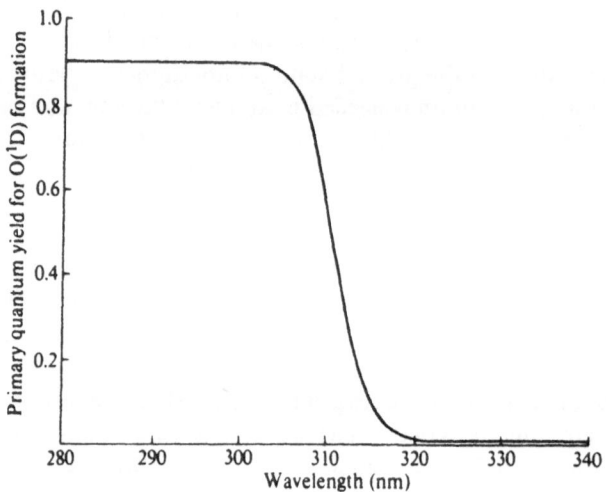

Fig. 5. Primary quantum yield for O(^{1}D) production in O$_3$ photolysis as a function of wavelength. Curve derived from quantum yields recommended by NASA [47]

The absorption spectrum of ozone is temperature dependent [37, 55, 59–63, 94, 95] (see pp. 15–16). Vibrational excitation of ozone seems to be responsible for the changes [42, 60, 94, 95]. The exact form of the $\phi(^1D) - \lambda$ curve is also temperature dependent [96–98], probably because vibrational excitation can contribute to the total energy available for dissociation of ozone. A later section (pp. 34–37) addresses the question of the photochemistry of vibrationally excited ozone, and further discussion is deferred until then. Electronic excitation of ozone might also play some part in modifying the $\phi(^1D) - \lambda$ relationship, thus providing further interest in the possibility of the existence of bound excited electronic states of ozone (pp. 18–19). If the states first populated dissociate at $\lambda > 310$ nm to yield O(^{1}D), or if collisional quenching of them produces "active" vibrationally excited ozone, then electronic excitation by visible radiation might enhance the overall rate of O(^{1}D) production in the atmosphere.

Moortgat and Kudszus [93] have cast the experimental data for both the wavelength and temperature dependences of $\phi(^1D)$ in an analytical form. With modification for the likely absolute values of $\phi(^1D)$, discussed in the next section, the expression is

$$\phi(\lambda,T) = A(\tau) \arctan[B(\tau)(\lambda - \lambda_0(\tau))] + C(\tau)$$

where $\tau = T - 230$ is a temperature function, with T given in Kelvin; λ is the wavelength in nm, and arctan is expressed in radians.

The coefficients $A(\tau)$, $B(\tau)$, $\lambda_0(\tau)$ and $C(\tau)$ are expressed as interpolation polynomials of the third order:

$$A(\tau) = 0.332 + 2.565 \times 10^{-4}\tau + 1.152 \times 10^{-5}\tau^2 + 2.313 \times 10^{-8}\tau^3$$
$$B(\tau) = -0.575 + 5.59 \times 10^{-3}\tau - 1.439 \times 10^{-5}\tau^2 - 3.27 \times 10^{-8}\tau^3$$
$$\lambda_0(\tau) = 308.20 + 4.4871 \times 10^{-2}\tau + 6.9380 \times 10^{-5}\tau^2 - 2.5452 \times 10^{-6}\tau^3$$
$$C(\tau) = 0.466 + 8.883 \times 10^{-4}\tau - 3.546 \times 10^{-5}\tau^2 + 3.519 \times 10^{-7}\tau^3.$$

In the limit where $\phi(\lambda, T) > 0.9$, the quantum yield is set $\phi = 0.9$, and similarly for $\phi(\lambda, T) < 0$, the quantum yield is set $\phi = 0$.
The mathematical expression is clearly convenient for atmospheric modelling purposes, and is, in fact, the basis for the current data evaluations [7, 47] as well as for Fig. 5 presented here.

Absolute Quantum Yield for O(^{1}D) Formation

Most measurements of the quantum yields for O(^{1}D) production at wavelengths around the fall-off region at the energetic threshold (i.e. ca. $\lambda = 310$ nm) have been made relative to those for shorter wavelengths, where the quantum yield has often been assumed to be unity on the basis of the spin-conservation arguments discussed earlier. Several pieces of information show that the O(^{1}D) + O$_2$($^1\Delta_g$) channel may not be the only one occurring for ozone photolysis at wavelengths shorter than the thermodynamic threshold, even though it may be the dominant one.

Two time-of-flight photofragment spectroscopy studies have indicated that the quantum yield for *ground* state, O(^{3}P), atom formation is about 0.1 at $\lambda = 274$ nm [99] and $\lambda = 266$ nm [100]. Photofragment spectroscopy is carried out by crossing a beam of radiation (photons) with a beam of reactant under collision-free conditions, and analysing the energy (and angular distribution) of release of the fragments. The two studies cited examined the atomic and the molecular fragments, respectively. Figure 6 shows some results of Sparks et al. [100] obtained by accumulating the molecular oxygen signal for 150 000 laser shots; the figure represents the centre-of-mass translational energy obtained from the photofragment time-of-flight spectrum. Most of the photofragments can be identified with formation of O$_2$($^1\Delta_g$) + O(^{1}D), the four peaks corresponding to v = 0, 1, 2, and 3 in the O$_2$; the broad high energy feature, however, can only result from dissociation in the O$_2$($^3\Sigma_g$) + O(^{3}P) channel, which must contribute roughly 10 percent to the total dissociation. Fairchild et al. [99] reached similar conclusions from their study of the atomic oxygen fragment.

Flash photolysis experiments tend to confirm that some of the dissociation proceeds via the triplet channel. Photolysis at $\lambda = 248$ nm [101, 102] and at

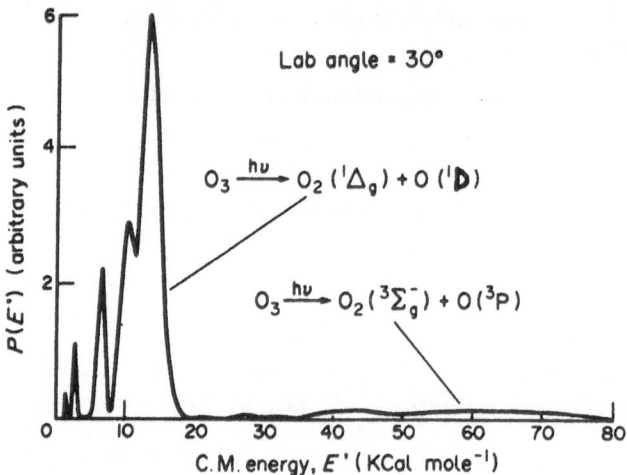

Fig. 6. Time-of-flight spectrum (centre of mass) for photofragmentation of ozone at $\lambda = 266$ nm. Reproduced, with permission, from Sparks et al. [100]

$\lambda = 266$ nm [103] show that some $O(^3P)$ is formed promptly after the photolytic flash (and thus presumably in the primary step), while additional $O(^3P)$ builds up subsequently as a result of the secondary reactions anticipated. Figure 7 displays the results of Greenblatt and Wiesenfeld [102] for photolysis at $\lambda = 308$ nm (Fig. 7a) and at $\lambda = 248$ nm (Fig. 7b). At the shorter wavelength, the $O(^3P)$ formed promptly after photolysis again accounts for about ten percent of the photolysis, in agreement with the time-of-flight experiments. Wine and Ravishankara [104] have measured $\phi(^1D)$ directly, and obtain a value of 0.9 at $\lambda = 248$ nm, in substantial agreement with the interpretation put on the measurements of $\phi(^3P)$. The longer wavelength at which $\phi(^3P)$ was determined (308 nm) is already in the "fall-off" region (cf. Fig. 5), so that the production of some $O(^3P)$ is expected; the experimental result suggests that $\phi(^1D)$ is 0.79. However, this result indicates another curious facet of the investigations, because combining it with the accepted $\phi(^1D) - \lambda$ relationship gives $\phi(^1D) = 0.96$ at $\lambda = 300$ nm. That is, the quantum yield is apparently higher at the threshold than it is at shorter wavelengths. Brock and Watson [87, 103] also find evidence that $\phi(^1D)$ decreases slightly between $\lambda = 304$ nm and $\lambda = 275$ nm. Such behaviour is not excluded on theoretical grounds, although it is not necessarily predicted, either. The formation of any triplet products in the ultraviolet region is presumably a result of a predissociation from the electronic states that more frequently yield the singlet fragments by direct dissociation. Figure 8 represents potential energy curves suggested [40] for ozone: the curves are cross sections through the potential energy surface in which one bond length alone is altered. The states are all of A' symmetry: that labelled B is the one first populated by absorption from the ground, X, state (pp. 13–15), and it correlates with the usual singlet fragments. However, the calculations [40] show that this state is crossed by a repulsive state, labelled R, that is also of A' symmetry, and that correlates with ground state products. Radiationless transition to this state

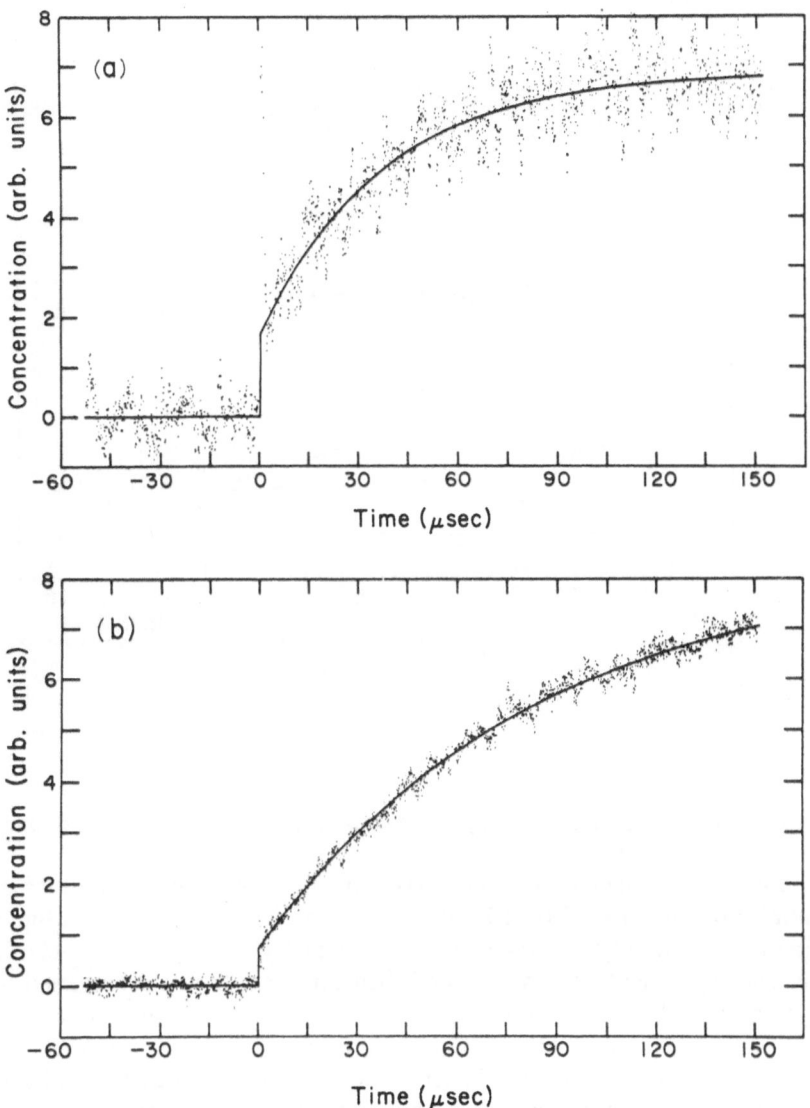

Fig. 7. Evolution of [O(^{3}P)] with time in the flash photolysis of ozone: (a) $\lambda = 308$ nm; (b) $\lambda = 248$ nm. Reproduced, with permission, from Greenblatt and Wiesenfeld [102]

will then permit the formation of O(^{3}P). The efficiency of crossing will be determined by the mixing of the electronic states and the velocity of approach to the intersection. The variation of this efficiency with wavelength will thus depend also on the exact and geometry of the surfaces. In particular, if crossing occurred at a much smaller value of R_2 than that indicated in Figure 8, so that the energy needed to reach it exceeded the binding energy of the A or B curves, then the triplet channel might be accessible only at wavelengths appreciably shorter than the threshold for

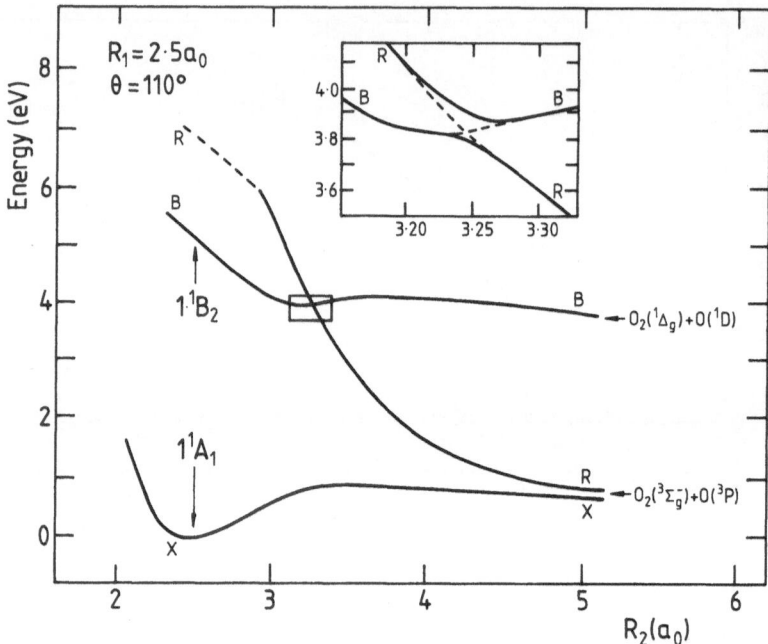

Fig. 8. Potential energy curves, obtained by *ab initio* calculation [40] for states of ozone involved in the photodissociation by ultraviolet radiation. The curves are cross sections through the potential energy surfaces in which one bond length ($R_1 = 2.5$ Bohr) and the bond angle ($\theta = 110°$) are held fixed

the singlets. That is, $\phi(^1D)$ would, for potential energy curves of this kind, decrease at wavelengths shorter than the threshold.

Experimental confirmation of the curve crossing process has been obtained by Valentini and co-workers [105, 106] who have examined rotational quantum state distributions in the $O_2(^1\Delta_g)$ fragment of the Hartley band photodissociation of ozone. The experiments themselves, and their interpretation, are discussed in the section on photodissociation dynamics. It suffices to say here that an anomalous distribution is observed, with an apparent propensity for even-J levels in the $O_2(^1\Delta_g)$. This result is explained in terms of a depopulation of the odd-J levels through the curve-crossing process described in the last paragraph. Not only do these experiments lend considerable weight to the curve-crossing hypothesis, but they also provide a way of determining the quantum yield for the triplet channel, by summing all the even-J and odd-J populations. Table 4 shows the quantum yields derived [106] (although data exist for $\lambda < 266$ nm, there are potential difficulties in interpretation that argue against inclusion of the quantum yields in the table). Two assumptions are made in the calculations: first, that the total dissociation quantum yield is unity at all wavelengths; and, secondly, that the *only* dissociation channels are to the singlet pair $O(^1D) + O_2(^1\Delta_g)$ or the triplet pair $O(^3P) + O_2(^3\Sigma_g^-)$. With these assumptions, the yield in the triplet channel is identical to $\phi(^3P)$. The variation in $\phi(^3P)$ over the wavelength range is stated [106] not to be statistically significant, since the accuracy of the measurements is ± 0.03. These results thus do *not* fit in

Table 4. Quantum yields for $O(^3P) + O_2(^3\Sigma_g^-)$ formation in the photodissociation of ozone derived from the rotational quantum state populations [106]

Wavelength nm	Quantum yield for triplet channel
266	0.15
280	0.08
291	0.11
293	0.14
300	0.06
305	0.10
309	0.12
311	0.11

with a value of $\phi(^1D)$ that is unity at wavelengths near the energetic threshold, but which decreases at shorter wavelengths. Rather, they argue for a constant branching ratio for the singlet channel of 0.89. In fact, the apparent constancy of the values of the triplet yield at the longest wavelengths is itself rather puzzling, since most other studies indicate values of $\phi(^3P)$ substantially above 0.11 for $\lambda > 308$ nm. One likely explanation is that there is an additional, spin-forbidden, dissociation channel in the long wavelength tail of the Hartley absorption band. This idea will be explored further in the next section.

Without a better knowledge of the identity and characteristics of the repulsive R surface, further speculation is probably pointless. However, the real values for the quantum yields at wavelengths near 300 nm are of vital importance for quantitative atmospheric modelling. The current NASA recommendation [47] is explicit in scaling the quantum yields to a limiting value of $\phi(^1D)$ of 0.9. Some modellers continue, nevertheless, to assume a limit of unity. There is clearly an urgent need to find out if there really is an increase in $\phi(^1D)$ from 0.9 to almost 1.0 as the threshold wavelength is approached from the *short* wavelength side.

The Molecular O_2 Fragment of O_3 Photolysis

Although there is enough energy for $O_2(^1\Delta_g)$ to be a product of the photolysis of ozone in the Chappuis region, all the available evidence indicates that excited products are not formed in the visible-region photolysis. That is, spin conservation (pp. 22–23) determines the pathway of photolysis. Static photolysis [79, 80] gives a quantum yield for ozone loss of about two, as expected if neither the atomic nor the molecular fragment is excited. Time-of-flight photofragment spectroscopy [99] shows no evidence of excitation for photolysis at $\lambda = 600$ nm, and Coherent Anti-Stokes Raman Spectroscopy (CARS: see p. 40) of the molecular product [107, 108] shows no detectable $O_2(^1\Delta_g)$ in the photolysis of ozone by visible radiation.

Considerable evidence from atmospheric studies had accumulated by the late 1960s to indicate that excited singlet molecular oxygen, $O_2(^1\Delta_g)$, is a product of the ultraviolet photolysis of ozone. Laboratory experiments, described in detail else-where [10, 15, 78], show explicitly that $O_2(^1\Delta_g)$ is the molecular fragment of ozone photolysis in the ultraviolet. Izod and Wayne [109] first detected emission at $\lambda = 1.27 \ \mu m$ from the products of ozone photolysis, although they believed that O_2 was needed as well to generate the singlet molecular oxygen. Huffman et al. [110] observed vacuum UV absorption bands, assigned to $O_2(^1\Delta_g)$, in ozone irradiated at $\lambda = 253.7$ nm. Gauthier and Snelling [111, 112] attributed optical emission at $\lambda = 1.27 \ \mu m$ seen on photolysis of ozone at $\lambda = 253.7$ nm to $O_2(^1\Delta_g)$ formed in the primary step. Although $O_2(^1\Sigma_g^+)$ could be produced at the photolytic wavelengths used (see Table 3), a limit of less than five percent is placed on its excitation efficiency [112]. Absolute calibration for $O_2(^1\Delta_g)$ emission intensities allowed Jones and Wayne to show [113] that the species was formed with an efficiency of 0.83 ± 0.11 at $\lambda = 253.7$ nm.

Other experiments confirm that $O_2(^1\hat{\Delta}_g)$ is a product of photolysis of ozone by ultraviolet radiation. The photofragment spectroscopy experiments [99, 100], described in the last section, show energy releases consistent only with the formation of the excited singlet oxygen. Furthermore, the coherent Raman studies [105, 106] directly examine the $O_2(^1\Delta_g)$ molecule, and are able to resolve detailed vibrational and rotational structure.

Both the photofragment [99, 100] and CARS [105, 106] results suggest, of course, not only that $\phi(O^1D)$ is less than unity, but also that the quantum yield for $O_2(^1\Delta_g)$ production will be less than one in the photolysis of ozone. Jones and Wayne [113] took their measurement of 0.83 ± 0.11 to indicate that $\phi(^1\Delta_g)$ approached unity, but it now seems that the result should be taken at face value. The significance of this conclusion for atmospheric studies is that the production rates of $O_2(^1\Delta_g)$ in the atmosphere must be scaled by the primary quantum yield. One application, for example, where a value of $\phi \simeq 0.8 - 0.9$ might have an impact is in the derivation of ozone concentrations from airglow intensities in the infrared atmospheric band, $O_2(^1\Delta_g \rightarrow {}^3\Sigma_g^-)$, as applied to data from the SME infrared instrument [2, 3].

Two experimental facts upset the neat picture of thermodynamically-controlled spin-conserved photolysis of ozone that was presented at the beginning of the discussion of photodissociation. The first concerns the absolute values of $\phi(O^1D)$ at wavelengths shorter than the $\lambda = 310$ nm threshold. The second of these results concerns the production of $O_2(^1\Delta_g)$ at wavelengths longer than the $\lambda = 310$ nm threshold. The evidence suggests [82, 83, 114] that $O_2(^1\Delta_g)$ formation remains efficient at $\lambda = 334$ nm, even though production of $O(^1D)$ no longer occurs. Line (c) in Fig. 4 shows some quantum yield data [82, 83] for photolysis at $\lambda = 334$ nm. The slope is near zero, but the intercept is just below four, almost the same as at the shorter wavelength experiments described by lines (a) and (b). Apparently, $O(^1D)$ is no longer formed at $\lambda = 334$ nm, but $O_2(^1\Delta_g)$ continues to be formed with an efficiency of about 0.9. That is, the dissociation appears to be spin forbidden in this weak absorption region.

Spin- "forbidden" processes are a consequence of spin–orbit coupling that makes S an imperfect description of the quantum state. Optical transitions can thus occur with $\Delta S \neq 0$, even though they may be much weaker than fully allowed transitions.

Spin-forbidden radiationless transitions are also possible, so that the rigour with which a (pre-)dissociation has to retain its spin state is relaxed. Figure 3 shows that the fragments $O(^3P) + O_2(^1\Delta_g)$ correlate with the 1^3B_1 and 2^3B_2 state of ozone, and the electronic states themselves are described in the section on electronic structure and excited states. Table 2 shows that the vertical excitation energy of the 2^3B_2 state of O_3 lies at about 3.27 eV, corresponding to an onset of absorption at $\lambda = 379$ nm. Direct, but forbidden, population of the 2^3B_1 state might therefore be achieved by the weak optical absorption at $\lambda = 334$ nm. An alternative route to the photolytic fragments that correlate with triplet states of ozone would, of course, be optical absorption in the singlet system, and (forbidden) crossing to the triplet.

Valentini et al. have considered [106] the rotational quantum state populations in the $O_2(^1\Delta_g)$ product of ozone photolysis at the long wavelength end of the Hartley continuum, as described at the end of the last section. They conclude that there is probably a dissociation channel additional to the $O(^1D) + O_2(^1\Delta_g)$ singlet pair. However, the CARS spectra show no evidence for the appearance of a $O(^3P) + O_2(^1\Delta_g)$ channel, which should be accompanied by changes in rotational and vibrational distributions at the longer wavelengths ($\lambda = 308–311$ nm in these experiments). An alternative possibility for the additional channel is that the O_2 is formed in the second, $b^1\Sigma_g^+$ state, for which, unfortunately, a search was not made. The quantum yield for a $O(^3P) + O_2(^1\Sigma_g^+)$ channel would have to be as high as 0.56 at $\lambda = 311$ nm to explain the observations [106]. The overall quantum yield measurements at $\lambda = 334$ nm [82, 83, 114] certainly do not distinguish between $O_2(a^1\Delta_g)$ and $O_2(b^1\Sigma_g^+)$ as the source of additional decomposition of ozone, but, equally, it does not necessarily follow that the same state is excited at $\lambda = 311$ nm and at $\lambda = 334$ nm. What is clear, however, is that there is experimental support for the idea of a singlet molecular fragment being formed along with a triplet atom in

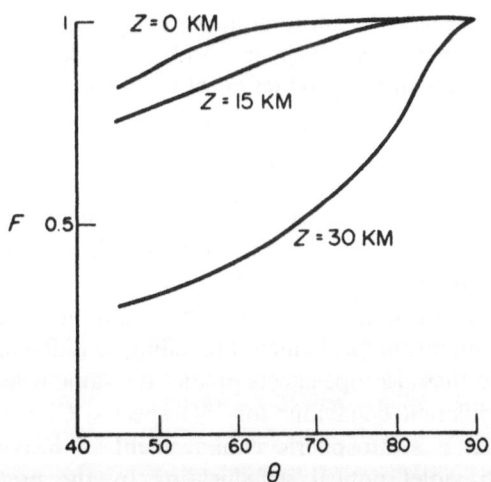

Fig. 9. Fraction F of $O_2(^1\Delta_g)$ that is produced by photolysis of ozone at wavelengths longer than 310 nm as a function of solar zenith angle. The values are shown at three different altitudes, and are calculated for Spring [26]

the tail of the Hartley band. From the point of view of atmospheric chemistry, it may not even matter which singlet molecular state is formed, since at altitudes below about 70 km quenching by N_2 is the most probable fate of $O_2(^1\Sigma_g^+)$, and the product of quenching is believed to be $O_2(^1\Delta_g)$ [10].

The direct or indirect production of $O_2(^1\Delta_g)$ at $\lambda > 310$ nm needs to be considered if IR Atmospheric band intensities are to be used to estimate ozone concentrations in the lower stratosphere and below [26]. Long-wavelength photolysis is likely to show the largest influence for large solar zenith angles and at low altitudes. Figure 9 gives one way of displaying the data: here, the fraction of $O_2(^1\Delta_g)$ produced in the "forbidden" region is shown as a function of solar zenith angle for three altitudes under irradiance typical of spring. Even at an altitude of 30 km, for zenith angles greater than about 70° half the $O_2(^1\Delta_g)$ is derived from the spectral region not usually included. It is evident that disregard of the contribution made by long wavelength photolysis is unjustifiable, even though the scaling of the $O_2(^1\Delta_g)$ yield in the "allowed" region, needed for the reasons presented at the beginning of this section, may partially compensate for the shortfall.

Isotope Effects in Ozone Photochemistry

The possible existence of stratospheric ozone enriched in the isotope ^{18}O was suggested [115] by Cicerone and McCrumb in 1980. Subsequent *in situ* balloon-borne mass-spectrometric measurements showed that stratospheric ozone is indeed enriched in ^{18}O. The mass 50 isotopomer of O_3 was enriched by up to 40% at altitudes of around 32 km [116, 117], although the enrichment was much less at higher or lower altitudes. Ground-based infrared spectroscopic observations [118] confirm an isotopic effect, with ^{18}O enrichments in O_3 of 1.05–1.11 averaged over the entire atmospheric ozone column. We consider in this section the relationship that laboratory studies have to interpretation of this atmospheric phenomenon.

A series of experiments by Thiemens and co-workers have demonstrated interesting isotope effects in the production of O_3 from O_2 by electric discharges [119–121], by ultraviolet light [122], and in the thermal or photolytic decomposition of O_3 [123]. Particularly noteworthy is the observation that the isotopic fractionation for ^{17}O and ^{18}O is similar ('mass independent') in all cases, apparently in accord with the latest stratospheric ozone measurements [117].

Normal chemical isotope effects result from changes in equilibrium positions and rate processes that are consequent on the change of molecular energy levels following isotopic substitution, although some dramatic kinetic isotope effects depend on differing quantum mechanical tunnelling of different mass isotopes. In no case, however, do these isotope effects predict the same behaviour for ^{17}O and ^{18}O species but a different behaviour for ^{16}O species. Cicerone and McCrumb [115] suggested that a stratospheric enhancement of heavy O_3 would be a consequence of ultraviolet optical self-shielding by the predominant $^{16}O^{16}O$ oxygen isotope. Preferential $^{16}O^{18}O$ dissociation might then lead to an enrichment of ^{18}O atoms, and thus of $^{50}O_3$. Such a shielding mechanism would presumably apply also to enhanced ^{17}O formation, and thus provide a mass-independent fractionation. However, the rate of isotopic exchange between ^{18}O (or ^{17}O) and $^{16}O^{16}O$ in the stratosphere was shown subsequently [124] to prevent any substantial enrichment of heavy ozone. Later interpretations of the laboratory experiments [120–123] have focused attention on different rates of ozone decomposition brought about by difference in the molecular symmetry of the ozone, although there has also been some discussion [125–127] of the influence of symmetry effects on the rate of formation of O_3 in the recombination of O with O_2.

None of the explanations presented so far for mass-independent enhancement of O_3 in the atmosphere or in the laboratory is entirely satisfactory. It is, therefore, worth examining a different hypothesis put forward by Valentini [128] that has its origins in an *observed* isotope effect in the photodissociation of ozone. As described briefly on p. 28, there is a propensity [105, 106] for even-J rotational quantum states in the $O_2(^1\Delta_g)$ fragment of photodissociation in the Hartley continuum, which is ascribed to a preferential depopulation of the odd-J levels. A fuller explanation of the phenomenon is deferred to the section on nascent product distribution. But note here that it is associated with the nuclear-spin statistics of ground state O_2 that exclude even-J levels. Since the nuclei in $^{18}O^{16}O$ or in $^{17}O^{16}O$ are not equivalent, there are no symmetry restrictions, in agreement with the observed absence [106] of a propensity for odd or even-J in $^{18}O^{16}O(^1\Delta_g)$. The arguments used to explain the propensity effects in photodissociation (pp. 40–48) can be extended to the collisional quenching of $O_2(^1\Delta_g)$, and can be used immediately to interpret the isotopic behaviour in the formation of O_3 from O_2 through which an electric discharge is passed. About 10% of $O_2(^1\Delta_g)$ is formed in the discharge [10] and non-adiabatic collisional relaxation will lead to mass-independent enrichment of ground state $^{17}O^{16}O$ and $^{18}O^{16}O$ and a concomitant depletion of the heavy isotopes in the $O_2(^1\Delta_g)$. Recombination of $O(^3P)$ with $O_2(^1\Delta_g)$ is likely to be inefficient since it is spin forbidden, so that only the heavy-isotope enriched ground state molecules are available for ozone formation. Detailed calculations [128] show that the experimental observations [120] of heavy-ozone enrichment can be quantitatively mimicked.

Heavy-ozone enrichments in the stratosphere might reflect a persistence of the isotopic fractionation between the $O_2(^1\Delta_g)$ and $O_2(^3\Sigma_g^-)$ fragments of ozone photodissociation, thus explaining the enrichment in the same way as the enrichment of the products of electric discharges. Unfortunately, the enrichment that could be achieved in the stratosphere in this way would be limited [128] to 10^{-5} in an oxygen-only mechanism. What is required is some way of capturing the isotopic fractionation generated by the selective heavy isotope depletions in $O_2(^1\Delta_g)$ generated photolytically from ozone. One suggestion put forward by Valentini [128] is that a small proportion of the $O_2(^1\Delta_g)$ could become incorporated in atmospheric CO_2, from which the return to other oxygen-containing species would be extremely slow in the stratosphere. A possible route starts with the process

$$O_2(^1\Delta_g) + HO_2 \rightarrow O_2(^3\Sigma_g^-) + HO_2 \qquad (25)$$

which is known [10] to be quite efficient. It is assumed by Valentini that 1% of the quenching of $O_2(^1\Delta_g)$ could involve H-atom transfer, and thus the formation of HO_2 with depletion in the heavier oxygen isotopes. Two further reactions exemplify the production of CO_2

$$HO_2 + NO \rightarrow OH + NO \tag{17}$$

$$OH + CO \rightarrow CO_2 + H \tag{21}$$

Overall, it is estimated [128] that about 1 in 10^7 of the $^{17}O, ^{18}O$-depleted $O_2(^1\Delta_g)$ molecules will produce CO_2, leading to an enrichment in the heavier isotopes in O_3. It is noted that the rapid interconversion of O, O_2, and O_3 ought to mean an enrichment in the heavier isotopes of all these species in the stratosphere, and a corresponding depletion of ^{17}O and ^{18}O in CO_2.

Vibrational Photochemistry of Ozone

Vibrationally excited ozone in the atmosphere is of interest for several reasons [129, 130]. Vibrationally excited molecules can produce a red shift in the ultraviolet absorption spectrum ozone, they can show enhanced rates of photodissociation and reaction with other molecules, and they can produce infrared airglow from the atmosphere. Chemical reactions of vibrationally excited ozone are likely only in the mesosphere and above [129], but changes in absorption cross section and in quantum yield for $O(^1D)$ formation might be important at lower altitudes, especially for photolysis in the threshold wavelength region.

The Hartley and Huggins absorption bands of ozone show distinct components due to vibrationally excited states, and hot bands can readily be identified in high resolution spectra [36, 60]. The temperature dependence of the absorption in the Huggins region, referred to in the section on absorption cross sections, results in part from the participation of these hot bands (as well as from rotational effects). Direct infrared laser excitation can lead to changes in the UV absorption spectrum [94, 131]; absorption of solar infrared radiation by atmospheric ozone could similarly affect the ultraviolet absorption. Ozone in the atmosphere is produced by the recombination of atomic oxygen with molecular oxygen

$$O + O_2 + M \rightarrow O_3 + M \tag{8}$$

We discussed on p. 19 the possibility that electronically excited states could be formed in this reaction. Most of the laboratory evidence, however, suggests that the newly formed ozone is vibrationally excited [132] and that it displays a modified UV spectrum [133]. Joens et al. [132] inferred initial formation of ozone with excitation in the stretching modes, followed by kinetically-limited collisional energy transfer to excite the bending mode.

The contribution of excited ozone to photolysis in the atmosphere will obviously depend on vibrational relaxation rates. There is, however, direct laboratory evidence that shows vibrational excitation to be effective in shifting the $\phi(^1D) - \lambda$ curve to longer wavelengths. A two-laser experiment was used [134] to demonstrate that, beyond the $\lambda = 310$ nm threshold, the cross section for $O(^1D)$ production increases by nearly two orders of magnitude on excitation of $O_3(\nu_3)$. The effect of vibrational excitation drops with increasing photon energy, a result expected because internal energy needs to make a smaller contribution to the total dissociation energy. The increased photodissociation cross section is not solely attributable to the increased absorption cross section: there seems to be a real enhancement of the quantum yield for $O(^1D)$ production by a factor of about 6.

Infrared chemiluminescence from vibrationally excited ozone formed in the recombination reaction was first observed by von Rosenberg and Trainor [72], who reported on the production and quenching of emission in the v_3, $v_1 + v_3$, and v_2 bands. The experiments were carried out at relatively high pressures at which extensive V→T and V→V intermode energy transfer could occur. More recently, Rawlins and co-workers [130, 135] have studied the dynamics of formation of vibrationally excited ozone in reaction (8) by using a low-pressure cryogenic reactor with the acronym COCHISE (cold chemiexcitation infrared stimulation experiment). Spectrally resolved infrared chemiluminescence at wavelengths near 10 μm (the v_3 fundamental band) was observed with a resolution of 0.027 μm. The spectra are consistent [130] with $\Delta v_3 = 1$ transitions from $(00v_3)$ and $(10v_3)$ vibrational levels with v_3 up to 5. Population of v_2 by collisional energy transfer does not appear to be significant, and levels in v_1 with $v_1 > 1$ are not observed. Thus, although the COCHISE experiments do not probe the true nascent distribution, they do indicate that the recombination-deactivation sequence is mode-selective. It seems [135] that the newly-formed ozone molecules are deactivated so rapidly that they retain a memory of the configuration of the approach of the O and O_2 reactants. Indeed, the absence of $v_3 > 5$ itself argues [135] for rapid deactivation of higher vibrational states, since the $v_3 = 5$ cutoff corresponds to about 58% of the recombination energy, and the O_3 potential well could accommodate up to eight or nine bound vibrational levels [130].

The presence in the upper atmosphere of vibrationally excited ozone has been established by several rocket-borne experiments that have observed infrared emission in the v_3 band [129]. For example, the HIRIS (high resolution interferometer spectrometer) has been used to obtain high resolution emission during an aurora [136], and the SPIRE (spectral infrared rocket experiment) circular-variable filter spectrometer has made possible observations of the diurnal variations of the emission in the quiescent atmosphere [137]. In these latter experiments, deductions were made of the local excited vibrational state concentrations as functions of altitude and solar illumination in the altitude region 70–105 km. Radiative and collisional energy transfer processes

$$O_3 + hv \rightarrow O_3(v_3 = 1) \tag{26}$$

$$O_3 + M \rightarrow O_3(v_3 = 1) + M \tag{27}$$

account [138] for much of the excitation, with the photons in reaction (26) being supplied by black-body emission from the Earth. However, the HIRIS and SPIRE data both exhibit significant non-thermal excitation of higher vibrational levels, and, in view of the evidence of the laboratory studies described in the last paragraph, it seems reasonable to ascribe the additional excitation of O_3 to the recombination of O with O_2 in reaction (8). The data from the rocket flights show [129, 136, 137] that the effective temperature of excitation of the levels with $v_3 > 1$

increases from ca. 500 K at night to ca. 2000 K in the daytime, and these results are shown [129, 137] to be consistent with the recombination mechanism. A dramatic enhancement of emission intensity was observed in the auroral measurements [136], and the relative vibrational population distribution was significantly more excited than in the non-auroral emission. Some additional processes that can excite $O_3(v_3)$ have been identified in the laboratory COCHISE studies [135]. These include V→V energy transfer from molecular oxygen

$$O_2(v=2)+O_3 \rightarrow O_2+O_3(v_3=2) \tag{28}$$

resonant E→V excitation from $O_2(b^1\Sigma_g^+)$

$$O_2(b^1\Sigma_g^+)+O_3 \rightarrow O_2(a^1\Delta_g)+O_3(v_3=5) \tag{29}$$

and possibly electron impact excitation

$$O_3+e \rightarrow O_3(v_3=1)+e \tag{30}$$

All three processes are plausible candidates for excitation of $O_3(v_3>0)$ in an aurora. The E→V process (29) is particularly interesting in that it can excite the highest vibrational level of O_3 observed in the laboratory experiments. It is necessary, to achieve resonant transfer, that $O_2(b^1\Sigma_g^+)$ be quenched to $O_2(a^1\Delta_g)$. This is exactly the behaviour demonstrated explicitly for the quenching of $O_2(b^1\Sigma_g^+)$ by many partners [139]. Although two-thirds of the interaction between $O_2(b^1\Sigma_g^+)$ and O_3 involves chemical change [10]

$$O_2(b^1\Sigma_g^+)+O_3 \rightarrow 2O_2+O \tag{31}$$

it is believed that the remainder proceeds via physical deactivation [140].

Solomon et al. [141] discuss further evidence for non-equilibrium excitation of the v_3 mode of mesospheric ozone, and some of the likely consequences. Enhanced chemical reactivity of vibrationally excited O_3 has been established in a number of cases that might be of atmospheric importance, such as interactions with O [135, 142], $O_2(^1\Delta_g)$ [135, 143], and NO [144–148]. Steinfeld et al. [35] review the information available. For the reaction with atomic oxygen [142], the rate constant for removal of vibrationally excited $O_3(100,001)$ is 1.5×10^{-11} cm^3 molecule^{-1}s^{-1}, or more than three orders of magnitude more rapid than the rate constant for *reaction* between O and O_3 at room temperature. West et al. [142] believe that more than 70% of the interaction with vibrationally excited ozone is non-reactive V→T deactivation. On the other hand, an analysis of the vibrational distributions observed in the COCHISE experiments [135] leads to the conclusion that chemical reaction

$$O+O_3(v_3>0) \rightarrow O_2+O_2 \tag{32}$$

itself possesses a rate constant in excess of 10^{-11} cm^3 molecule^{-1}s^{-1}, and that the rate constant increases with increasing v_3. Rawlins et al. point out [135] that this reaction could be energetic enough to populate metastable excited states of O_2 such as the $A^3\Sigma_u^+$, $A'^3\Delta_u$, and $c^1\Sigma_u$, for which atmospheric sources are not yet clearly identified [10, 15]. The mechanism and product channels of this reaction are clearly worthy of further attention. Rawlins et al. also find indirect evidence for the enhancement of the reaction between $O_2(^1\Delta_g)$ and vibrationally excited ozone

$$O_2(^1\Delta_g) + O_3(v_3 > 0) \rightarrow 2O_2 + O \tag{33}$$

in the anomalously large loss rates they observe for $O_3(v_3 \geqslant 2)$ in the COCHISE experiments. Kurylo et al. [143] have demonstrated directly that vibrational excitation in O_3 enhances the rate of the reaction. A CW CO_2 laser operating on the P(30), $\bar{v} = 1034 \text{ cm}^{-1}$, line was used to excite the v_3 mode in O_3. Although it was believed probable that energy redistribution would occur before reaction, this excitation was shown to lead to increased atomic oxygen production in the reaction. Rate enhancements of 38 ± 20 were obtained, which correspond to effectively 100% utilization of vibrational energy in the active modes. Such efficient use of one vibrational quantum is consistent with a stepwise mechanism for decomposition of ozone that first involves E–E energy transfer [143], probably to excite the $^3B_2$ state of O_3 [10]

$$O_2(^1\Delta_g) + O_3(v_3 > 0) \rightarrow O_3(^3B_2) + O_2 \tag{34}$$

followed by dissociation of the electronically excited O_3. Parker [149] has proposed a similar scheme for the reaction with *unexcited* O_3 (reaction (24)) to account for the detailed time evolution of $O_2(^1\Delta_g, v' = 0,1)$ in laser-pumping experiments.

This section will have shown that there are several outstanding problems in the chemistry of vibrationally excited ozone. The experiments of Rawlins et al. [130, 135] are specifically designed to help in the interpretation of the mesospheric and thermospheric processes that involve the species, and testing of the concepts and models that they have presented is eagerly awaited [135].

Photodissociation Dynamics

The Goals of Chemical Dynamics

A recurring topic in this chapter has been the formation of electronically excited oxygen atoms and molecules in the photodissociation of ozone. These photolytic products are clearly not in thermal equilibrium with their surroundings. In a more general way, the vibrational, rotational, and even translational energies and distributions of the photofragments may be in disequilibrium. The energy disposal in the products of reaction, and the alignment or orientation of the products themselves, provide valuable clues about the detailed dynamics of the interaction. In turn, an understanding of the chemical dynamics at the molecular level provides the key to interpretation of the macroscopic behaviour of the bulk system [150]. A knowledge of the microscopic, molecular mechanism then allows rationalization of the observed bulk photochemistry, and also permits informed prediction of the properties of the system for conditions that have not been tested experimentally. The present section of this article provides an outline of the progress that has been made in elucidating the dynamics of the photodissociation of ozone.

Reactants turn into products in a chemical reaction through an evolving intermediate species [6]. Chemical dynamics aims to characterize the forces bringing about chemical change by investigating reaction paths in which both the initial and the final conditions are as well-defined as possible [150]. In a full reactive collision, the initial conditions can sometimes be arranged so that the energy spread and partitioning in the reactants is under the experimentalist's control. Even so, a wide range of impact parameters and of orientations in the reaction partners is usually unavoidable. Photodissociation, on the other hand, represents a "half-collision" in which the dynamics are started off within the interaction region, and the fragments are observed once the interaction is over, or, better still, the intermediate is investigated as it evolves into the fragments. The absorption of a photon causes the transference of the ground-state vibrational wave-function, which is often well characterized, to the excited-state potential energy surface. Thus the dynamic study of photofragmentation is particularly fruitful, because the ability is gained to select a subset of initial conditions that are as well-defined as possible, within the limits of the Uncertainty Principle [6].

Figure 10 illustrates what happens after the ground-state wave function is transferred to a dissociating upper electronic state [6]. The wave function is not an eigenfunction of the excited state, so that it becomes a moving wave packet on the

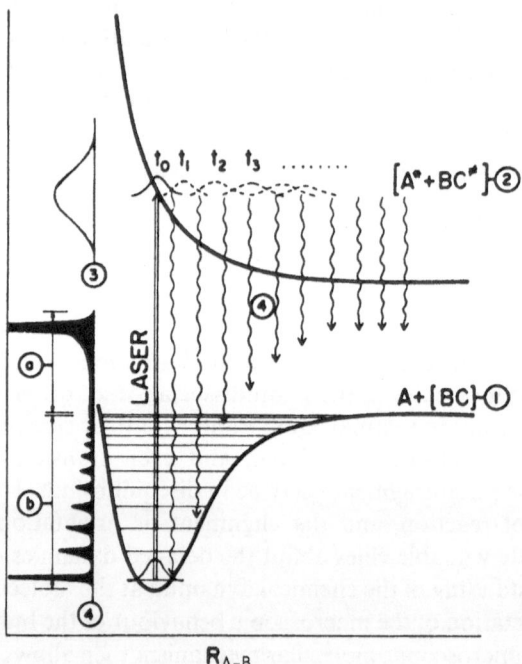

Fig. 10. Generalized view of photodissociation. The ground-state wave function is transferred to the upper surface at $t = 0$, for example by a laser pulse. Dissociation proceeds *via* the evolving packets shown at $t = 1$, $t = 2$, etc. The numbers in circles show the types of experiment, discussed in the text, that can be used to probe the dissociation. Reproduced, with permission, from Imre et al. [6]

upper surface whose evolution in time is the dynamic process of interest. The numbers in circles on the figure show the types of experiment that can be used to probe the dissociation dynamics. Long-term dynamics (on the time scale of tens of femtoseconds or more) can be probed by investigating the structure and spectroscopy of the final products (circle 1), to give information about the shape of the potential at infinite internuclear separation. More detail is provided by the velocities, internal states, and angular distributions of the products (circle 2), which reflect the cumulative history of the system. Absorption spectroscopy (circle 3) yields information about the behaviour in the 'Franck–Condon' region (corresponding to very short times of the order of femtoseconds), while emission spectroscopy (circle 4) can probe the dynamics as the fragments separate, but while they are still close enough for there to be appreciable interactions between them. All these techniques have been applied to the problem of ozone photochemistry. Absorption spectroscopy is essential in understanding the excited state first reached. The question of the electronic states of the products has been the subject of earlier sections and we have seen how more detail about distribution between the different internal and translational modes can be provided by time-of-flight photofragment spectroscopy. One problem in studying dynamics is that intermolecular collisions redistribute initial energies, driving the system ultimately towards thermal equilibrium. What is of interest, from a dynamic point of view, is the nascent energy distribution. Experiments on the microsecond time-scale are able to probe at least the electronic excitation in the fragments of ozone photolysis, because electronic deactivation is *relatively* inefficient. But to achieve a more intimate picture of the disposal of energy in the nascent products, experiments must be performed in collision-free conditions, either by using low enough pressures (as in the photofragment spectroscopy experiments), or by sufficiently reducing the time scales of the probe.

Nascent Product Distributions

Even relatively modest experiments have been used to look at the products on a time-scale short enough that vibrational relaxation has not wiped out more detailed information. Klais et al. [151] were able to use flash photolysis, followed by absorption in the vacuum ultraviolet region, to study vibrationally excited $O_2(^1\Delta_g)$. The experiments not only provide further confirmation that $O_2(^1\Delta_g)$ is a primary product of photolysis, but also give an idea of the partitioning into the accessible vibrational levels. Time-of-flight photofragment spectroscopy provides even more detailed information. For example, Sparks et al. [100] use the product energy information in conjunction with polarization dependence data to show that the relative yields of photolysis at $\lambda = 266$ nm into $v' = 0$, 1,2 and 3 of $O_2(^1\Delta_g)$ are 0.57, 0.24, 0.12, and 0.07. The exact position of the $v' = 0$ peak indicates that about 17 percent of the energy remaining after production of this level is deposited in rotational excitation, a result in accord with a model in which energy is released inpulsively from a geometry near that of the ground state of ozone (see next section). Fairchild et al. [99] investigated, in addition to the total energy release, the angular distribution of the fragment species, for dissociation both in the ultraviolet and the red regions of the spectrum. They show how the experimental data are consistent with the orientation of the transition dipoles in the dissociating molecule with respect to the electric vector of the photolytic laser radiation. The data are also used to explain the width of the time-of-flight spectra according to the model of dissociation.

Coherent Anti-Stokes Raman Spectroscopy (CARS) has been used in studies by Valentini and co-workers on the dissociation of ozone in the Chappuis [39, 107] and Hartley [105, 106] absorption regions. CARS is a multiphoton spectroscopy dependent on non-linear effects resulting from the use of high intensity lasers. Very high time resolution is inherent in the technique, because the time scale is that of the Raman process itself, and the gathering of data under collision free conditions is much simpler than with absorption or infrared emission spectroscopy [107]. In the experiments described, total pressures are about 2 torr, and the mean time between collisions is thus about 50 ns. The probe laser pulses are delayed from the photolysis pulse by only 4 ns [39] or 1 ns [106], so that they probe the true nascent molecular oxygen photofragments. The CARS technique is of quite general applicability, since optically accessible electronic states are not necessary in Raman spectroscopy. That means that ground-state molecular oxygen, $O_2(^3\Sigma_g)$, can be investigated in CARS studies of ozone photolysis as well as $O_2(^1\Delta_g)$.

The dissociation by radiation in the visible region of the spectrum is a good starting point for our discussion, since complications associated with the photo-fragments in more than one electronic state are absent. Figure 11 shows [39] the vibrational Q-branch of the CARS spectrum of O_2 formed from the photolysis of O_3 at nine wavelengths between 560 and 638 nm. Each spectrum shows five bands associated with $v = 0, 1, 2, 3,$ and 4 in $O_2(X^3\Sigma_g^-)$ at 1530, 1510, 1490, 1465, and 1450 cm^{-1}. There is no evidence of CARS transitions due to $O_2(a^1\Delta_g)$, and the yield of $^1\Delta_g$ oxygen in the Chappuis absorption region can thus be shown to be less than 0.5% (see pp. 29–32).

The photofragment quantum flux distributions, normalized to $v = 0$, are shown in Fig. 12 for photolysis at $\lambda = 572$ nm. Almost identical distributions are found at the other wavelengths, as the similarity of the spectra in Fig. 11 indicates. Figure 13 illustrates this point by plotting the vibrational flux distributions for all nine photolysis wavelengths. The distribution remains virtually the same, even though at $\lambda = 638$ nm $v = 4$ is the highest accessible vibrational level, while at $\lambda = 560$ nm, $v = 6$ could be populated. Note the small population inversion between $v = 3$ and $v = 2$ at all wavelengths. The constancy of the distribution indicates that the dissociation is vibrationally adiabatic, with the exit channel dynamics being involved only in the partitioning of energy between rotational and translational modes. Simple kinematics explains [39] the vibrational adiabaticity. Because the equilibrium bond angle in O_3 is quite small (ca. 117°: see note b, Table 1), the repulsive impulse between the photofragments has a small projection along the O–O bond in the O_2 product.

Only about 5% of the available energy could appear as vibration even if the interaction were purely impulsive [107]. Appreciable rotational excitation is expected to follow repulsion of O from O_2 because of the small bond angle, and

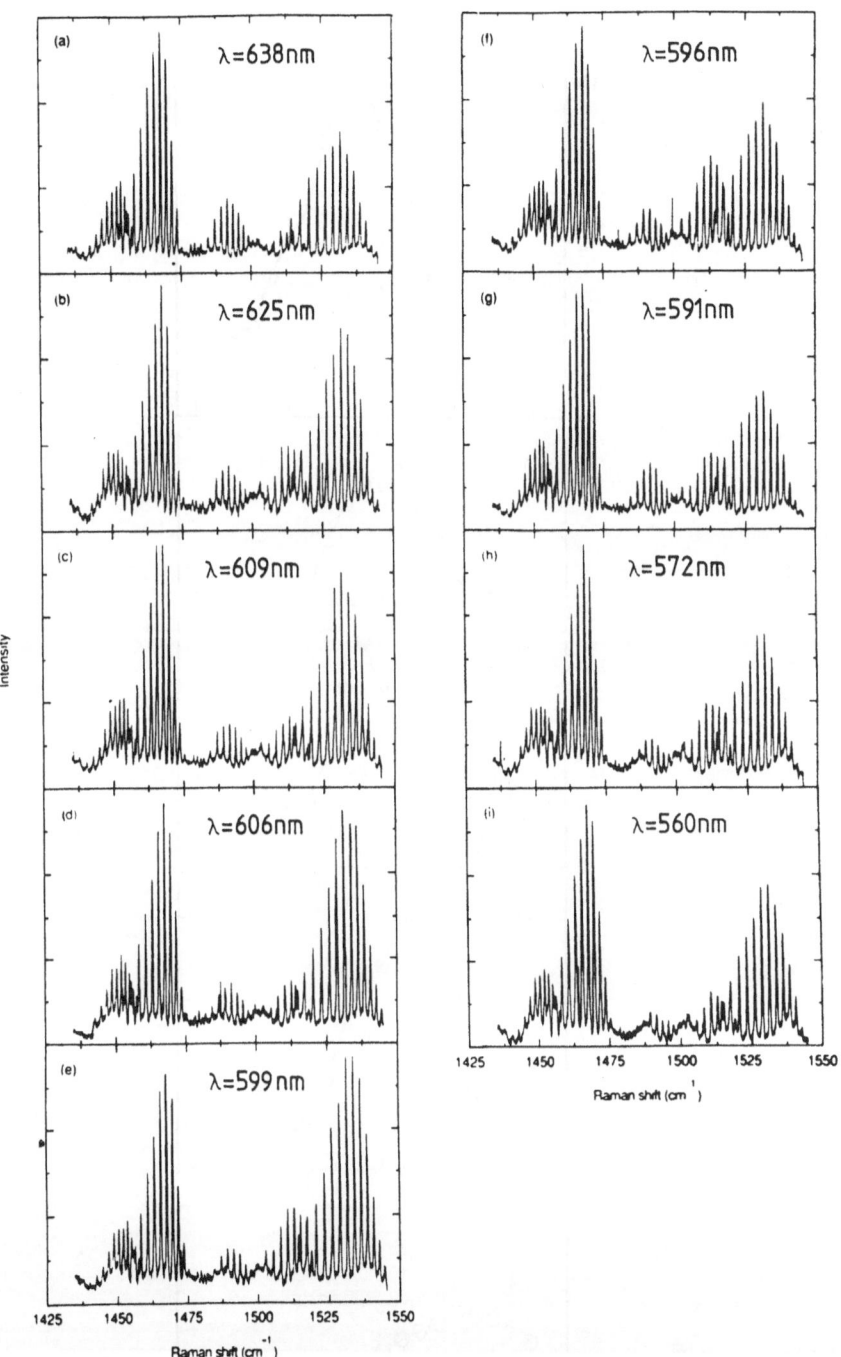

Fig. 11. Coherent Raman anti-Stokes spectrum (vibrational Q branch) of $O_2(^3\Sigma_g^-)$ produced in the photolysis of ozone by visible radiation: (a) $\lambda = 638$ nm; (b) $\lambda = 625$ nm; (c) $\lambda = 609$ nm: (d) $\lambda = 606$ nm; (e) $\lambda = 599$ nm; (f) $\lambda = 596$ nm; (g) $\lambda = 591$ nm; (h) $\lambda = 572$ nm; (i) $\lambda = 560$ nm. Reproduced, with permission, from [39]

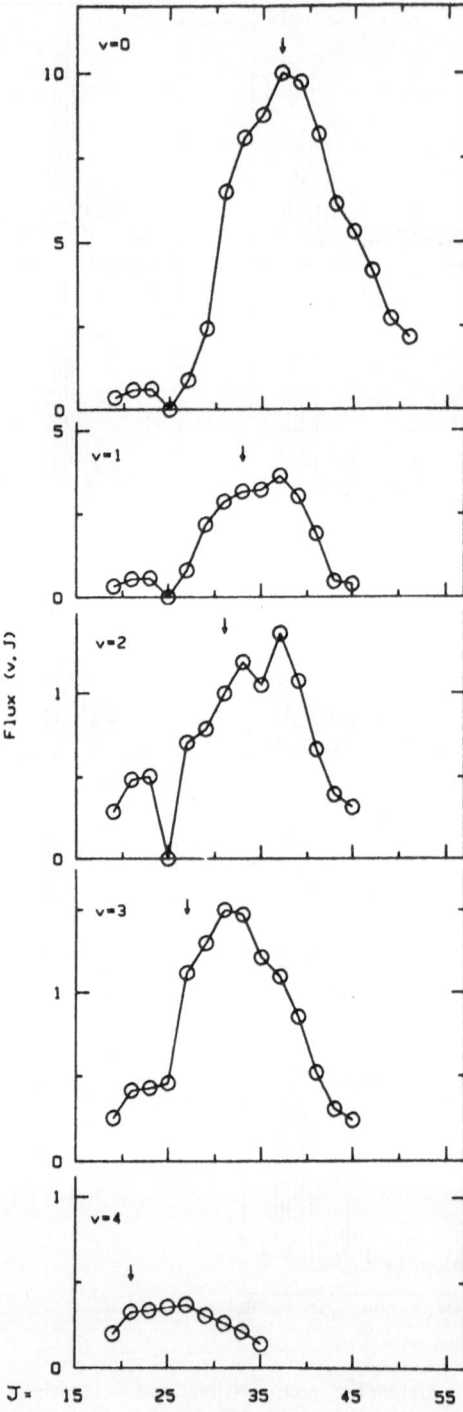

Fig. 12. Quantum state flux distributions, normalized to a maximum of 10, of the $O_2(^3\Sigma_g^-)$ fragment in the photolysis of ozone at $\lambda = 572$ nm. Reproduced, with permission, from [39]

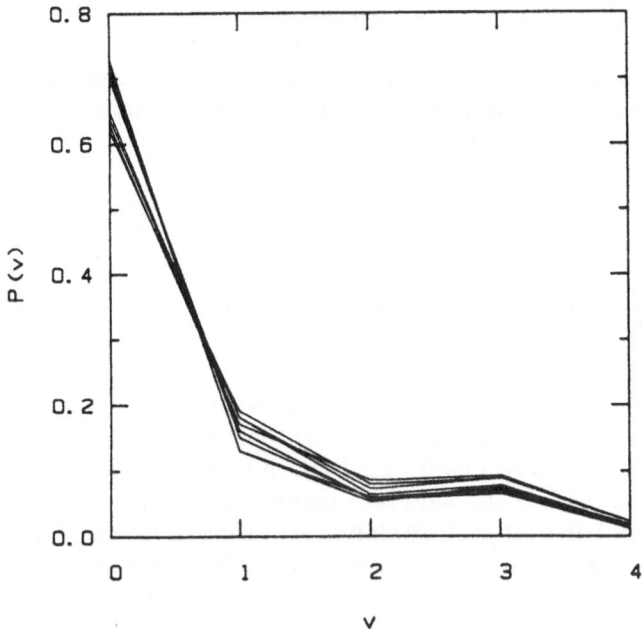

Fig. 13. Vibrational state flux distributions (summed over rotational states) of the $O_2(^3\Sigma_g^-)$ fragment in the photolysis of ozone at nine wavelengths in the visible region. Reproduced, with permission, from [39]

such excitation is, indeed, seen. For $v = 0$, the peak of the distribution in J lies between 35 and 40, which is the most probable rotational quantum number for an equilibrium temperature of around 6500 K. A rotationally impulsive, vibrationally adiabatic, fragmentation would place 21% of the available energy in rotation. The arrows in Fig. 12 show the values of J predicted to be the most probable on this basis, neglecting thermal rotational and zero-point vibrational energy of the O_3 before dissociation. The agreement between prediction and observation is good, the deviations being accounted for by the neglected parent motions.

A more detailed calculation of the rotational distribution reveals a further interesting feature. Final rotational states of O_2 are excluded if the angular momentum release is anti-parallel to the in-plane angular momentum of the initial O_3. A vector correlation [152–154] thus selects the lower orbital momentum state from two energy-degenerate final states. The two degenerate states correspond to breaking of one or other of the two bonds that are equivalent in the C_{2v} ozone molecule. In-plane rotation effectively lifts the degeneracy, because the fragment rotational and orbital angular momenta and recoil linear momenta depend on which bond is broken. The bond selectivity in the photodissociation may arise because the fragment state of minimum momentum gives the larger Franck–Condon overlap between the initial bound state and the final continuum state in the photoexcitation.

Levene et al. [39] rationalize their results in terms of a potential energy surface for ozone in the excited $^1B_1$ state. This surface is also able to explain the structure of the Chappuis absorption band. The gradient of potential on this surface is very nearly parallel to the dissociative bond, except for a small region at precise C_{2v} geometry, so that vibrationally-adiabatic dissociation dynamics are expected. The O–O bond length in the ground state of O_3 is almost identical with the O–O bond length at the outer turning point for ground state O_2, $v = 0$ [39], so that low vibrational excitation is predicted. The precise way in which the Franck–Condon overlaps decrease with increasing v can explain the observed population inversions between $v = 2$ and $v = 3$.

CARS studies of the photodissociation of ozone in the ultraviolet region [105, 106] show the formation of $O_2(^1\Delta_g)$ in different vibrational levels. Although in some respects the results are broadly the same as those found for the photolysis in the Chappuis bands, there is one remarkable difference. Unexpectedly, the rotational structure in the $O_2(^1\Delta_g)$ shows an anomalous propensity for even-J, as indicated in Fig. 14. We have already noted some of the consequences of this propensity in the section on isotope effects and we shall return shortly to an interpretation of the phenomenon. CARS spectra have been obtained [106] at 17 different wavelengths in the Hartley band. At each photolysis wavelength, all energetically accessible vibrational states are populated (with v up to 7 for photolysis at $\lambda = 230$ nm), in

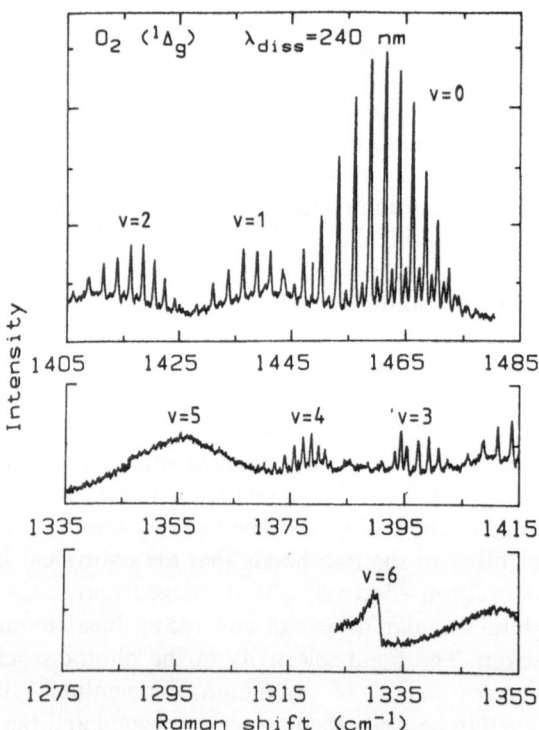

Fig. 14. Coherent Raman anti-Stokes spectrum (vibrational Q branch) of $O_2(^1\Delta_g)$ produced in the photolysis of ozone at $\lambda = 240$ nm. Note the alternation in intensities between neighbouring J levels. Reproduced, with permission, from [106]

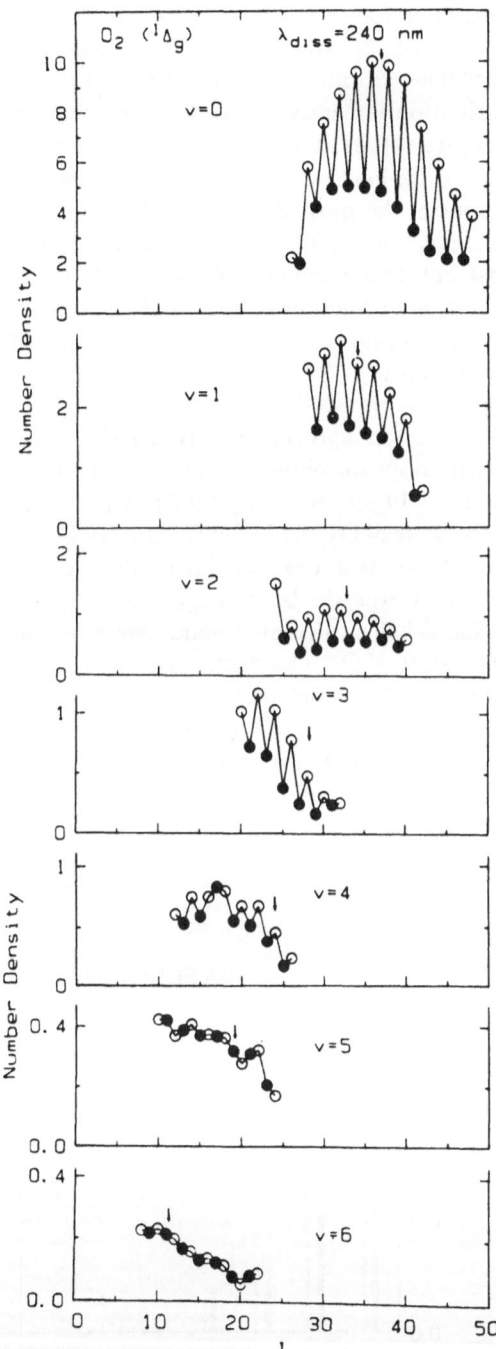

Fig. 15. Quantum state flux distributions, normalized to a maximum of 10, of the $O_2(^1\Delta_g)$ fragment in the photolysis of ozone at $\lambda = 240$ nm. Open circles represent even-J states, and closed circles odd-J states. Reproduced, with permission, from [106]

contrast to the behaviour for photolysis by visible radiation described above. Nevertheless, the most probable vibrational state is always v = 0, and, apart from the population of the higher vibrational levels of $O_2(^1\Delta_g)$ at the shorter wavelengths, the vibrational distribution is nearly independent of photolysis wavelength. The range of rotational states within each vibrational level is quite narrow, and high-J values are populated for the lower values of v, as indicated in Fig. 15 for photolysis at $\lambda = 240$ nm, Of course, the most dramatic difference between Fig. 15 and its Chappuis-band analogue, Fig. 12, is the alternating populations for even and odd J. Even though high-J rotational states are produced, conservation of angular momentum prevents a large fraction of the available energy from appearing as rotation. Thus, since v = 0 predominates in the vibrational distribution, most of the energy difference between the photon energy and that needed to form $O(^1D) + O_2(^1\Delta_g)$ appears as translation.

Figure 16 shows the good agreement between the CARS measurements of vibrational state distributions for photolysis at $\lambda = 266$ nm and those obtained by photofragment time-of-flight spectroscopy [100] at the same wavelength. Initial bond lengths in the $^1A_1$ state of O_3 and in the $^1\Delta_g$ state of O_2 are not much different, so that vibrational excitation is not expected from the Franck–Condon overlap of reactant and product. However, the bond length in the upper state of the Hartley system (1^1B_2: see Table 1) has a calculated bond length of 1.405 Å [40], as against 1.271 Å for the ground state of O_3. The relative lack of vibrational excitation must therefore mean that the gradients on the potential energy surface are sufficient to drive the dissociating system out of the Franck–Condon region so rapidly that there is not time to excite the preserved O–O bond.

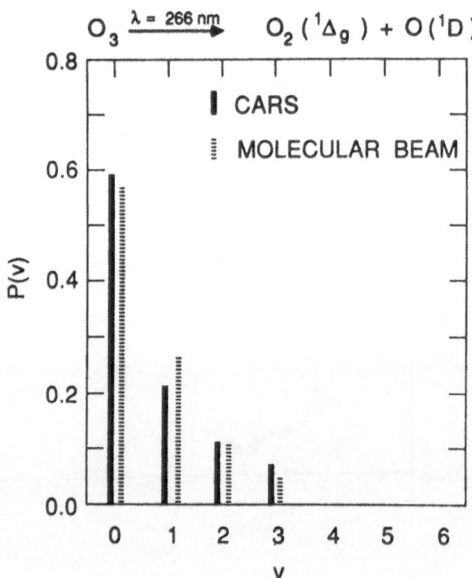

Fig. 16. Vibrational state distributions in the $O_2(^1\Delta_g)$ fragment of photolysis of ozone at $\lambda = 266$ nm obtained by CARS [106] and time-of-flight [100] techniques. Reproduced, with permission, from [106]

It is clear that vibrationally adiabatic, rotationally impulsive energy release must dominate the dissociation, just as it did in the case of photolysis in the visible region. However, since there are changes of vibrational and rotational populations with wavelength, the energy release cannot be perfectly impulsive, even for rotational energy, in the photolysis by ultraviolet radiation. The agreement between prediction of the most populated rotational level by the impulsive model (arrows in Fig. 15) and experiment is not as good for the photolysis in the Hartley band as it is for the longer-wavelength photolysis, even when allowance is made for the rotation and zero-point vibrational energy of the initial ozone reactant. Nevertheless, the width and shape of the rotational distributions can be well matched by the calculations. As in the case of the visible-region dynamics, it is necessary to incorporate a vector correlation that requires the rotational angular momentum of the O_2 fragment derived from the in-plane rotation of the O_3 to be parallel to the angular momentum induced by the impulsive release of dissociation energy.

We come now to a consideration of the anomalous alternation in rotational populations observed in the ultraviolet photolysis [105, 106]. Intensity alternations of this kind cannot be due to spectroscopic effects or nuclear degeneracy differences [105]. Rather, a dynamical bias in the photodissociation must be behind the alternations. Nuclear exchange symmetry restrictions allow only odd-J levels in ground state $^{16}O^{16}O$ ($^3\Sigma_g^-$), because the nuclei have zero spin; in $O_2(^1\Delta_g)$ the situation is complicated by the existence of Λ-doubling, but again there are restrictions, even-J levels being associated with the Δ^+ components, and odd-J with the Δ^- components. Nuclear spin statistical effects might thus be a good place to look for an interpretation of the photodissociative behaviour. Valentini et al. [106] have carried out CARS photofragment spectroscopy on $O_2(^1\Delta_g)$ formed in the dissociation of isotopically substituted ozone. Both $^{16}O^{16}O$ and $^{18}O^{18}O$ products show the same propensity for even-J rotational states, but the heteronuclear isotopomer $^{16}O^{18}O$ does not. Comparisons of the relative populations for $^{16}O^{18}O$ and $^{16}O^{16}O$ rotational levels demonstrate clearly that it is the odd-J levels in the homonuclear isotope that are depleted relative to the even-J levels, rather than the even-J levels that are preferentially populated. This result thus excludes an explanation that calls for a preference in the photodissociation for the Δ^+ Λ-doublet components of $O_2(^1\Delta_g)$. The data are, however, consistent with a preferential removal of the odd-J levels in the curve-crossing process discussed in the section on absolute quantum yields. In that section, it was pointed out that a considerable body of evidence suggests that about 10% of the photolysis of ozone in the Hartley region results in the formation of two triplet products, $O(^3P) + O_2(^3\Sigma_g^-)$, by means of a radiationless energy transfer process. Figure 8 was used to illustrate the principle with reference to diatomic-like potential energy curves calculated for the ozone molecule [40]. Crossing is assumed to occur at large enough internuclear distances that the system behaves as separate atomic and diatomic species. Radiationless transitions require that the parity of the two states involved should be the same, and thus that J remain unchanged, or change by an even number of quanta. Because the level to which transfer is to occur, effectively $O_2(^3\Sigma_g^-)$, has only odd-J levels available, transfer is possible only from odd-J levels of $O_2(^1\Delta_g)$. Relative to these odd levels, therefore, there is an apparent propensity for the even-J levels in the $O_2(^1\Delta_g)$ channel that remains. The chemical and atmospheric consequences of this phenomenon have been explored in the section on isotope effects but here we observe a striking effect on nascent populations that arises from the dynamics of photodissociation coupled with quantum-spin effects.

The Dissociating Molecule

The studies described in the previous two sections have revealed much about the nascent products of ozone photodissociation and hence about the nature of the dissociative process. Yet more detailed information about the dynamics of dissociation requires a study of the ozone molecule while it is in the process of falling apart: that is, on a time scale of a few femtoseconds. 'Pump-and-probe' experiments of the kind used for time resolutions up to picoseconds seem doomed to failure for shorter times, because the Uncertainty Principle would appear to limit the available spectral resolution. One way out of this problem is to conduct the experiments in *frequency*, rather than *time*, space [5, 6]. Molecules that photodissociate, such as ozone, are not normally thought of as fluorescent, because spontaneous emission is orders of magnitude slower than dissociation. However, with a high enough intensity of illumination, the few photons that are emitted can be detected. The emission process can be considered either as fluorescence or as resonantly enhanced Raman scattering; the point of consequence is that because the photons that are emitted refer to the shortest possible time-scale, while the two photofragments are still close to each other, they probe the unbound potential surfaces on which dissociation occurs. Emission from the upper (dissociating) electronic state to discrete vibrational levels of the lower state reflects the projection of the time-evolving reaction onto the known vibrational wave functions of the ground electronic state. The frequencies of the fluorescence spectra are characteristic of the vibrational spacing of the electronic state from which absorption arises, while the intensities reflect the dynamics on the upper surface. Because the molecule is dissociating, it passes through infinite displacements, thus allowing effective Franck–Condon overlap over a wide range of lower-state vibrational levels.

The intensities of the vibrational emission bands, which depend on the Franck–Condon overlaps, are sensitive to small changes in nuclear geometry at the sub-Angstrom level, and thus provide information about the evolution of the structure on the requisite time scale. In other words [6], the intensities of the lines emitted during dissociation constitute a 'photograph' of the motion of the molecule on the upper surface. For a diatomic species, the principle is illustrated in Fig. 10, circle 4. The initial wave packet is excited at t_0 onto the upper surface, where it evolves with time (wave packets at t_1, t_2, etc). The equivalent emission spectrum shown alongside the lower state curve possesses wing emission (4(a)), and discrete emission (4(b)) into discrete, well-characterized, vibrational eigenstates. Although the Uncertainty Principle must still apply to such experiments, the linewidths of the transitions are constrained by the need to conserve energy, and the uncertainty broadening appears only in the variation of scattering cross-section with probe frequency [6].

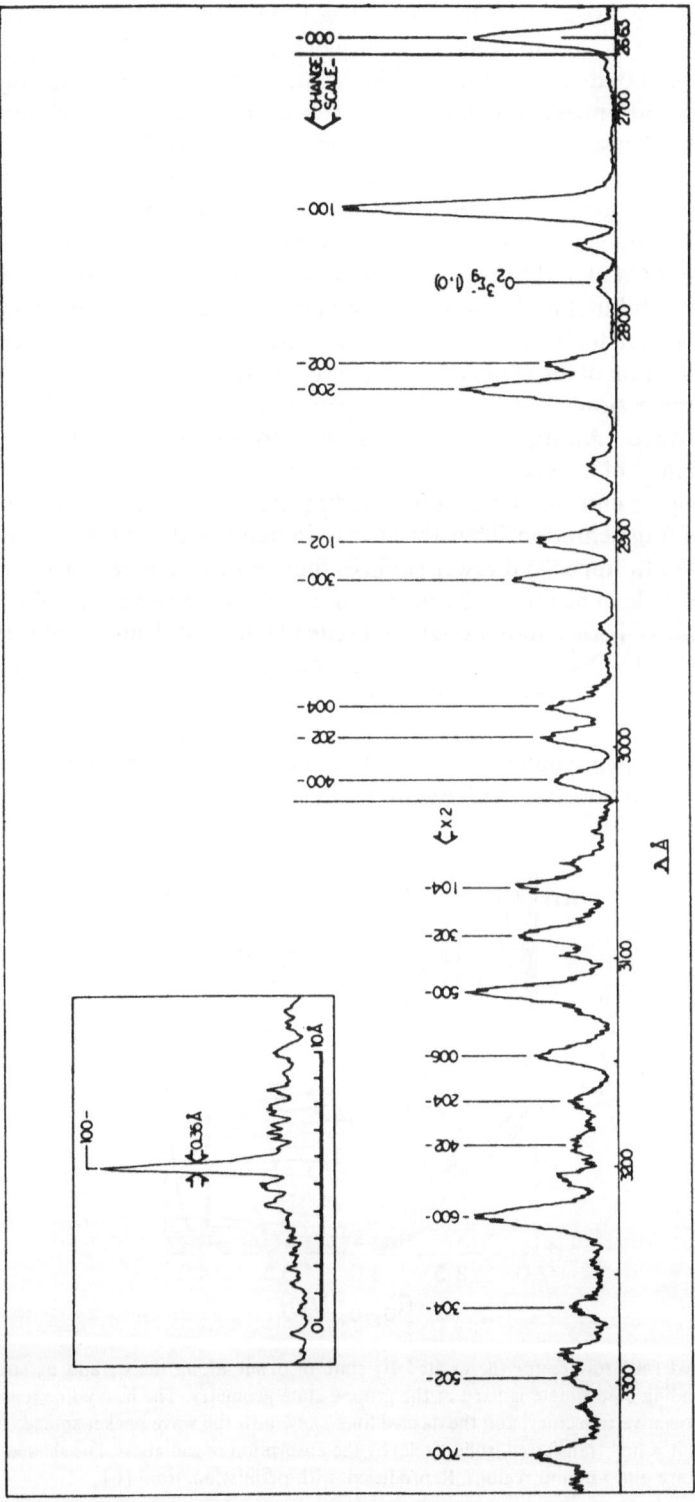

Fig. 17. Photoemission spectrum of O_3 excited at $\lambda = 266$ nm (resolution $= 0.8$ nm FWHM). The inset shows the (100) peak under higher resolution (0.035 nm FWHM). The Raman fundamental of ground state O_2 appears weakly, and is labelled on the figure. Reproduced, with permission, from [6]

The dynamics of dissociation of ozone are more complicated than described for a diatomic molecule as a result of the symmetry of O_3, because the initial equivalence of the two O–O bonds means that the dissociation involves the spreading of the wave packet on the upper surface instead of force-dominated motion. Nevertheless; the emission spectroscopy of dissociating ozone has permitted characterization of the repulsive upper potential energy surface and of the dynamics of dissociation [5, 6], as well as allowing investigation [46], via excitation spectroscopy, of the Huggins absorption spectrum. Figure 17 reproduces the photoemission spectrum obtained by Imre et al. [5] for excitation at $\lambda = 266$ nm. The spectrum consists of overtones and combination bands in v_1 (symmetric stretch) and even quanta in v_3 (antisymmetric stretch). No bands with $v_2 > 0$ (bending) are evident. As anticipated from the explanation of the emission given in the preceding paragraphs, the vibrational progression in the fluorescence spectrum is unusually long. Vibrational levels of the ground, $^1A_1$, state of ozone are seen up to (700) at $v = 7523$ cm^{-1}, which is within 500 cm^{-1} of the energy of dissociation to $O(^3P) + O_2(^3\Sigma_g^-)$. The absence of v_2 activity suggests immediately that the bending mode is not much involved in the early stages of fragmentation. Thus the change in bond angle with excitation must be small, and both upper and lower surfaces must possess similar curvature with respect to the angle of bending. The relative intensities of the (100) and (002) bands yields quantitative information about the excited state, and demonstrates that the upper state potential has a maximum rather than a minimum with respect to the normal coordinate q_3 for the v_3 mode. The upper state is therefore better described as possessing C_s rather than C_{2v} symmetry. Calculations by Hay et al. [40] for the upper states of $^1A'$ (C_s) symmetry show that the 1^1B_2 (C_{2v}) state has this form, while the 2^1A_1 state has a potential well. Figure 18 illustrates the type of potential surface

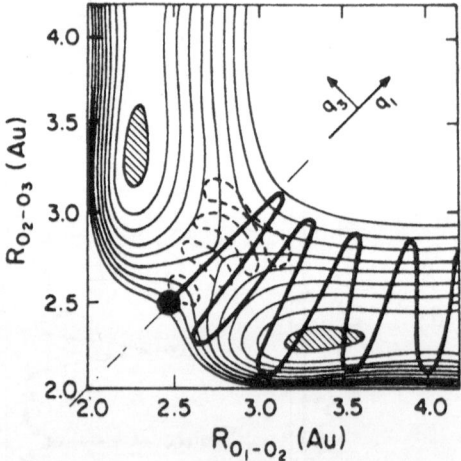

Fig. 18. Calculated [40] surface for the excited B_2 state of ozone along the q_1 and q_3 stretch co-ordinates: the bending co-ordinate is fixed at the ground-state geometry. The heavy line represents a typical photodissociative trajectory, and the dashed lines show how the wave packet spreads from the position to which it is first transferred (solid circle) by the absorption of radiation. The shaded areas in the exit channels are quasi-bound regions. Reproduced, with permission, from [6]

envisaged. The relatively large intensity in the v_3 mode is interesting in this context, because the force along q_3 vanishes at the initial configuration of the wave packet, and a low intensity in this mode might be expected. However, rapid spreading of the wave packet along the v_3 mode, indicated in Fig. 18 by dashed lines, permits the overtone transitions in this mode. The spreading motions maintain symmetry about the C_{2v} axis. Odd quanta in v_3 are antisymmetric, and therefore vanish, leaving the observed even quanta. Figure 19 emphasizes the point by showing the spreading of the wave packet over a section through the potential surface along q_3 (for q_1 and q_2 at the ground-state equilibrium geometry).

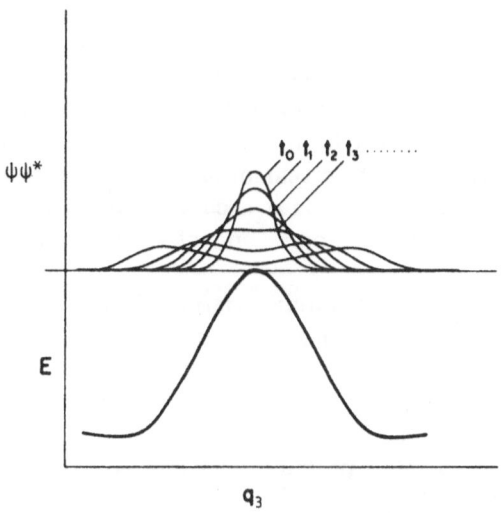

Fig. 19. Cross section through the surface shown in Figure 18 along the q_3 co-ordinate (lower portion) and the spreading of the wave packet over it (upper portion). Reproduced, with permission, from [6]

Photodissociation itself starts by the transference of the wave packet to the excited state near the top of a saddle point along the C_{2v} axis, represented by the closed circle in Fig. 18. The initial displacement from the energy minimum along q_1 leads to a rapid acceleration along the q_1 coordinate, and the generation of the long v_1 progression in emission. Bifurcation of the wave packet spreading along q_3 leads to product formation. The heavy line in Fig. 18 shows one photodissociative trajectory. The absorption spectrum of ozone shows only very weak fine structure, which indicates that the packet does not revisit the Franck–Condon region many times before the irreversible motion in the q_3 coordinate. Some information is thus provided about the rate of spread relative to the rate of motion along the symmetric stretch.

Detailed analyses of the dynamics may ultimately provide the solutions to the remaining problems of ozone photochemistry that both physical chemists and atmospheric scientists await. Further developments that are anticipated [6] include wavelength tuning to obtain further time resolution, higher-resolution spectroscopy, and accurate calculations of the motions of the wave packets on adjustable model surfaces to generate intensity distributions for comparison with experiment. Full deconvolution of the upper and lower surface information will demand a combination of clever experiment with clever theory.

Conclusion

In the ultraviolet region, photolysis of ozone leads to the production of energy-rich $O(^1D)$ that itself drives much atmospheric chemistry, including that of the hydroxyl radical. An understanding of the dependence on wavelength of the quantum yield for $O(^1D)$ formation is recognized as essential in quantitative modelling of atmospheric chemical behaviour. The molecular fragments of the ultraviolet photolysis can also be excited. Such excited molecules not only contribute to the natural airglow, but emissions from them are now routinely used to derive atmospheric concentrations of the precursor ozone. Once again, correct calculation of concentrations depends critically on an accurate knowledge of the efficiency with which the excited molecules are born, and of how that efficiency varies with wavelength.

A good outline of the ozone photodissociation phenomenon has emerged from laboratory studies, but there remain some disquieting features, especially from the point of view of applications to atmospheric models. The principal thesis of this Review is that the laboratory studies already cast doubt on the way in which the photochemical data are used. Those studies afford ample evidence that the idea of a transition from singlet to triplet products, *each formed with unity efficiency beyond the threshold region*, is an oversimplification. First, at wavelengths well short (240–270nm) of the threshold ($\simeq$ 310nm), the quantum yield for $O(^1D)$ formation does not exceed 0.9. The value *might* be unity at wavelengths nearer 300–310nm, which are atmospherically more important, but, on the available evidence, it also might remain at 0.9. Secondly, the same arguments may apply to the efficiency of $O_2(^1\Delta_g)$ formation, so that assumption of unity production efficiency for all wavelengths shorter than 310nm may lead to an underestimate of atmospheric ozone concentrations derived from airglow intensities. Thirdly, and in contrast, the experimental evidence also shows that $O_2(^1\Delta_g)$ continues to be formed with high efficiency at wavelengths at least as long as 334nm, a situation excluded by the generally-accepted view of spin-conserved photolysis. Production of $O_2(^1\Delta_g)$ in the longer wavelength region would mean an additional contribution to the Infrared Atmospheric Band intensity beyond that otherwise expected.

Some of the conclusions set out above might be strengthened by further experiments designed to look expressly for the "unexpected" atomic and molecular products, and to assess the efficiency of their production in the threshold region. One major obstacle to further progress is the evidently limited understanding of the photodissociation process. The elements of the physical chemistry involved are clear enough, but interpretation of the details of dissociation is still awaited. *Why* dissociation should proceed in a certain way is one of the questions addressed by the study of photodissociation dynamics. Sophisticated dynamical experiments and theories might not at first sight seem relevant to atmospheric studies, but they most surely are since successful interpretation at the more detailed level implies an improved understanding of the overall photolytic process. This review therefore outlines the kinds of experiment currently in progress, and the part the results could play in extending the theory of dissociation.

References

1. Wayne R P (1985) Chemistry of atmospheres, Oxford University Press, Oxford
2. Thomas R J, Barth C A, Rottman G J, Rusch D W, Mount G H, Lawrence G M, Sanders R W, Thomas G E, Clemens L E (1983) Geophys. Res. Letts. 10: 245
3. Thomas R J, Barth C A, Rusch D W, Sanders R W (1984) J. Geophys. Res. (D) 89: 9569
4. Simons J P (1984) J. Phys. Chem. 88: 1287
5. Imre D, Kinsey J L, Sinha A, Krenos J (1984) J. Phys. Chem. 88: 3956
6. Imre D G, Kinsey J L, Field R W, Katayama D H (1982) J. Phys. Chem. 86: 2564
7. World Meteorological Organization (1986) Atmospheric Ozone 1985. Global Ozone Research and Monitoring Project, Report Number 16
8. Angeletti G, Restelli G (eds) (1987) Fourth European Symposium on the Physico-Chemical Behaviour of Atmospheric Pollutants, Reidel, Dordrecht
9. Watson R T, Geller M A, Stolarski R S, Hampson R F (1986) NASA reference publns. No. 1162
10. Wayne R P (1985) In: Frimer A A (ed) Singlet molecular oxygen, CRC, Boca Raton, FL, p 81
11. Wayne R P (1988) Chem. Br. 24: 225
12. Levine J S (1985) In: Levine J S (ed) The photochemistry of atmospheres, Academic, Orlando, FL, chap 2
13. Lovelock J E (1979) Gaia: a new look at life on Earth, Oxford University Press, Oxford
14. Berkner L V, Marshall L C (1965) J. Atmos. Sci. 22: 225
15. Wayne R P (1984) J. Photochem. 25: 345
16. Noxon J F, Vallance-Jones A (1962) Nature 196: 157
17. Wraight P C, Gadsden M (1975) J. Atmos. Terr. Phys. 37: 717
18. Noxon J F (1982) Planet. Space Sci. 30: 545
19. Pick D R, Llewellyn E J, Vallance-Jones A (1971) Canad. J. Phys. 49: 897
20. Wood H C, Evans W F J, Llewellyn E J, Vallance-Jones A (1970) Canad. J. Phys. 48: 862
21. Llewellyn E J, Evans W F J, Wood H C (1973) In: McCormack B M (ed) Physics and chemistry of upper atmospheres, Reidel, Dordrecht, p 193
22. Vallance-Jones A (1973) Space Sci. Rev. 15: 355
23. Wayne R P (1967) Q. J. Roy. Met. Soc. 93: 69
24. Evans W F J, Hunten D M, Llewellyn E J, Vallance-Jones A (1968) J. Geophys. Res. 73: 2885
25. Evans W F J, Llewellyn E J (1970) Ann. Geophys. 26: 167
26. Crutzen P J, Jones I T N, Wayne R P (1971) J. Geophys. Res. 76: 1490
27. Tachibana K, Phelps A V (1981) J. Chem. Phys. 75: 3315
28. Borrell P, Borrell P M, Pedley M D (1977) Chem. Phys. Lett. 51: 300
29. Leiss A, Schurath U, Becker K-H, Fink E H (1978) J. Photochem. 8: 211
30. Noxon J F, Traub W A, Carleton N P, Connes P (1976) Astrophys. J. 207: 1025
31. Traub W A, Carleton N P, Connes P, Noxon J F (1979) Astrophys. J. 229: 846
32. Connes P, Noxon J F, Traub W A, Carleton N P (1979) Astrophys J. 233: L29
33. Kong T Y, McElroy M B (1977) Planet. Space Sci. 25: 839
34. Trauger J T, Lunine J I (1983) Icarus 55: 272
35. Steinfeld J I, Adler-Golden S M, Gallagher J W (1987) JILA Data Center Report No. 31; J. Phys. Chem. Ref. Data 16: 911
36. Katayama D H (1979) J. Chem. Phys. 71: 815; (1986) J. Chem. Phys. 85: 6809

37. Bass A M, Paur R J (1985) In: Zerefos C S, Ghazi A (eds) Atmospheric ozone, Proceedings of the Quadrennial Ozone Symposium, Reidel, Dordrecht, p 606
38. Hay P J, Dunning T H Jr (1977) J. Chem. Phys. 67: 2290
39. Levene H B, Nieh J -C, Valentini J J (1987) J. Chem. Phys. 87: 2583
40. Hay P J, Pack R T, Walker R B, Heller E J (1982) J. Phys. Chem. 86: 862
41. Sheppard M G, Walker R B (1983) J. Chem. Phys. 78: 7191
42. Adler-Golden S M (1983) J. Quant. Spectroscop. Radiat. Transfer. 30: 175
43. Atabek O, Bourgeois M T, Jacon M (1986) J. Chem. Phys. 84: 6699
44. Devaquet A, Ryan J (1973) Chem. Phys. Lett. 22: 269
45. Brand J C D, Cross K J, Hoy A R (1978) Canad. J. Phys. 56: 327
46. Sinha A, Imre D, Goble J H, Jr, Kinsey J L (1986) J. Chem. Phys. 84: 6108
47. DeMore W B, Molina M J, Sander S P, Golden D M, Hampson R F, Kurylo M J, Howard C J, Ravishankara A R (1987) Chemical Kinetics and Photochemical Data for use in Stratospheric Modeling. Evaluation Number 8 JPL Publication 87–41.
48. Vigroux E (1953) Ann. Phys. (Paris) 8: 709
49. Inn E C Y, Tanaka Y (1953) J. Opt. Soc. Am. 43: 870
50. Inn E C Y, Tanaka Y (1959) Adv. Chem. Ser. 21: 263
51. Griggs M (1968) J. Chem. Phys. 49: 857
52. Watanabe K, Zelikoff M, Inn E C Y (1953) Geophysical Research Paper No. 21, Air Force Cambridge Research Laboratories, Bedford, MA
53. Ackerman M (1971) In: Fiocco G (ed.) Mesospheric models and related experiments, Reidel, Dordrecht, p 149
54. Molina L T, Molina M J (1986) J. Geophys. Res. 91: 14501
55. Bass A M, Paur R J (1986) Unpublished data reported in reference 7
56. Mauersberger K, Barnes J, Hanson D, Morton J (1986) Geophys. Res. Lett. 13: 671
57. Mauersberger K, Hanson D, Barnes J, Morton J (1987) J. Geophys. Res. 92: 8480
58. Hearn A G (1961) Proc. Phys. Soc. London 78: 932
59. Yoshino K, Freeman D E, Esmond J R, Parkinson W H (1988) Planet. Space Sci (in press)
60. Simons J W, Paur R J, Webster H A, III, Bair E J (1973) J. Chem. Phys. 59: 1203
61. Brion J, Daumont D, Malicet J, Marche P J (1985) J. Phys. (Paris) Lett. 46: L105
62. Freeman D E, Yoshino K, Esmond J R, Parkinson W H (1984) Planet. Space Sci. 32: 239
63. Freeman D E, Yoshino K, Esmond J R, Parkinson W H (1984) In: Zerefos C S, Ghazi A (eds) Atmospheric ozone, Proceedings of the Quadrennial Ozone Symposium, Reidel, Dordrecht, p 622
64. Thunemann K H, Peyerimhoff S D, Buenker R J (1978) J. Mol. Spect. 70: 432
65. Wilson C W Jr, Hopper D G (1981) J. Chem. Phys. 74: 595
66. Jones R O (1984) Phys. Rev. Lett. 52: 2002
67. Jones R O (1985) J. Chem. Phys. 82: 325
68. Morin M, Foti A E, Salahub D R (1985) Canad. J. Chem. 63: 1982
69. Tanaka T, Morino Y (1970) J. Mol. Spect. 33: 538
70. Swanson N, Celotta R J (1975) Phys. Rev. Lett. 35: 783
71. McGrath W D, Maguire J M, Thompson A, Trocha-Grimshaw J (1983) Chem. Phys. Lett. 102: 59; McGrath W D, Thompson A, Trocha-Grimshaw J (1986) Planet. Space Sci. 34: 1147
72. von Rosenberg C W Jr, Trainor D W (1974) J. Chem. Phys. 61: 2442; (1975) J. Chem. Phys. 63: 5348
73. Wraight P C (1977) Planet. Space Sci. 25: 1177
74. Kleindienst T, Locker J R, Bair E J (1980) J. Photochem. 12: 67; Locker J R, Joens J A, Bair E J (1987) J. Photochem. 36: 235
75. Ramirez J E, Bera R K, Hanrahan R J (1984) Radiat. Phys. Chem. 23: 685
76. Norrish R G W, Wayne R P (1965) Proc. Roy. Soc. Lond. Ser. A 288: 200
77. Norrish R G W, Wayne R P (1965) Proc. Roy. Soc. Lond. Ser. A 288: 361
78. Wayne R P (1972) Faraday Disc. Chem. Soc. 53: 172
79. Castellano E, Schumacher H J (1962) Z. phys. Chem. NF34: 198
80. Tkachenko S N, Zhuraviev V E, Popovich M P, Zhitnev Y N, Filippov Y V (1980) Russian. J. Phys. Chem. 54: 1304
81. Lissi E A, Heicklen J (1972) J. Photochem. 1: 39
82. Jones I T N, Wayne R P (1969) J. Chem. Phys. 51: 3617

83. Jones I T N, Wayne R P (1970) Proc. Roy. Soc. Lond. Ser. A 319: 273
84. Lee L C, Black G, Sharpless R L, Slanger T G (1980) J. Chem. Phys. 73: 256
85. Taherian M R, Slanger T G (1985) J. Chem. Phys. 83: 6246
86. Arnold I, Comes F J, Moortgat G K (1977) Chem. Phys. 24: 211
87. Brock J C, Watson R T (1980) Chem. Phys. 46: 477
88. Cobos C, Castellano E, Schumacher H J (1983) J. Photochem. 21: 291
89. Fairchild P W, Lee E K C (1978) Chem. Phys. Lett. 60: 36
90. Kajimoto O, Cvetanovic R J (1979) Int. J. Chem. Kin. 11: 605
91. Moortgat G K, Warneck P (1975) Z. Naturforsch. 30a: 835
92. Philen D L, Watson R T, Davis D D (1977) J. Chem. Phys. 67: 3316
93. Moortgat G K, Kudszus E (1978) Geophys. Res. Lett. 5: 191
94. Alder-Golden S M, Schweitzer E L, Steinfeld J I (1982) J. Chem. Phys. 76: 2201
95. Astholz D C, Croce A E, Troe J (1982) J. Phys. Chem. 86: 696
96. Kajimoto O, Cvetanovic R J (1976) Chem. Phys. Lett. 37: 533
97. Kuis S, Simonaitis R, Heicklen J (1975) J. Geophys. Res. 80: 1328
98. Moortgat G K, Kudszus E, Warneck P (1977) J. Chem. Soc. Faraday Trans. 2, 73: 1216
99. Fairchild C E, Stone E J, Lawrence G M (1978) J. Chem. Phys. 69: 3632
100. Sparks R K, Carlson L R, Shobatake K, Kowalczyk M L, Lee Y T (1980) J. Chem. Phys. 72: 1401
101. Amimoto S T, Force A P, Wiesenfeld J R, Young R H (1980) J. Chem. Phys. 73: 1244
102. Greenblatt G D, Wiesenfeld J R (1983) J. Chem. Phys. 78: 4924
103. Brock J C, Watson R T (1980) Chem. Phys. Lett. 71: 371
104. Wine P H, Ravishankara A R (1982) Chem. Phys. 69: 365
105. Valentini J J (1983) Chem. Phys. Lett. 96: 395
106. Valentini J J, Gerrity D R, Phillips D L, Nieh J -C, Tabor K D (1987) J. Chem. Phys. 86: 6745
107. Moore D S, Bomse D S, Valentini J J (1983) J. Chem. Phys. 79: 1745
108. Levene H B, Nieh J -C, Valentini J J (1987) J. Chem. Phys. 87: 2583
109. Izod T P J, Wayne R P (1968) Nature 217: 947
110. Huffman R E, Larrabee J C, Barsley V C (1969) J. Chem. Phys. 50: 4594
111. Gauthier M, Snelling D R (1970) Chem. Phys. Lett. 5: 93
112. Gauthier M, Snelling D R (1971) J. Chem. Phys. 54: 4317
113. Jones I T N, Wayne R P (1971) Proc. Roy. Soc. Lond. Ser. A 321: 409
114. Castellano E, Schumacher H J (1972) Chem. Phys. Lett. 13: 625
115. Cicerone R J, McCrumb J L (1980) Geophys. Res. Lett. 7: 251
116. Mauersberger K (1981) Geophys. Res. Lett. 8: 935
117. Mauersberger K (1987) Geophys. Res. Lett. 14: 80
118. Rinsland C P, Malathy Devi V, Flaud J -M, Camy-Peyret C, Smith M A H, Stokes G M (1985) J. Geophys. Res. 90: 10719
119. Thiemens M H, Heidenrich J E, III (1983) Science 219: 1073
120. Heidenrich J E III, Thiemens M H (1986) J. Chem. Phys. 84: 2129
121. Bains-Sahota S K, Thiemens M H (1987) J. Phys. Chem. 91: 4370
122. Thiemens M H, Jackson T (1987) Geophys. Res. Lett. 14: 624
123. Bhattacharya S K, Thiemens M H (1988) Geophys. Res. Lett. 15: 9
124. Kaye J A, Strobel D F (1983) J. Geophys. Res. 88: 8447
125. Kaye J A (1986) J. Chem. Phys. 91: 7865
126. Bates D R (1986) Geophys. Res. Lett. 13: 664
127. Anderson S, Kaye J A (1987) Geophys. Res. Lett. 14: 91
128. Valentini J J (1987) J. Chem. Phys. 86: 6757
129. Rawlins W T (1985) J. Geophys. Res. 90: 12283
130. Rawlins W T, Armstrong R A (1987) J. Chem. Phys. 87: 5202
131. McDade I C, McGrath W D (1980) Chem. Phys. Lett. 73: 413
132. Joens J A, Burkholder J B, Bair E J (1982) J. Chem. Phys. 76: 5902
133. Kleindienst T, Burkholder J B, Bair E J (1980) Chem. Phys. Lett. 70: 117
134. Zittel P F, Little D D (1980) J. Chem. Phys. 72: 5900
135. Rawlins W T, Caledonia G E, Armstrong R A (1987) J. Chem. Phys. 87: 5209
136. Rawlins W T, Caledonia G E, Gibson J J, Stair A T Jr (1985) J. Geophys. Res. 90: 2896

137. Green B D, Rawlins W T, Nadile R M (1986) J. Geophys. Res. 91: 311
138. Ogawa T (1976) Planet. Space Sci. 24: 749
139. Wildt J, Bednarek G, Fink E H, Wayne R P (1988) Chem. Phys. 122: 463
140. Slanger T G, Black G (1979) J. Chem. Phys. 70: 3434
141. Solomon S, Kiehl J T, Kerridge B J, Remsberg E E, Russell J M, III (1986) J. Geophys. Res. 91: 9865
142. West G A, Weston R E Jr, Flynn G W (1978) Chem. Phys. Lett. 56: 429
143. Kurylo M J, Braun W, Kaldor A, Freund S M, Wayne R P (1974) J. Photochem. 3: 71
144. Gordon R J, Lin M C (1976) J. Chem. Phys. 64: 1058
145. Moy J, Bar-Ziv E, Gordon R J (1977) J. Chem. Phys. 66: 5439
146. Stephenson J C, Freund S M (1976) J. Chem. Phys. 65: 1893
147. Bar-Ziv E, Moy J, Gordon R J (1978) J. Chem. Phys. 68: 1013
148. Hui K K, Cool T A (1978) J. Chem. Phys. 68: 1022
149. Parker J G (1977) J. Chem. Phys. 67: 5352
150. Levine R D, Bernstein R B (1987) Molecular reaction dynamics and chemical reactivity, Oxford University Press, Oxford
151. Klais O, Laufer A H, Kurylo M J (1980) J. Chem. Phys. 73: 2696
152. Bernstein R B, Herschbach D, Levine R D (1987) J. Phys. Chem. 91: 5365
153. Simons J P (1987) J. Phys. Chem., 91: 5378
154. Houston P L (1987) J Phys. Chem., 91: 5388

Nonenzymatic Biomimetic Oxidation Systems: Theory and Application to Transformation Studies of Environmental Chemicals

B. L. Worobey

Food Research Division, Bureau of Chemical Safety, Food Directorate, Health Protection Branch, Health and Welfare Canada, Ottawa, Ontario K1A 0L2 CANADA

Summary

Nonenzymatic oxidation systems are used to mimic enzymatic oxidation under *in vitro* conditions. Such systems are termed model systems or biomimetic systems. Model systems can provide an elegant means to study the oxidative transformations of environmental chemicals. An overview is presented of model systems which have been studied, their unique advantages, reaction mechanisms, applications to environmental chemicals, and a discussion of which system(s) most faithfully models enzymatic oxidation.

The chapter describes the usefulness of these systems and elucidates their functional details. Novel systems and applications are presented for several groups of environmental chemicals such as herbicides, insecticides, polycyclic aromatic hydrocarbons, substituted anilines, halogenated hydrcarbons, and other environmental contaminants and transformation products.

Future trends and developments include: optimizing reaction conditions, controlling product formation (isomer ratios and specific compounds) and yields, using these systems to replace animals in metabolism studies, studying mechanisms of enzymatic reactions, elucidating stereochemical effects of enantiomers and diastereomers on oxidative transformation products, and studying oxidative products as they occur when two or more contaminants are competing for oxidative reagents.

The review is divided according to the system described and within each system according to class of environmental contaminant. Chemical and mechanistic aspects of biomimetic systems are not discussed in detail, however, references are given for further information. Many systems were discussed in a previous review [1]; especially pertaining to pesticides. Newer, and novel systems (e.g. the metal-loprophyrin biomimetic system) are discussed as are applications of all systems to environmental contaminants in general.

Introduction

Many biological enzymatic reactions may be mimicked by model systems, e.g. reduction, oxidation, hydrolysis and transamination. Biomimetic chemistry is concerned with building chemical models that imitate enzyme activity and specificity. Chemical oxygenation of organic compounds as a model for the unspecific hydroxylation *in vivo* has frequently been investigated for a variety of substrates. Many different oxidants of various complexity have been used, leading to a more or less successful biomimetic oxygenation. How successful the model reaction is depends on the criteria used to assess the outcome of the reaction. A system that yields an acceptable product pattern could, e.g. show a kinetic isotope effect which is far removed from that displayed by enzymatic systems.

Generally, a satisfactory model reaction should yield the main products found in biotic systems. How good such an agreement is must be an arbitrary decision for there is no such thing as *the* biotic product pattern of a compound. Many factors can influence the product distribution such as species, age, sex, mode of application, organ studied etc., all of which may show profound influence on the product pattern.

A second concern in our simulation studies is that only a limited number of reagents should be necessary for a large variety of substrates. Therefore, our aim cannot be to find the optimal oxidant for each substrate but to identify a few simple systems which have an oxygenating effect on a broad range of environmentally relevant compounds. It is clear that under those circumstances a perfect fit of biotic and chemically simulated products can neither be attained nor should it be expected.

The metabolic or degradative fate (transformations) of a xenobiotic *in vivo* has many pathways (Fig. 1.). Model systems are especially useful to study any one of these areas of xenobiotic oxidation reactions (shown above), mechanisms, products and end effects. As a complement to biological studies on metabolism of environmental contaminants, it is possible to use simple chemical reactions to simulate biotic pathways. An absolute correspondence between biotic transformation and chemical simulation can, of course, never be expected since the reaction conditions are widely different, but still, a correspondence is observed.

Pollution or environmental contamination is the introduction by man (usually) of substances (or energy) into the environment that are capable of being hazardous to human health or harmful to living organisms. Environmental contaminants are a major class of xenobiotics. Environmental contaminants [2–4] studied in biomimetic systems include those listed in Table 1.

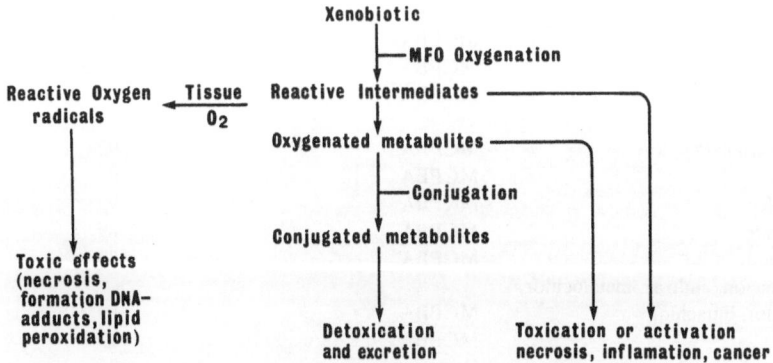

Fig. 1. Xenobiotic transformations and interactions

Table 1. Environmental Contaminants Studied with Biomimetic Systems

Compound Class	Biomimetic System	Ref.
Pesticides		
matacil	AAOS	289
parathion	AAOS	256, 290
	PTFAA	291
	MCPBA	85
paraoxon	AAOS	290
	MCPBA	85
aldrin	titanous Cl, peroxyacetic acid, AAOS, Fenton, mercaptobenzoic acid	14, 77, 292
aldicarb	AAOS	293
carbofuran	AAOS	294
schradan	AAOS	295
	MCPBA	75
carbaryl	AAOS	51, 296
coumarin	AAOS, Fenton	297, 298
o-coumaric acid	AAOS, Fenton	297
linuron	AAOS	50
malathion	PTFAA	291
	MCPBA	85
2,2-dime 2,3-dehydrobenzofuranyl 7-N-me N-(dimethoxy phosphinothionyl) carbamate	MCPBA	299
N-(dimethoxy phosphinothionyl) carbamate	MCPBA	299
fonofos (dyfonate)	MCPBA	87, 88
including chiral isomers	MCPBA	95
methamidophos	MCPBA	85
diazinon	MCPBA	85
deltamethrin, resmethrin, methfuroxam	peroxyacids	275
EPN (o-ethoxy-(p-nitrophenyl) phenylphosphonothioate	MCPBA	85
phosmet	MCPBA	85
monuron	AAOS, Fenton	300
tetramethrin	MCPBA	75 R
allethrin	MCPBA	75 R
sulprofos	MCPBA	75 R
profenofos	MCPBA	75 R
metribrzin	MCPBA	75 R
EPTC		
diallate, triallate	MCPBA	92
nitrofen	MCPBA	92
aciflurofen	MCPBA	92
glyphosate	MCPBA	89
fluvalinate	MCPBA	89
chlordimefon, amitraz, metalochlor, alachlor, butachlor	MCPBA	89
	MCPBA	93
S-triazine herbicides	Fenton	36
triazine herbicides	t-butylhydroperoxide/Co/Salen	231

Table 1. (*continued*).

Compound Class	Biomimetic System	Ref.
methfuroxam (fungicide)	MCPBA	90
organothiophosphorous	sodium hypochlorite,	301
insecticides	MCPBA, hV	94
daminozide	MCPBA, H_2O_2, NaOCl	302
Pesticide analogs		
30 organophosphates	MCPBA	85
($-$)-menthylmethylphenyl		
phophinothioate	MCPBA	303
Synergists		
WARF, (N,N-dibutyl-p-chloro benzene		
sulfonamide)	AAOS	256
piperonyl butoxide	AAOS	256
sesamex	AAOS	256
MGK 264 (N-(2-ethexyl)		
5-norbornene-2,3-dicarboximide)	AAOS	256
6-nitro-1,3-benzodioxole	Fenton	77
Nitrosamines		
diethylnitrosamine	AAOS	52–56
N-nitrosopiperidine	AAOS	52–56
dimethylnitrosamine	AAOS	52–56
N-nitrosomorpholine	Fenton	44
several nitrosamines	AAOS, Fenton	58–62, 301
Polyaromatic Hydrocarbons		
	2-mercaptobenzoate/Fe(II)/O_2	273
	Fe(III) TPPCl/Iodosylbenzene	143
naphthalene	PFTAA; hV/pyridine-N-oxides	252
	trifluoroacetic acid/Cu(III)	251
	Fenton and aqueous H_2O_2/UV	38
7-methylbenz(a)anthracene	AAOS	69
9,10-dialkylanthracene	tris (phenanthroline) tris	
	(hexafluorophosphate) iron	18
several PAHs including		
benzo(a)pyrene, etc.	AAOS	39, 64, 65, 67, 70
	Fe(III) TPPCl/Iodosylbenzene	143
	Fenton, TFA/H_2O_2	66
	Fentons	39, 40
anthracene	Iodosylbenzene/BF_3	147
	Fe(III) TPPCl/Iodosylbenzene	143
Aliphatic & Aromatic Amines		
1- and 2-naphthylamine	AAOS	63, 304–306
secondary and tertiary N-alkylated	Fe(III)/TPPCl	307
tertiary alkylated amines	Fe(III)/porphyrin/iodosoxylene	184
aniline	heminthiol	144–146, 308
	peroxomonophosphoric acid	216
	peroxyacetic acid	81
aniline	Fe(II)/cysteine methyl SO_4 or	
	thiosalicylic acid, cystein methyl	
	SO_4/hemin	209

Table 1. (*continued*).

Compound Class	Biomimetic System	Ref.
N-dimethylaniline et al.	iodosylbenzene/vanadium	
	acetylacetonate	212, 213
N-dimethylaniline	di (2-ethyl-hexyl) sulfosuccinic	
	acid/O_2 or cumene hydroperoxide	258
	TTP Fe(III)Cl/iodosoxylene	176
aniline, acetanilide	hemoglobin/methylene blue/NADH	34, 214
tertiary amines such as:		
3 & 4 substituted	iodosylbenzene/Fe(II)TPP	195
N,N-dimethylbenzylamines	or Mn(III) TPP	
p-substituted dimethylanilines	electrochemical oxidation,	
	metalloporphyrin systems	207
substituted N-alkylnitroanilines	hV	215
p-nitroaniline	peroxomonophosphoric acid	216
	peroxyacetic acid	81
2,3-disubstitued indoles	MCPBA	309
N-benzylamines, thioamides		
and ureas	potassium superoxide/crown ether	238
o-toluidine	hemin	146
p-toluidine	peroxyacetic acid	81
	Fe(II) or hemin/thiols	154, 155
4-chloroaniline	peroxyacetic acid	81
aniline	mercaptopropionic acid or	
	α-mercaptopropionylglycine, Feıı or	
	hemin, and pyridine	141
	hemin/cystein	148, 149
aniline	AAOS	310
aniline	haemoglobin/cumene hydroperoxide	
	or riboflavin or NADH or ascorbic	
	acid or dihydroxyfumarate	102, 210, 211
aliphatic and alicyclic hydrocarbons		
	2-mercaptobenzoate/Fe(II)/O_2	273
	PFTAA, hv/pyridine-N-oxide	252
cyclohexane	iodosylbenzene/MnIV porphyrins	311
	iron porphine/iodosylbenzene	173
	trifluoroacetic acid/Cu(III)	251
several hydrocarbons	irons/porphyrin	180
	TFPAA	221
halogenated alkanes	AAOS	71
1,2-dichloropropane	AAOS	72
olefins (stilbenes,	Fe/hemoprotein/H_2O_2	312
non-1-ene, styrene,	TPP/O_2	179
cyclohexene etc)	Mn(II) porphyrin/NaOCl	194, 191
	Iron porphyrins/PhIO, MCPBA,	175, 253
	Fe $(ClO_4)/H_2O_2$, tris (acetylacetonato)	
	Fe(III)/BuO$_2$H, KO$_2$/BuBr,	
	Biacetyl/O_2/hV	
	PFTAA, hV/pyridine-N-oxides	252
	iodosylbenzene/Fe porphyrins,	177, 178, 143
	pentafluoroidosylbenzene/Fe(III) tetra	313
	bis (2,6-dichlorophenyl)porphyrin Cl,	
	metalloporphyrins,	142, 281, 314

Table 1. (*continued*).

Compound Class	Biomimetic System	Ref.
	lithium hypochloride/MnTPPCl	315
	iron porphyrin/p-CN dimethylaniline	
	N-oxide	174
	Mn(III)/porphyrin/ascorbate	183
alkanes	Mn(II)/porphyrin/ascorbate	183
alkanes, cycloalkanes	trifluoroacetic acid/H_2O_2	316
	iodosylbenzene/MnIVporphyrins	311
	iron porphyrin/p-CN	
	dimethylaniline N-oxide	174
aliphatic acids	Fenton	317
methyllinolenate,	AAOS	220, 287
linoleic acid	Fe(III)/Cysteine	288
cinnamic acid	AAOS	298
cyclohexene	tetraphenylporphyrinato/MnIII	318
	acetate-sodium hypochlorite	
Other Environmental contaminants		
phthalate esters	Fenton-type, MCPBA	96
	Fenton and aqueous H_2O_2/UV	38
p-toluidine		
	hemoglobin/methylene blue/NADH	34, 214
7-hydroxycoumarin		
	Fenton	155, 223
	Hamilton, Fenton, AAOS	240, 241
	PFTAA, hV/pyridine-N-oxide	252
halogenated benzenes, xylene	trifluoroacetic acid/Cu(III)	251
xylene	TFPAA	221
4-nitrophenol	H_2O_2/Cu or Fe Phenanthroline	247
	hemoglobin/methylene blue/NADH	34, 214
m-hydroxybenzoate	phenzine methosulphate/NADH/O_2	226
aromatic acids	Fenton and aqueous H_2O_2/UV	38
eg. salicylate,	AAOS	319, 73, 298
benzoate etc.	phenazine methosulfate/NADH/O_2	320–323
p-chlorotoluene	hemin/thiol,	159, 160
	AAOS, Fenton	
phenol	peroxyacetic acid	255
	electrochemical	208
	PFTAA, hV/pyridine-N-oxide	252
anisole	hemin/thioglycolate ethyl ester	150, 153
	or cysteine ethyl ester	
	Hamilton, Fenton	240, 241
	electrochemical	208
	trifluoroacetic acid/Cu(III)	251
	AAOS	74, 240, 241
	Fe(III) TPPCl/Iodosylbenzene	143
alkylphenols	trifluoroperoxyacetic acid	324
aromatic acids	Fenton	317
phenols	Fenton	317
	electrochemical	325, 326
	TFPAA	221
	trifluoroacetic acid/Cu(III)	251

Table 1. (*continued*).

Compound Class	Biomimetic System	Ref.
	peroxydisulfate/M^x	242
toluene	catecholamine/H_2O_2/O_2	233
	Fe(III) TPPCl/Iodosylbenzene	143
	Mn(III)/porphyrin/ascorbate	183
toluene	2-mercaptobenzoate/Fe^{2+}/O_2	273
	Fenton	327, 223
	PFTAA, hV/pyridine-N-oxide	252
benzene	trifluoroacetic acid/Cu(III)	251, 328
	peroxydisulfate/M^x	242
	Hamilton, Fenton, AAOS	240, 241
	Fenton	223, 327
phenylacetic acid	peroxydisulfate/M^x	242
	trifluoroacetic acid/Cu(III)	251, 328
p-nitroanisole	iron porphyrin/cumylhydro-	
7-ethoxycoumarin	peroxide or t-buhydroperoxide	182
benzylalcohol	Mn(III)/porphyrin/ascorbate	183
aromatic aldehydes	metallotetraphenylporphyrins	329
nitrobenzene	Fenton	155
	AAOS	298
	Hamilton, Fenton, AAOS	240, 241
alkyl and phenyl substituted	peroxydisulfate,	43
alcohols and phenyl ethers	Fenton	
PCBs	anodic electrochemistry	203
toluene, anisole,	Fe compounds/H_2O_2	330
benzene	[Fe(III) bis tris (acetylacetonato);	
	Fe(III) tris (acetylacetonato);	
	tetraphenylporphinatoiron III chloride;	
	ferrocene; pentacarbonyliron,	
	Fe(III) chloride,]	
trimethylbenzene	TFPAA	221

Many nonenzymatic chemical *in vitro* model systems (reductive and especially oxidative), have been shown to produce similar metabolic and or degradation products as *in vivo* systems (enzymes or whole organisms). The advantages of these *in vitro* model systems are as follows:

a) Less time consuming and less costly compared with the tedious separation of the microsomal fraction of a tissue homogenate, or whole organism studies.

b) Less time required for cleanup and separation of products.

c) Larger amounts of a test compound can be used, which could facilitate product identification and isolation.

d) Production of high yields of metabolites or degradation products facilitates ease of cleanup and separation, as well as, providing products which are unavailable or difficult to synthesize.

e) The use of appropriate model systems is one way to simplify the problem of examining the possible biological modes of degradation of such complex mixtures as toxaphene, where the toxic components of the mixture are not currently available in amounts adequate for routine studies on their metabolic and environmental fate.

f) Elucidation of biochemical mechanisms (re. activation or detoxification) may be aided by the use of simpler model nonenzymatic systems.

g) To screen for compounds with synergistic, inhibitory, potentiation, agonistic or antagonistic effects using a model substrate and determining the action of such compound(s) on substrate oxidation by a given biomimetic system.

Enzymatic Metabolism of Environmental Chemicals

Before proceeding to discuss biomimetic systems, let us first consider some of the basic enzymatic metabolic pathways related to environmental chemicals. The four most important types of *in vivo* chemical changes that occur are oxidation, reduction, hydrolysis, and conjugation. Each group of reactions may be subdivided into a number of different metabolic activities depending on the type of substrate involved.

There are two major steps of detoxification of xenobiotics [5, 6]: the primary phase involving oxidative, hydrolytic, or other enzymatic processes to produce polar end-products (nonsynthetic process), and the secondary phase producing water-soluble conjugates (synthetic process). Usually, this secondary phase is most active after the primary phase has occurred; it is the primary phase which model systems simulate. Microsomal oxidases are dominant in the nonsynthetic (primary phase) metabolism of xenobiotics. To begin let us make an explicit distinction between *oxygenation* where an oxidant oxygen atom is incorporated into the organic substrate and *oxidation* where one or more electrons are removed from the organic substate and transferred to the oxidant, which acts as an electron acceptor.

Microsomal oxidases, are characterized by the requirement of NADPH, microsomes, and oxygen *in vitro* for degradation of their substrates, as well as, by their sensitivities toward methylenedioxy derivatives, i. e., synergists such as sesamex and piperonyl butoxide. By far the most important biochemical reactions involving the initial stages of metabolism are the NADPH-requiring general oxidation system and the hydrolysis of esters. Both of these reactions are catalysed by enzymes found in the microsomes of the liver [7, 8] and so it is with these enzymes we shall be most concerned.

Oxidation: Mixed-Function Oxidase (MFO) System

There are several different types of enzymes that catalyze hydroxylation and epoxidation reactions. The most ubiquitous is a system isolated from microsomes which has a heme-containing enzyme as the actual oxygenase. Because of its unusual spectrum in the presence of carbon monoxide and reducing agents, the terminal oxygenase is usually referred to as cytochrome P-450.

The NADPH-requiring general oxidation system, commonly referred to as the MFO (or oxygenase) system is located in the microsomal portions of various tissues. It is characterized by (i) requiring NADPH as a cofactor, (ii) involving an electron transport system with cytochrome P-450, and (iii) being capable of oxidizing many different kinds of substrates, i.e., by substrate nonspecificity. The MFO system is important in the metabolism of xenobiotics in all living organisms [10]. The system is located in microsomes, and is composed primarily of material derived from the cellular endoplasmic reticulum. It is usually precipitated by high-speed centrifugation of postmitochondrial supernatant fractions of either plant or animal tissue [11]. We now know that most xenobiotics are metabolized by this system and that the pathways are important in detoxification, toxification (activation), and in other functions (e.g. the resistance of insects to insecticides [7, 10]).

The complete mechanism of the microsomal electron transport system as proposed by Casida [12] is shown in Figure 2. In short, it involves cytochrome P-450 (P-450) and cytochrome b. The final process of oxidation (or hydroxylation) of xenobiotics involves a reduced cytochrome P-450-oxygen complex, which, on oxidation of the substrate, becomes a stable oxidized form itself. Within the microsomal particles, NADPH-requiring oxidation activity is generally concentrated in microsomes isolated from the smooth endoplasmic reticulum although in some cases the activity can be evenly distributed among smooth and rough fractions [13]. More will be said about the mechanism shown in Figure 2 in a later discussion of nonenzymatic versus enzymatic mechanisms.

The reactions catalyzed by the MFO system include deamination, demethylation, dealkylation, aromatic ring hydroxylation, alkyl and N-hydroxylation, hydrolysis of esters, epoxidation, oxidation of sulfides to sulfoxides and sulfones, conversion of phosphorothioates to phosphates (desulphuration), conversion of methylenedioxyphenyls to catechols, and oxidation of alcohols and aldehydes to acids. For an overview of the metabolism of xenobiotic functional groups the reader is referred to an excellent review by Jakoby et. al. [14].

P-450 monooxygenases catalyze the reductive activation of dioxygen by NADPH and the insertion of one oxygen atom into compounds, e.g. for S = substrate:

$$S + O_2 \text{ NADPH} \xrightarrow[\text{P-450}]{} SO + H_2O$$

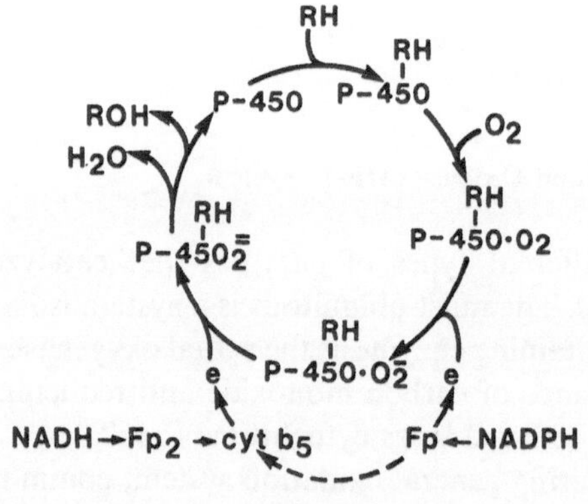

Fig. 2. Microsomal electron transfer reactions occurring during xenobiotic or environmental chemical metabolism

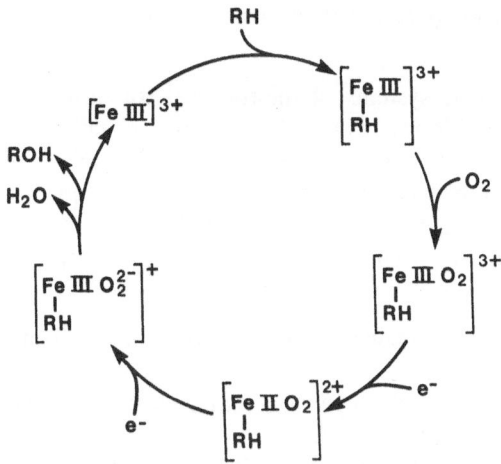

Fig. 3. Mechanism of P-450 catalytic cycle

Oxidative metabolism by P-450 is followed by conjugation and detoxication. If no conjugation occurs then activated species can react with vital intracellular macromolecules, resulting in necrosis, redox cycling, oxygen radical formation, and mutations [15].

According to recent data [16–19] the mechanism of the P-450 catalytic cycle involves the steps shown in Fig. 3 (Fe represents the iron porphyrin nucleus of P-450). Complexation of O_2 at iron (II) porphyrin is followed by electron transfer, acylation, and heterolytic cleavage of the O-O bond resulting in formation of an oxyferryl particle capable of hydroxylating the CH bond [17]. Hamilton [20] has reviewed enzymatic and model oxidation reactions for various classes of enzymes and groups of organic compounds as substrates. Oxidases are enzymes that catalyze the transfer of hydrogen from some substrate to O_2. Depending on the specific enzyme, either H_2O_2 or H_2O may be the product of O_2 reduction in these cases. The term oxygenase is reserved for enzymes that catalyze the incorporation of oxygen from O_2 into at least one other substrate; i.e., in these reactions at least one of the original oxygen atoms of O_2 ends up in a product other than H_2O_2 or H_2O. Oxygenases are further subdivided into two groups: (a) monooxygenases, which catalyze the incorporation of one of the oxygen atoms of O_2 into some substrate and the reduction of the other oxygen atom to H_2O, and (2) di-oxygenases which catalyze the incorporation of both oxygen atoms of O_2 into some substrate(s). Most redox enzymes utilize O_2 as a reactant; peroxidase and catalase utilize H_2O_2 as the oxidant.

In the oxidation of a substrate one assumes that the active site of the cytochrome is in close proximity to the heme iron because it is the iron atom that binds the molecular oxygen in forming a ternary iron-oxy-substrate complex prior to monooxygenation [21].

Nonenzymatic Model Systems (Biomimetic Systems)

The oxidation enzyme systems of microsomal fractions are responsible for the metabolism of xenobiotics. The MFO system includes an electron donor, an electron transfer protein (flavoprotein) and P-450 as terminal oxidase. These systems may be studied *in-vitro*, however, P-450 systems are unstable and their preparation is often expensive; the cost is further increased by the quantity of the cofactor NADPH. Research dealing with P-450 catalysis can be approached in several ways:

1) by stabilization of the enzyme system through immobilization of microsomes or of isolated components ;
2) by substitution of the electron donor and electron transfer components with simpler molecules and
3) by preparation of model systems exhibiting P-450-like properties.

Details of the chemicals used in most model systems has been previously reviewed (1). Table 1 lists the systems and references to environmental chemicals studied for each system. This Table contains many references to compounds (studied in these systems) which are not necessarily discussed in the text.

Fenton Oxidation System

Fenton's reagent [22, 23] has been widely employed as a model in investigations of the mechanisms of enzymatic hydroxylations [24, 25] and has served as a useful model for the microsomal mixed-function oxidases. The products resulting from aromatic hydroxylation using Fenton's reagent (employing a combination of $FeSO_4$, H_2O_2) have been shown to be similar to those produced enzymatically [26]. The Fenton-type of reaction (nonenzymatic hydroxylation) can take place *in vivo* [27–30]. Hydroxyl radical formation in the brain, via a Fenton-type reaction, has been described [31]. Details concerning the elucidation of oxidizing species were recently reviewed [32].

The Fenton process is probably the earliest chemical means of generating hydroxyl radicals. Fenton [22] oxidized tartaric acid by the action of a mixture of ferrous ion and H_2O_2. Subsequent work by Haber and Weiss [33] suggested that the oxidant was the hydroxyl radical. The overall reaction in the Fenton system is $Fe(II) + H_2O_2 \rightarrow Fe(III) + \cdot OH + OH^-$. Chelating the metal with EDTA allows reactions to be carried out over a much wider pH range. The formation of hydroxyl radicals in aqueous media was recently reviewed [34]. Fenton oxidation may proceed with other metals: $Mn(II) < Cu(I) < Co(II) < Fe(II)$. The addition of ascorbic acid increased the activity of the system but altered the reactivity $Mn(II) < Fe(II) < Co(II) < Cu(I)$. Chemical action of hydroxyl radicals may involve hydrogen abstraction $(RH + \cdot OH \rightarrow R\cdot + H_2O)$ or addition reactions with π electron systems. With aromatic rings hydroxylation occurs via a hydroxycyclohexa-dienyl radical.

Electron transfer reactions occur with metal ions and are a complicating factor with Fenton's reagent:

$$Fe(II) + OH\cdot \rightarrow Fe(III) + OH^-$$
$$2OH\cdot \rightarrow H_2O_2$$
$$OH\cdot \rightarrow H^+ + O^{\cdot-}$$

The $O^{\cdot-}$ radical anion has different properties compared with the $OH\cdot$ undergoing hydrogen abstraction rather than addition to π cloud systems. Oxidation with Fenton's reagent is sensitive to experimental technique; for example the products formed can depend on the order of addition of the reagents [35]. The interaction of hydroxyl radicals on various functional groups or class of organic compound has been reviewed [35].

Pesticides

S-triazine pesticide dealkylation by free radical generating systems was reported [36] for atrazine, simazine and propazine. Similar reactions to *in vivo* metabolism were found using the Fenton system. Phenoxyacetic acids and halogenated substituted phenoxyacids were studied in the Fenton system resulting in the following oxidative products: CO_2, formaldehyde, phenols and 1,2-diphenoxyethanes [37].

Aromatic Hydrocarbons

Polycyclic aromatic hydrocarbons (PAH) have been studied in several model nonenzymatic systems (including Fenton's system) and the products formed compared with *in vivo* oxidative products or *in vivo* enzymatic oxidative products (e.g. naphthalene, [38]). For a series of alkylated benzopyrenes the products formed in the Fenton system were identical to those formed in a horseradish peroxidase system [39]. Nagata et. al. [40], demonstrated a potentiation of benzo(a)pyrene carcinogenicity when reacted with Fenton's reagent. The attack by Fenton's reagent on benzo(a)pyrene probably resulted in the formation of hydroxylated epoxy-benzo(a)pyrene thus yielding the ultimate carcinogen. Benzene and toluene were oxidized in the Fenton's system forming o-, m- and p-hydroxylated products [41]. The mechanisms of these reactions involving the hydroxy radical are well understood, and have been reviewed (395, 398). When Cu(II) was replaced for Fe(III) in the Fenton's system [41], hydroxylation yields for benzene or toluene were changed. The yield of different o-, m-, p-isomers were different and dependent upon the concentration of the transition metal.

Other Environmental Chemicals

Aminophenols were oxidized with $Fe(III) + H_2O_2 + EDTA +$ phosphate buffer and compared with $Fe(III) + H_2O_2$, however, no products were found for the latter system [42]. Snook and Hamilton [9] studied the effect of various substituents on phenylalkanols on the yield of oxidation products formed in the Fenton system. They also reported the effect of pH and competitive oxidation of two substrates [43].

Most recently Jarman and Manson [44] studied a MFO and Fenton's model system for the oxidation of N-nitrosomorpholine. Both systems formed similar products e.g. acetaldehyde, glyoxal, N-nitroso-2-hydroxy morpholine *interalia*. They concluded that the Fenton's system and other biomimetic models are sometimes considered unsatisfactory simulations mechanistically of biological oxidation. Nevertheless, they resemble microsomal systems in producing electrophilic oxidizing species under mild conditions, as well as, similar oxidation products. Nitrosamines have also been studied in the Fenton system as discussed in section 2.2.2.

The Udenfriend (or Ascorbic Acid) Oxidation System

In 1952 Udenfriend and Cooper [45] showed that a water soluble system obtained from rat liver could, in the presence of NADP and oxygen, catalyze the conversion of L-, but not D-phenylalanine to tyrosine. Subsequent studies showed that ascorbic acid increased the efficiency of the conversion of tyramine to hydroxytyramine by adrenal medulla homogenates. Udenfriend et. al. [46] then found that the adrenal tissue was unnecessary, and that the hydroxylation and other compounds could be effected at 37° C by oxygen in the presence of ferrous sulphate, ascorbic acid, and EDTA in phospate buffer. This system became known as the Udenfriend or ascorbic acid oxidation system (AAOS).

The AAOS has been shown to be an especially good model for biological and photochemical radical oxidation [1, 46–50]. Recently, it was reported that the ascorbic acid oxidation system was an excellent model for microsomal oxidation of alcohols and the data supported a mechanism consistent with the concept of hydroxy radical formation [51].

Pesticides

Recently Worobey [50] reported the oxidation products from the herbicide linuron in the AAOS. The results obtained show that the AAOS is capable of effecting demethoxylation, demethylation, hydroxylation of the aromatic ring, dechlorination, and cleavage of the amide functional group to a substituted phenylisocyanate. Most of these reactions are the same as those observed using microsomal fractions. It therefore appears that the products formed by interaction of linuron with hydroxy free radicals generated by the AAOS closely resemble metabolites formed *in vivo* in plants and animals or formed by photochemical decomposition.

Locke [93] investigated the metabolites produced from carbaryl by rat liver; upon acid hydrolysis of the water soluble glycones (conjugates) he identified N-hydroxycarbaryl. N-hydroxycarbaryl was also formed by a mammalian enzymatic system and by the AAOS.

Nitrosamines

The AAOS has been used to study the activation of nitrosamines to mutagenic oxidation products. Hsieh et. al. [52] used this system with diethylnitrosamine and nitrosopiperidine: the obtained products were identified and tested using the bacterial Ames test. Other reports using the AAOS with nitrosamine substrates are described in the literature [53–56]. Archer and Eng [57] reported the chemical activation of nitrosodiethylamine and nitrosopiperidine in the AAOS; after oxidation both nitrosamines became potent alkylating agents. They concluded that the Udenfriend model system may prove useful in elucidating the mechanism of carcinogenic and mutagenic alkylating agents and for detecting alkylating agents in environmental samples.

Suzuki et. al., [58] studied the oxidation products formed from N,N-dibutylnitrosamine (DON) in AAOS, Fenton's and MFO systems. Three hydroxyl derivatives were formed in all systems indicating the faithful replication of oxidation products by these biomimetic systems with enzymatic products; the pattern of oxidation products (regioselectivity) and the position(s) oxidized (stereoselectivity) were similar to MFO oxidation. Similar studies have been reported for other nitrosamines [59–62].

Aromatic Amines

N-Hydroxyaromatic compuonds are of great interest because of their potential mutagenic and carcinogenic effects [51]. Mayer [63] showed that N-hydroxylated derivatives of 1-naphthylamine and 2-naphthylamine produced by the AAOS are mutagenic. Thus, in addition to producing typical oxidation products found with *in vivo* systems, the AAOS may be used to screen for the possible formation of N-hydroxy metabolites. This information would enable one to predict their presence in the food supply which may pose a serious toxicological hazard to man, as well as, other non-target organisms.

The carcinogenic amines are recognised as requiring metabolic activation for these compounds to exert their carcinogenic potential. Aromatic amines are metabolically converted by hydroxylation of the amino nitrogen or ring carbon atoms, and oxidative dealkylations to more carcinogenic compounds. Use of the AAOS may serve as a useful technique to evaluate the possible genotoxic effects of metabolic breakdown products of compounds formed *in vivo* by hydroxylating mechanisms.

Polycyclic Aromatic Hydrocarbons

Tierney et. al. [64] reported a rapid simple method of obtaining small amounts of trans-dihydrols of PAHs. They required these metabolites of benz(a)anthracene and 7,12-dimethylbenz(a)anthracene for use as reference compounds for metabolite analysis in *in vivo* studies and to help elucidate the mechanisms of reactive metabolites formed *in vivo*. They used the AAOS to successfully prepare many trans-dihydrol oxidative metabolites which would have been more difficult to prepare using unambiguous multistage synthetic routes. The major dihydrols, trans 5, 6 and trans 8, 9 were prevalent as is the case for *in vivo* prpearation using rat liver microsomal preparations.

Previously, the AAOS was used to generate dihydrodiols to prove or provide evidence for the hypothesis that diol-epoxides were involved in the metabolic activation of PAHs and are derived from non K-region dihydrodiols, and that the formation of reactive epoxides within the bay region of hydrocarbons is of biological importance. Sufficient material of K-region and non K-region dihydro-diols was obtained for use in PAH metabolism studies, as well as, genotoxicity studies [64].

Other PAHs were also studied in the AAOS and the products formed compared with those from liver microsomes [65]. The chemical and enzymatic oxidation of benzo(a)pyrenes to carcinogenic forms (e.g. 7,8-diol-9, 10-epoxide) was also reported [66].

Peroxidizing polyunsaturated fatty acids can convert benzo(a)pyrene to hydroxylated oxidized products when catalyzed by Fe(II)/ascorbic acid in the absence of any enzyme [67]. The ultimate carcinogenic benzo(a)pyrene 7,8-diol was formed in the presence of polyunsaturated fatty acids undergoing peroxidation catalyzed by Fe(II)/ascorbic acid in the absence of any enzyme (The biomimetic system contained: ferrous sulfate, ascorbate, arachidonic acid or docosahexaneoic acid in acetone, and phospate buffer at pH 7.4). Diols and tetrols were detected using this model system, as well as, benzo(a)pyrene 7,8-diol, 9,10-epoxide which exists in two stereo isomeric forms-diol epoxide I and II [68]. Tierney et. al. [69] reported the preparation of dihydrols from 7-methylbenz(a)anthracene with the AAOS in which all five possible dihydrodiols were formed. Preparation of oxidative metabolites by the AAOS permitted correlations between structure and biological activity; particular dihydrodiols involved in metabolic activation of carcinogenic PAHs could thus be confirmed.

Various mono and dihydroxy oxidation products of PAHs were formed and were identical to PAH hydroxylation studied using the AAOS, rat liver microsomes and whole animals [70]. The PAHs studied were: phenanthrene, benz(a)anthracene, naphtalene, 1- and 2-naphthol, 1,2-dihydroxynaphthol, antharcene and pyrene.

Other Environmental Chemicals

Jones and Walsh [71] compared the oxidation products of 1-bromopropane in the AAOS and in the rat: oxidation products (3-hydroxypropionate, 1-bromo-2-hydroxypropane, and 1,2-epoxypropane) were the same in both cases. AAOS oxidation of PAHs produced some products (e.g. 11, 12-diols) not found in microsomal oxidation [39]. Jones and Gibbon [72] used the AAOS with 1,2-dichloropropane and identifed propane-1,2-diol the same *in vivo* metabolite found in the rat. Salicylate reaction in the AAOS resulted in 2,3 and 2,5-dihydroxy benzoates while benzoic acid gave o-, m- and p-hydroxybenzoates [73].

Anisole hydroxylation was studied in the AAOS and 8 products were identified [74]; oxidation was improved by addition of metallic iron (Fe°) for continuous elimination of oxalic acid since this by-product of the oxidation of ascorbic acid displaces EDTA and inhibits selective hydroxylation.

Peroxy Acid Oxidation Systems

Casida and Ruzo [75] recently reviewed peroxyacid oxidations as possible biomimetic models of nonenzymatic oxidation. They found that reactive intermediates generated by metabolic epoxidation, N-oxidation and S-oxidation are often identical to those obtained on peroxyacid oxidation as illustrated with a variety of pesticides. Peroxyacid systems are not always suitable biomimetic models for N-oxidation but epoxidations, sulphoxidations and reactive intermediates formed are the same as those formed by microsomal oxidases. Their review considers reactive intermediates formed by epoxidation, N-oxidation, and S-oxidation reactions that proved difficult to study without the use of biomimetic systems. Advantages of peracid systems as biomimetic models for xenobiotic metabolism include:

1. The rate or sites of oxidation is governed in part by oxidative strength of the peracid and the reaction conditions. Commonly used acids: peroxyacetic, peroxybenzoic, peroxyphthalic, peroxytrifluoroacetic, and especially m-chloroperoxybenzoic acid (MCPBA) are preferred for reactivity, yield of products and ease of storage and use. Product yields are improved and isolation was facilitated by addition of potassium fluoride to the MCPBA reactions or by binding the peroxyacid to insoluble polymers to decrease the time of exposure to the organic acid.
2. Products are often obtained in large amounts because many xenobiotics and system reactants readily dissolve in organic solvents.
3. Temperature of the reaction can be varied over a wide range which is an advantage when studying unstable products.
4. ^{18}O-MCPBA is useful in mechanistic studies with mass spectrometric analysis.
5. It is often possible to use nuclear magnetic resonance to directly monitor reactions.

The peroxyacid model systems are mechanistically different from P-450 reactions where a formal perferryl oxygen species $[Fe(v)=O)III]$ exists. Peroxyacid oxidations are generally due to $+OH$ species. Reaction patterns may therefore differ and functional groups that are readily metabolized in the MFO system appear unreactive to peroxyacids. Sometimes difficulties result from incorporation of the acyl fragment from the peracid into the oxidation product. Consequently applicability of the model is greatest in cases of epoxidation, amine oxidation and S-oxidation when the MFO and peroxyacid oxidizing species lead to the same ultimate products and reactive intermediate.

These biomimetic systems have been helpful in elucidating activation mechanisms involving: epoxidation of pyrethroids and juvenoids (tetramethrin, allethrin, tetrallethrin), P-oxidation of dimethylphosphoramides (e.g. the insecticide schradan), N-oxidation of nitrophenylether herbicides (e.g. nitrofen, acifluorofen, oxyfluorofen) and S-oxidation (metribuzin, EPTC, diallate and triallate herbicides, sulprofos, profenofos organophosphorous insecticides). The oxidation of these environmental chemicals will be discussed in detail later.

The epoxidation of a variety of unsaturated compounds by peroxyacids is well established and the mechanism is considered to occur through a concerted reaction involving the OH^+ ion [24, 26, 76, 77]. Peroxytrifluoroacetic acid (PTFAA) is an oxidizing reagent which has been used to study the mechanisms of the mixed function oxidase catalyzed hydroxylation of aromatic substrates [26, 76, 78, 79].

The observation of intramolecular migration of aryl ring substituents (a 1,2-shift known as the NIH shift) during both enzymatic and nonenzymatic hydroxylation using PTFAA [80] verifies the usefulness of this compound as a model catalyst for the MFO catalyzed metabolism of aromatic compounds.

Peroxyacid oxidations are commonly carried out in organic solvents. However, aniline oxidation with peroxyacetic acid was found to proceed.twenty-five times faster in water than in ethanol [81]. Aniline was oxidized to phenylhydroxylamine, nitrosobenzene, azobenzene and azoxybenzene. Oxidations with PTFAA include: hydroxylation of aromatic rings, epoxidation of alkenes, oxidation of unsaturated ketones, oxidation of amines and oximes to nitro compounds, ketone conversion to esters, alkenes to ketones, and oxidation of alkanes [82]. Zinburg and Ballschmitter [83] reported the oxidation of hexachlorocyclohexanes (e.g. the insecticide lindane) to various chlorosubstituted phenols using NiO_2 peroxy systems.

m-Chloroperoxybenzoic Acid (MCPBA) Oxidation System

The general usefulness of MCPBA may be seen from the reactions reported for the following classes of compounds [84]: olefins to epoxides, cyclopropenes to α, β-unsaturated aldehydes and/or ketones, α, β-unsaturated ketones and esters to epoxides, ketones to esters, acid chlorides to alcohols, acids to alcohols, primary alkylamines to nitroalkanes, primary aromatic amines to aromatic nitroso compounds, secondary amines to nitroxide radicals, tertiary amines to N-oxides, and sulfides to sulfoxides or sulfones.

Pesticides

Peracids in organic solvents provide many of the desired features for a model chemical oxidation system. Bellet and Casida [85] investigated the products of MCPBA oxidation of several organothiophosphorus insecticides in relation to those formed by microsomal oxidase metabolism and by photo-decomposition. They reported that the MCPBA oxidation system generates many oxidative products similar to those obtained via enzymatic and photochemical oxidation. They also suggest that an understanding of MCPBA oxidation of a variety of organothiophosphorus compounds can provide further insight into the mechanisms by which they undergo enzymatic metabolism and photodecomposition. The MCPBA system provided a means of preparation of possible metabolites and photoalteration products of thionophosphorus compounds not readily accessible by other means or for investigations on the biochemical toxicology and environmental fate of these pesticides [86–88].

Biomimetic oxidation of the herbicide glyphosate and the insecticide fluvalinate may involve transformation at the α-carbon by hydroxylation or by N-hydroxylation [89]. The latter process was examined with MCPBA oxidation as a possible biomimetic system. Several products were isolated and were identical to those obtained by *in vivo* metabolism; trapping experiments also supported the presence of imines as reactive intermediates in glyphosate degradation. The authors [89] rationalized their results in terms of free carboxyl involvement in the reactions and initial N–hydroxylation followed by degradation to an imine which can undergo hydrolysis or further oxidation; yields were 6–10%. Thus oxidation of these compounds in the MCPBA system led to many of the same products formed during *in vivo* metabolism and environmental degradations.

The identification of pesticide reactive intermediates is important in understanding pesticide degradation pathways and their possible toxicological role [90]. *In vitro* model chemical oxidations are used to achieve these transformations. MCPBA readily converted the pesticides, resmethrin and methfuroxam, to unsaturated ketoaldehydes or diketones [90]. These products are also obtained by photolysis or in *in vivo* metabolism. Epoxide intermediates formed in these reactions were established by direct NMR observation and mechanistic studies with ^{18}O-MCPBA.

The possible metabolic formation of disubstituted anilines and nitrosobenzenes from formamidine insecticides (chlordimeform and amitraz) and chloroacetamide herbicides (metalachlor, alachlor and butachlor) was studied by MCPBA oxidation and compared with *in vivo* (rat) and hepatic MFO metabolism [91]. *In vivo* metabolism of 2,4-, and 2,6-disubstituted anilines occurred with conversion to nitrosobenzenes. The authors prepared nitrosobenzenes with MCPBA to obtain standards for genotoxicity testing and found 2,6-dialkynitrosobenzenes to be direct-acting mutagens. Alkylamines where oxidized to various derivatives including: hyroxylamino-, nitroso-, nitro- and azoxy- derivatives. Alachlor was oxidized to 17 products which permitted comprehensive toxicological profiles of "typical" oxidative metabolites to be established [91].

Ruzo et. al. [92] also studied the MCPBA oxidation of several nitrodiphenyl herbicides (nitrofen, acifluorfen, oxyfluorfen) and found several mutagenic (Ames test) oxidative degradation products related to nitrosobenzenes and hydroxylamino intermediates, which may contribute to their toxicological profiles.

The formation of the sulfoxides of S-triazine herbicides using MCPBA has also been reported [93]; biomimetic oxidations were compared with microsomal oxidation products and similarities were discussed. Miyamoto with Yamamoto [211] studied the transformation of an organophosphate insecticide [S(2-acetylaminoethyl) dimethylphosphorodithioate] under various oxidative conditions. MCPBA oxidation in CH_2Cl_2, a rat liver microsome NADPH system, and photodecomposition (sunlight irradiation on glass or bean leaves) yielded similar products from all systems.

Oxidation of (S)p-fonofos by MCPBA took place stereoselectively [95], resulting predominantly in (R)p-fonofos oxon and (R)p-phenyl ethyl (ethoxy) phosphinyl disulfide. (R)p-fonofos gave (S)p-oxon and (S)p-disulfide; oxidation to the oxon proceeded with retention, while oxidation to the disulfide occurred with inversion. Mechanisms of isomer product formation for phosphorothionate insecticides, were proposed on the basis of these studies.

Simulation of the biotic oxidative transformation of di(2-ethyl-hexyl)phthalate [DEHP] and dioctylphthalate by abiotic means was studied by Brodsky et. al. [96]. DEHP is a high volume industrial chemical used as a plasticizer for PVC; it is ubiquitous in the environment and can cause tumors in bioassays of rats and mice. The authors studied oxidation with Fenton's reagent (in CCl_4 + cupric perchlorate or ferric perchlorate $+30\%H_2O_2$) with no products detected; however, with MCPBA in CH_4Cl_2 there were several products (ketoalcohols and diols) formed involving hydrolysis to the monoester followed by hydroxylation of the ester alkyl group and further oxidation to the corresponding ketone or an ω-alcohol to a carboxylic acid. MCPBA oxidation proceeds to the ketone stage most likely through further oxidation of an initially produced alcohol compared with trifluoroperacetic acid (TFPAA) which gives alcohols as their trifluoroacetates; also there is exclusive aliphatic oxidation with MCPBA whereas, TFPAA hydroxylates aromatic rings.

Iron Porphyrin Oxidation System

Iron salts in the presence of H_2O_2 can catalyze the oxidation (dealkylation and hydroxylation) of various chemicals by affecting the interaction of O_2^- and H_2O_2 after chelation with EDTA with production of more reactive O_2 intermediates. The catalytic power of iron greatly increases when it is coordinated with the ring of protoporphyrin IX and, especially, as a prosthetic group of P-450. The structure and reactions of iron porphyrins were recently reviewed [97, 98].

Numerous biomimetic systems have explored the chemistry of catalysis by P-450 and related enzymes. Of those, the closest model is the system which uses iron porphyrin to catalyze the insertion of oxygen from iodosylbenzene into a substrate [99]. The catalases and most peroxidases are haemoproteins in which the essential metal cofactor is present in the native enzymes as an iron (III) porphyrin (ferriheme). In these species 2/3 of the octahedral (or near-octahedral) first coordination sphere of iron is occupied by the porphyrin ligand (commonly called protoporphyrin IX). The remaining coordination positions are probably occupied by ligands derived from the protein, solvent or solutes. Protein-free water soluble ferrihemes exhibit both "catalatic" and "peroxidatic" properties [100].

Various iron porphyrins (IP) of natural and synthetic origin have been used as models of P-450. Sakurai [101] has reviewed studies on the bioinorganic chemistry of P-450 models (metalloporphyrins). Changes of the porphyrin ring substituents generally have only a secondary influence on the reactions. Mansuy [102] has detailed the similarities of IP biomimetic systems to P-450 enzyme properties.

The design of heme models able to mimic the reactions catalyzed by P-450 dependent monooxygenases has escalated in recent years. They have proven especially useful not only for understanding the structure of the active oxygen P-450 complex, which is largely unknown, but also for the preparation of substantial amounts of metabolites of xenobiotics. Their close similarity to P-450 makes the IP system an excellent model for determining the mechanism involved in the formation of reactive metabolites of xenobiotics.

Systems associating an IP with a thiol and a base in large excess in the presence of dioxygen have been used to oxidize a wide range of xenobiotics (see Table 1). Since it is difficult to mimic the dioxygen activation cycle of P-450 (Fig. 1.), where the arrival of dioxygen and electrons are determined by the protein components of the monooxygenase, it appears easier, initially, to mimic the peroxy reactions of

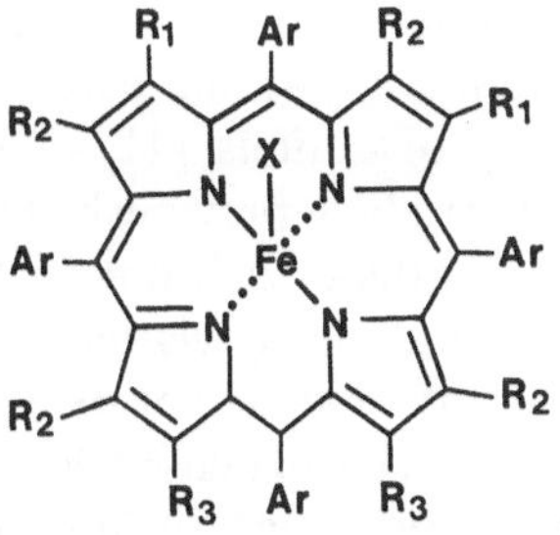

Fig. 4. Structural diagram of iron porphyrin nucleus and Table of Analogues

P-450. And because of the rapid, irreversible oxidation occuring between thiolates and any peroxy compound, research has focused on systems that associate an iron-porphyrin and a peroxy compound in organic solvents. For details the reader is referred to the review of Mansuy [102].

The study of metalloporphyrin properties has proven very useful in understanding the reaction mechanisms of peroxidase, catalase and cytochrome-catalyzed reactions and has been reviewed previously [102, 157, 158, 210]. IP and oxidizing agents such as iodosylarenes or alkylhydroperoxides have been shown to hydroxylate various substances including inert hydrocarbons.

IP model systems basically involve an iron porphyrin substituted nucleus, as shown in Fig 4:

Formula of IPs

Name	R_1	R_2	R_3	Ar
Iron protoporphyrin IX (Fe PPIX) [hemin, Cl salt]	–vinyl	–CH_3	–$CH_2CH_2CO_2H$	–
Dimethylester of Fe PP IX (Fe PP IX DME)	-vinyl	–CH_3	–CH_3	–
Iron deutero porphyrin DME (Fe DP DME)	-H	–CH_3	–CH_3	–
Iron octaethylporphyrin (Fe OEP)	–C_2H_5	C_2H_5	–C_2H_5	–
Iron tetraphenylprophyrin (Fe TPP)	–H	–H	–H	–C_6H_5
Iron tetraparatolylporphyrin (Fe TTP)	–H	–H	–H	–para–CH_3–C_6H_4

All P-450 enzymes contain iron protoporphyrin IX as the prothetic group and have a common reaction cycle with four well-characterized states:

1) a ferric resting state,
2) reduction to a ferrous form,
3) binding of this to dioxygen (for the subsequent hydroxylation step) and
4) a hydroxylation step.

Metalloporphyrins have utilized oxidants such as iodosylbenzene [107–109], hypochlorite [110–113], persulfate [114] and peroxide [115, 116] and have been reported to catalyze many O_2 oxidations [117–121]. The redox properties of metalloporphyrins have been studied previously [122–124, 126]. They are of interest as biomimetic models of the cytochromes which are enzymes containing iron porphyrins, and function catalytically via the Fe(II) $\rightleftarrows$ Fe(III) couple and of hemoglobin (an Fe(II) porphyrin) which is not oxidized by oxygen to Fe(III) but reversibly binds oxygen at the Fe(II) oxidation level. Dolphin et al. [126] concluded, as a result of model studies using various IPs, that the electron tranfer which accounts for the overall Fe(II) $\rightarrow$ Fe(III) couple is unlikely to involve direct electron transfer at the iron but rather electron transfer occurs initially at the porphyrin periphery followed by an internal electron transfer i.e.

$$Fe(II) \underset{+e^-}{\overset{-e^-}{\rightleftarrows}} [Fe(II)heme]\cdot + \rightleftarrows Fe(III)heme$$

Taylor and Xw [125] recently reported a biomimetic model for the enzyme catalase; experiments were reported which revealed the mechanisms of each of the steps of the model system employing hydrogen peroxide and hydroperoxides with iron (II) prophyrins. Reactions that have been studied include: aromatic hydroxylation [127–129], aliphatic oxidation [130], oxidative dealkylation [131], N-demethylation [132] and aromatic methyl migration [133].

Sukurai [134] has studied models for P-450 under physiological conditions. He used hemin-thiol (cysteine or cysteine methyl ester) in the presence of a base such as pyridine, imidazol and histidine and showed that under exposure to carbon monoxide a typical P-450-CO type complex was formed. He reported that cysteine binds to ferric porphyrin to form axial S-Fe-N coordination in the presence of base and is easily reduced to ferrous porphyrin by exess thiol. Spectral properties (VIS, EPR and magnetic circular dichromism) of the complex were also similar to that for the P-450-CO complex [135–140]. Optical and EPR spectra of P-450 were also similar to those for the biomimetic oxidation system of α-mercaptopropionylglycine or mercaptopropionic acid with Fe(II) or hemin and pyridine [141]. Detailed mechamistic studies have been reported for metalloporphyrin biomimetic systems [142, 143].

Aniline

Sukurai et al. [144] also studied aniline hydroxylation using heminthioglycolate ester and various reducing agents. Hydroxylation (o, m and p) was enhanced upon addition of a thiol as a reducing agent suggesting an involvement of a similar reaction for P-450 monooxygenases requiring NADPH as a reducing agent. Adams et. al. [145] found that hemin NADH or NADPH system catalyzed p-hydroxylation of aniline with an efficiency approaching that of the enzymic system. Adams et al. [146] investigated the mode of interaction between aniline, and other substituted anilines, in the hemin (ferriprotoporphyrin IX) system. The interaction resembled P-450 reactions with regiospecifity (p-hydroxylation) and competitive inhibition by hydroxy radical scavengers [147]. Recent studies indicate that hemin plays an important biological role as the prosthetic group of proteins involved in the *in vivo* transport and metabolism of molecular oxygen. That is, iron porphyrins and their complexes may be potential models for some of the oxygen activation/insertion reactions mediated by P-450.

Hemin provided higher hydroxylation activity than Fe(II) for aniline hydroxylation in a thiol/hemin or Fe(II) system [141]. Also α-mercatopropionylglycine was more effective than mercaptopropionate in aniline hydroxylation. Hydroxylation yield of the α-MPG/hemin system was approximately 1.5% compared with that by rat liver microsomes. Ortho/para ratios of hydroxylation were lower for the model system compared with microsomes. They concluded that the biomimetic system used for hydroxylation was influenced by some factors such as solvent effect or polarity, tertiary structure of the thiol and steric interactions; however the biomimetic system was still a useful model and formed no meta isomer as in the biological system providing excellent stereoselectivity.

Aniline hydroxylation with a hemin-cysteine system was studied as a model for P-450 enzymes by Sakurai [148,149]. The reaction was characterized by use of various inhibitors such as hydroxyl scavengers KI, CH_3OH, DMSO or formic acid, a singlet oxygen trapping agent 1,3-diphenylisobenzofuran, SOD and catalase. This system was compared with the AAOS and Fenton's oxidation system. Possible mechanisms for hydroxylation by the hemin/cysteine system were proposed. Aniline hydroxylation was not inhibited strongly by any inhibitor and therefore does not involve ·OH, O_2^- $^1O_2$ or H_2O_2, and may have a different reactive intermediate; other systems were inhibited; and optimum pH and reaction times were established. Sakurai et. al. [150–153] also studied other thiols in a hemin model system using p-methylanisole as a substrate; several ring hydroxylation products were identified.

Aromatic Hydrocarbons

Para-toluidine oxidation was studied in several hemin systems using different thiols [154, 222]. The hemin/thiol system was a good model for P-450 liver microsomal MFO oxidation. The yield of methyl group migration (for p-toluidine) and hydroxylation was dependent upon the type of thiol, chemical form of iron [Fe(II) or hemin], pH of the reaction mixture, reaction time, concentration of the substrate and/or thiol. Thiols studied included: β-mercaptopropionic acid, thiosalicylate, 2,3-dimercaptopropanol, cysteine, 0-aminobenzenethiol and thiophenol. 2,3-Dimercaptopropanol with Fe(II) gave the highest yields of hydroxylation products followed by cysteine with hemin.

DDT has been shown [156] to be degraded rapidly by the P-450 model system containing hemin/cystein. Products identified are shown below (Fig 5.): (see p. 80).

In the presence of a designed 24-residue DDT-binding peptide [157] the rate of DDT degradation was increased suggesting that the peptide-hemin-cysteine mixture might represent an early stage in the evolution of an enzyme. Horseradish peroxidase catalyzes the oxidation of reduced glutathione; this reaction is accompanied by light emission which is attributed to the generation of singlet oxygen. The

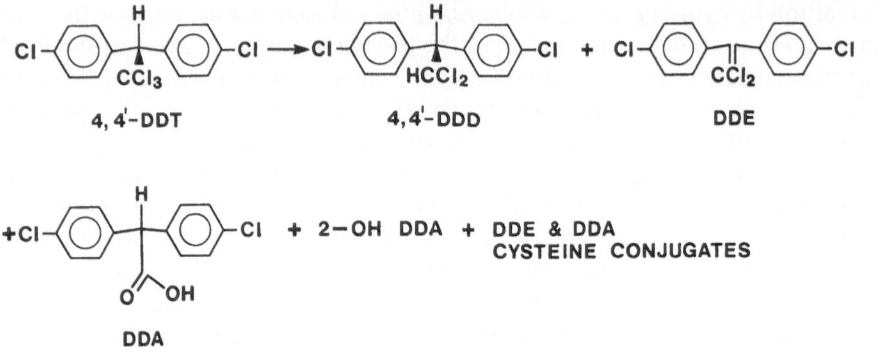

Fig. 5. DDT oxidation in the hemin/cystein biomimetic system

chemiluminescene was shown to be directly related to thiyl radical formation. Replacement of horseradish peroxidase by hemin was also accompanied by chemiluminescene [158] and so may involve a similar mechanism in the hemin/thiol biomimetic systems.

Sukurai et al. [159] compared the hydroxylation of p-chlorotoluene in several model systems; the hemin-thiol system was found to be the best model of P-450 monooxygenases and the AAOS was better than Fentons system. The hydroxylation of p-chlorotoluene with hemin-thiol complexes [160] resulted in five hydroxylation of products; methyl NIH shift was demonstrated and cysteine ethyl ester was the best thiol. Product formation was compared as were product ratios for various thiols and for other model systems. The hemin-thiol biomimetic system appeared to be an excellent model for P-450 monooxygenases.

Alkanes and Alkenes

Recently, model systems have been described [103, 161–172] that mimic the enzymatic oxidation of alkanes and olefins. These systems consist of a synthetic metallo(III)porphyrin and a singlet oxygen donor such as iodosylbenzene. Grove et. al. [173] studied the stereoselectivity of iron porphyrin oxidation model systems on cis and trans stilbine. Alkenes and alkanes were also oxidized in a system containing: FeTPPCl, CH_2Cl_2, and p-cyanodimethylaniline N-oxide [174].

Alkene epoxidation was regioselective (styrene, stilbenes etc.) using an IP system [175]. The 2-iodoso-m-xylene oxidation of tertiary aromatic amines catalyzed by iron (III) chloro-α, β, γ, δ-tetraphenylporphinato (Fe(III) TPPCl) was reported for N,N-dimethylaniline [176]; N-demethylation and N-methoxylation were prominent.

Olefin oxidation was studied using iodosylbenzene with various synthetic IPs [177, 178]. The wide range of oxidative transfers catalyzed by P-450 suggested that metalloporphyrin complexes should also catalyze such reactions under suitable conditions. Olefin oxidation was studied using iodosylbenzene with various synthetic iron porphyrins [177,178]. Tetraphenylporphinato iron (III) catalyzed the reaction of several olefins (cyclohexene, biphenyls) with molecular oxygen (179); oxidized products were isolated in high yields.

Hydrocarbons

Oxygenation reactions of hydrocarbons (epoxidation and hydroxylation) catalyzed by synthetic metalloporphyrins may be associated with various oxidants, e.g.
single oxygen donors such as iodosylbenzene, sodium hypochlorite, alkylhydroperoxides, hydrogen peroxide, amine N-oxides, potassium hydrogen persulfate or
molecular oxygen in the presence of an electron source. The metalloporphyrin-
catalysed oxygenation of hydrocarbons was recently reviewed [180].

Synergists

Iron porphyrin-carbene complexes were prepared [181] for models of P-450
complexes and their effects upon metabolic oxidation on the insecticide synergists
of the 1,3-benzodioxole series studies.

Mansuy et. al. [181] studied the high stability of the iron-metabolite bond in the
iron porphyrin-benzodioxole-carbene complexes. The high stability of the complexes is probably at the origin of synergistic action of the benzodioxole derivatives. It has been proposed that the iron ligand present in these complexes is the
1,3,-benzodioxole-2-carbene formed by oxidation of the methylene group of 1,3-
benzodioxole. They isolated and characterized an iron (II) porphyrin complex
which brings indirect evidence of the presence of this carbene as a ligand in the
benzodioxole-derived P-450–iron (II)complexes.

The formation of the P-450-benzodioxole-derived carbene complex is the first
example of the involvement of an iron-carbene bond after the NADPH- and O_2
dependent oxidation of a substrate by P-450. This should be an important point in
the understanding of the mechanism of oxygen activation and oxidation of some
substrates by P-450.

Mansuy et al. [266] prepared an iron porphyrin-carbene complex as a model for
P-450 complexes obtained upon metabolic oxidation of 1,3-benzodioxole insecticide synergists. It has been proposed [183] that the iron ligand present in these
latter complexes is the 1,3-benzodioxole-2-carbene formed by oxidation of the
methylene group of 1,3-benzodioxole. Mansuy et. al. [266] also described the
isolation and characterization of an IP-1,3-benzodioxol-2-carbene complex, which
provides evidence for the presence of this carbene as a ligand in the benzodioxole-
derived P-450-iron(II)complexes. Visible spectra of the microsomal P-450(II)-
benzodioxole-derived metabolite complex compared with that for Fe(TPP)Cl just
after addition of n-BuS⁻ to both systems resulted in very similar spectra indicating
that the biomimetic system was a good model for the P-450-iron(II)-benzodioxole-
metabolite complex, and strongly supporting the 1,3-benzodioxole-2-carbene
nature of this metabolite. Because of the strength of the iron-carbene bond in
several previously described iron porphyrin-carbene complexes these data could
explain inactivation of P-450 by the benzodioxole carbene metabolite and thus the
severe inhibition of the detoxifying monooxygenases of the insect since various
derivatives of 1,3-benzodioxole are well known insecticide synergists. Their use in
combination with an insecticide results in marked increases in toxicity presumably
because of their ability to inhibit the enzymes responsible for insecticide detoxification.

Other Environmental Chemicals

Miyata et. al. [184] studied the mechanism of N-dealkylation of tertiary alkylated amines using a iron (III) porphyrin biomimetic system; they correlated this to P-450 catalysis. Mansuy et. al. [182] studied the IP catalysis of oxidative dealkylation 4-nitroanisole and 7-ethoxycoumarin by cumylhydroperoxide or t-butylhydroperoxide.

P-450 enzyme contains a protoporphyrin IX group which catalyzes the monooxygenation and oxidative dealkylation of many organic compounds. These enzymes are known to cleave the oxy-oxy bond of molecular oxygen by a two electron uptake followed by elimination of one oxygen atom as H_2O. The other oxygen atom which is left on the iron as a carbene like oxene $[FeO]^{3+}$ is thought to be the key oxidizing species [185].

The catalytic cycle of P-450 is believed to involve reductive activation of dioxygen at the heme center and subsequent peroxy bond cleavage to give a ferryl ion species as the active oxygen transfer agent. Support for an iron-oxo species (I)

$$\text{P-450} \cdot \text{Fe(III)} \xrightarrow[\substack{2e^- \\ 2H^+}]{O_2} \text{P-450} \cdot \text{Fe(III)} \cdot H_2O_2 \rightarrow \underset{I}{\text{P-450} \cdot\cdot \text{FeO(III)}} + H_2O$$

is derived from the fact that a number of single oxygen donors, hydroperoxides, peroxyacids and iodosylbenzene, effect oxygen transfer in a manner similar in many respects to the fully reconstituted enzyme system [186–188]. Groves et al. [82] studied cyclohexane oxidation in the IP systems: chloro-tetraphenylporphinato iron (III) (a) and chlorodimethylferriprotoporphyrin IX (b). The importance of the catalyst was illustrated with cis and trans stilbine (173); with (a) as catalyst, the corresponding cis and trans stilbine oxides with complete retention of configuration was obtained; (b) catalyzed the conversion of cis to the oxide while trans was inert. Trans isomers were formed from both cis/trans stilbine using a tris (acetylacetonato) iron(III)/hydrogen peroxide system. A mechanism was proposed for these differences based on steric consideration for various catalysts and substrate.

Mansuy et. al. [182] studied the IP catalysis of oxidative dealkylation of p-nitroanisole and 7-ethoxycoumarin by cumylhydroperoxide or t-butylhydroperoxide.

Although H_2O_2, ascorbic acid, Cu and Fe transition metals are integral components of biological systems, it is because of the direct similarity of iron porphyrins to P-450 that the most faithful reproduction (regioselectively and stereoselectively) results and is of prime importance in mechanistic and metabolic profile studies. IPs act as iron catalysts for oxygen transfer in oxidative reactions [97, 98].

Other Metal Porhyrin Systems

The desire to understand the mechanism of P-450 catalyzed reactions and the need for selective and effective synthetic oxygenation or reduction catalysts has spurred interest to study heme model systems (transition metal–porphyrin complexes). Fontecave and Mansuy [189] compared extensively the metalloporphyrin

(Mn/TPP)-C_6H_5-IO system to the Mn/TPP-O_2-ascorbate system. The major problem of the Mn(TPP) (Cl)-O_2-ascorbate system was its low yield of oxidation products. The active species was mainly consumed by reduction by ascorbate. This is a general problem for chemical models of monooxygenases which produce a highly oxidizing species in a medium where high concentrations of a reducing agent are required for reductive dioxygen activation. However, it does oxidize alkenes and olefins in very mild and simple conditions (room temperature, pH 8.5) which are similar to those for monooxygenases and it leads to a completely regioselective oxidation of olefins with the exclusive formation of epoxides differing from P-450 and most other metalloporphyrin based models. Data strongly suggested different oxidizing species in these two systems.

The MnTPP (manganese III tetraphenylporphinato acetate) sodium hypochlorite system has been reported to epoxidize olefins [112,113], e.g. epoxidation of cyclohexene, styrene, methylstyrene, cyclooctene, 2-methylheptene etc.

Backvall et al. [190] recently reported the use of a macrocyclic metal complex with p-benzoquinone/hydroquinone as an electron transfer system for oxidation reactions. Three redox couples were used for catalysis involving molecular oxygen oxidation. A multistep electron transfer with Pd(II)/Pd(0), p-benzoquinone/hydroquinone [TPP]/Co(TPP) as redox catalyst couples were used to study the oxidation of 1,3-cyclohexadiene. This oxidation system resulted in high product yields (89% 1,4-diacetoxy-2-cyclohexene) and has similarities with biochemical processes where an oxidation becomes very mild and selective when several redox couples with falling redox potentials are interacting. This is the first model system of a selective oxidation using triple catalysis with oxygen as an ultimate oxidant.

Mansuy et al. [183] studied the selective epoxidation of olefins and hydroxylation of alkanes under mild conditions using a monooxygenase-like dioxygen activation by an Mn(III) porphyrin-ascorbate system. The model reaction proceeds by oxygen atom transfer to form an oxo-metal compound, which is able to remove hydrogen and "capture" the resulting carbon radical with the other product, a metal-bound hydroxyl group. Manganese porphyrins have also been shown to catalyze oxygen insertions.

Sodium hypochlorite and manganese porphyrin with pyridine was found to be a suitable model for understanding the role of P-450 and peroxidase enzymes [191]. Several transition metal ligands were screened. NaOCl was found to exhibit chemoselectivity, stereoselectivity, greater reaction rate, and was highly efficient for epoxidation of di, tri and tetra substituted olefins with 60–70% yields for alicyclic and aliphatic olefins. NaOCl was the best donor of one oxygen atom and especially in the presence of catalytic amounts of manganese porphyrin. Olefin and alkane oxidations were catalyzed by Mn porphyrins in the presence of various oxidants such as O_2, ascorbic acid and iodosobenzene; mechanisms were elucidated and discussed [192].

Cobalt and manganese porphyrins are known to catalyze the oxidation of phenols to quinones by molecular oxygen [193]; these oxidations have similarities with biochemical processes.

Meunier et. al. [194] developed the Mn(III) porphyrin, NaOCl two phase H_2O/CH_2CL_2 system which upon addition of 25% methanol resulted in rate increases of 3 times and changed the reaction order from zero to first order, without changes in product distribution. Iodosylbenzene/Mn(III)chloride oxidation of tertiary amines proceeded by an initial one electron transfer process while t-butyl hydroperoxide oxidation, e.g., was initiated by hydrogen atom abstraction [195].

To clarify the active site, structure and functions of P-450, several hydrophilic and lipophilic model compounds were constructed [101] using metalloporphyrins and various thiol-containing ligands; olefin oxidation was detailed (e.g. cyclohexene by biological systems and model systems). Several model systems were studied and resulted in the same five oxidation products, in the same proportions as isolated from hepatocytes. The various metals studied included Fe, Cu, Co, Mn and vanadate ion. The oxidation of acetanilide was studied in eighteen biomimetic systems. The o, m, p ratios for the hydroxylated product were described, as well as yields, rates of conversion and product ratio yields. Hemin systems gave the most similar product ratios.

Electrochemical Oxidation Systems

Historically, synthetic electrochemistry made its debut in 1801 with the electrochemical oxidation of alcohol [196]. Aromatic compounds are oxidized electrochemically (anodic oxidation reactions) by three well recognized routes [197–199]:

1. Direct electron transfer to form a cationic species (cation radical or dication). The electron is lost from the highest occupied molecular orbital to the anode electrode.
2. Reaction of the aromatic compound with an electrogenerated oxidizing agent. Depending on the sytem, oxidizing agents such as hydroxy radicals, anode oxides, supporting electrolyte radicals, or oxidizing agents generated from redox couples [e.g. Co(III)/Co(II)] may be employed.
3. Dehydrogenative chemisorption. This is a fuel cell-type oxidation in which the adsorbed aromatic dissociates into chemisorbed hydrogen atoms and organic residue in the primary electrochemical step.

The actual oxidation route of a particular compound will depend on many factors including:

1) the solvent-supporting electrolyte system,
2) the stability of the electrogenerated reactive species (e.g. cation radical, radical, or cation),
3) the role of the anode material,
4) specific adsorption of various species,
5) the electrode (working electrode) potential (e.g. as the potential increases the yield increases).

Hydroxylation in an aqueous media can be altered preferentially by addition of a second solvent, e.g. acetone has a moderating effect on the oxidation of toluene since acetone is oxidized, which, in effect, protects the product, benzaldehyde, from further oxidation. Hydroxy radicals result from discharge of water adsorbed on the working electrode.

$$H_2O \rightarrow \cdot OH + H^+ + e^-$$

Some electrochemical oxidation reactions include: naphthalene hydroxylation and quinone formation; o-, m-, p- acetoxylation of many halobenzenes, alkyl-substituted benzenes and PAHs, hydroxylation of dibenzo-p-dioxin, furans, aromatic amines; oxidation including hydroxylation in aqueous media of olefinic and acetylenic compounds, phenols, alcohols, saturated hydrocarbons and S-containing compounds. Reaction products varied and are given for various anode materials and mediums used or various experimental conditions [198, 199].

Amines

In some electrochemical systems oxidation product distributions closely parallel those obtained chemically, while in others they are entirely different. Whereas the decomposition reactions of aliphatic amine cation radicals are characterized by fragmentation and hydrolysis reactions, the electrochemistry of aromatic amines is dominated by various coupling reactions through the aromatic rings. For example, 2,4-dimethylaniline in H_2O, H_2SO_4 (Pt or PbO_2) was oxidized to 2,4-dimethyl-4-OH-2,5-cyclohexadienoneimine, 2,4-dimethyl-4-OH-2,5-cyclohexadienone, 3-OH-2,6-dimethyl-p-benzoquinone, and 2,5-dimethylhydroquinone [198].

The oxidation of anilines (alkyl, nitro, halogenated and halo/alkyl substituted) can reach yields of 30 to >90%. The electrooxidation of aliphatic amines, aromatic amines, amides and carbamates was recently reviewed by Shono [200]. Oxidation products were obtained in good yields and often with regioselectivity.

Polychlorinated Biphenyls (PCBs)

Anodic oxidation of two PCBs (4,4'-dichlorobiphenyl and 2,5, 2',5'-tetrachlorobiphenyl) and Aroclors 1232, 1242, 1254, 1260 commercial mixtures using very high anoidic potentials has been reported [201]. Hydroxychlorobiphenyls were formed and isolated. Many products were similar to known metabolic or photochemical degradation products. This indicated that electrochemistry is a convenient way to study the degradation of persistent environmental contaminants, as well as, a synthetic route for certain key degradation products. Most biological significant hydroxylations involve reaction with molecular oxygen, oxygen transition metal complexes, hydroxyl or perhydroxyl radicals or nucleophilic addition of water to electron deficient centers. Degradation of PCBs as studied by oxidative electrochemistry [202] resulted in hydrolysis, hydroxylated compounds, quinone formation and oxidative coupling. Fenn et al. [203] reported the anodic oxidation of two polychlorinated biphenyls and a PCB Aroclor mixture. PCBs are very resistant to environmental oxidative reactions; oxidation at high anodic potentials at a Pt electrode (+1.2 to +2.5V) for oxidation of the ring was required. They found hydroxychlorobiphenyls and benzoquinones similar to known metabolic or photochemical degradation products.

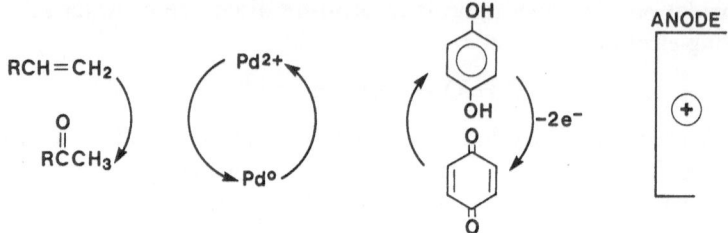

Fig. 6. Pd catalyzed electrochemical oxidation

Hydrocarbons

The formation of p-quinones as the major electrolysis products from the electro-lytic oxidation of aromatic hydrocarbons is well known [203] and are formed from hydroxy intermediates; up to 10% yield of a hydroxychlorobiphenyl through a single-pass flow-through cell was found. They concluded that electrochemistry was a convenient method to study the degradation, as well as, a convenient synthetic route for certain key degradation products. Tsuji and Mirato [331] used anodic electrochemical oxidation to oxidize olefins (styrene, cyclopentene, and cyclohex-ene) to ketones by catalysis with $Pd(OAc)_2$ and benzoquinone in high yields. Pd(II) was reduced to Pd° and benzoquinone oxidizes Pd° to Pd(II) as shown below (Fig 6).

Electrochemical Models of MFOs

An electrochemical model of hydroxylase has been developed on a hemin-modified cathode in an aqueous solution for aniline oxidation to p-aminophenol [204, 205]. An electrochemical model of alkane oxidation by P-450 was proposed recently by Khenkin and Shilov [206]. Catalytic oxidation of alkanes to alcohols and ketones took place in an electrochemical cell with iron porphyrin deposited on a graphite electrode. The oxidation mechanism was assumed to be the same as P-450 action. In monooxygenase based P-450 the active center contains iron porphyrin and the electrons are transferred from NADH or NADPH by the electron transport chains.

The oxidation, and mechanism thereof, for p-substituted dimethylanilines by P-450 was studied using electrochemical and metalloporphyrin models by MacDonald et al. [207]; they compared the model reactions to those occuring *in vitro* using rat liver P-450 MFOs. Electrochemical reductant replacement of ascorbic acid in the Udenfriend (AAOS) system was used [208] to study the hydroxylation of phenol and anisole and resulted in enhanced oxidation.

A complete chemical model of the P-450 cycle was constructed [17] by com-bining the iron-containing activating center with a source of electrons which continuously reduces the iron to the state capable of reacting with O_2 and activating it to further reaction; an iron porphyrin complex placed on the cathode of a electrochemical cell was used. These authors [17] used an iron tetraphenylporphyrin chloride coated electrode to oxidize cyclohexane to cylohexanol and hexane to 2- and 3-ols.

Hemin ($C_{34}H_{32}N_4O_4FeCl$, Teichmann's crystals) is an IP obtained by heating haemeglobin with acetic acid and sodium chloride. Sikurai and Ogawa [209] reported the oxidation of aniline using a cysteine methylester-hemin system. Hydroxylation to o- and p- aminophenols was 5–9 greater than systems not containing hemin.

Other Biomimetic Oxidation Systems

Many other biomimetic systems have been studied. Most of these have only recently received attention.

Amines

Haemoglobin contains four peptide chains each bound to a heme or IP molecule. Cumene hydroperoxide-dependent aniline hydroxylation with haemoglobin participation has been reported [210]. The results showed that hydroxylation of aromatics by hydroperoxides can be catalyzed not only by P-450 but also by haemoglobin. The mechanisms and kinetics of this oxidation were also reported [210]. Haemoglobin is the respiratory protein of the red blood cells (M.W. 65×10^3, 94% protein, 6% heme) and like catalases, peroxidases and P-450 hydrolases is a protoporphyrin IX hemoprotein. Aniline hydroxylation to 4-aminophenol was also studied with haemoglobin [102] in which NADH was substituted by cheaper election donors such as ascorbic acid and dihydroxylfumaric acid. The haemoglobin was immobilized by crosslinking with glutaraldehyde as insoluble particles. This gave hydroxylase activity similar to P-450 and a peroxidative mechanism was postulated. The same researchers studied another system using hemoglobin/riboflavin/NADH [244].

The oxidation of aromatic amines (aniline and N-dimethylaniline e.g.) by iodosylbenzene/vanadium acetylacetonate in benzene was recently reported [212, 213] as an oxidative model of P-450 catalyzed oxidations. A P-450 model system contained hemoglobin, methylene blue and NADH was employed to oxidize several xenobiotics (see Table 1). Regioselectivity (position of hydroxylation) for aniline and phenol was virtually the same as enzymatic oxidation of these contaminants. Generally, hemoglobin is involved in reactions that can mimic catalysis by hemoprotein enzymes such as peroxidase, catalase and P-450. Ascorbic acid can also be used as an electron donor in lieu of NADH; a peroxidatic-type mechanism was shown to occur [214]. Ascorbic acid gave twice the yield of hydroxylation products compared with NADH. Ortho-, meta- and para- hydroxy isomers were all detected for aniline in the model system with ortho and para hydroxylation preferred compared with para hydroxylation for red blood cells. In the NADH and ascorbate systems, haemoglobin gives rise to aromatic hydroxylations with a specificity towards aniline and phenol which are hydroxylated, whereas hydroxylations of acetanilide and p-toluidine did not seem to be increased by haemoglobin.

A model for xanthine oxidase was proposed by McMahnon et al. [215]; they studied photolysis of N-alkyl or-nitroanilines to N-dealkylated compounds, which were similar to dealkylated products of similar compounds formed enzymatically. Peroxomonophosphoric acid oxidation of aniline and p-nitroaniline to azobenzene, azoxybenzene, p-aminophenol and p-benzoquinone has also been reported [216].

Other oxidants, although receiving far less attention, appear to be promising for future studies as biomimetic systems. Ruthenium tetroxide was reported to mimic biological oxidation involving N-demethylation and N-oxidation of alicyclic amines [217]. Perrone et al. [218] also studied the oxidation of alicyclic amines regarding α-C oxidation using the same oxidant; this type of endo- and exo- cyclic regioselective α-C oxidation is one of the most important in vivo metabolic pathways for xenobiotics containing amino or ether functional groups. Only the carbon atom next to the heteroatom was oxidized for heterocyclics indicating high regioselectivity [218].

Nonaqueous oxidation with $CuCl/O_2$ has been demonstrated for 3,3',4,4'-tetrachloroazobenzene formation [219] from 3,4-dichloroaniline and for 3,3'-dichloroazobenzene from 3-chloroaniline [220]. The formation of other azobenzenes from primary aroamtic amines was also reported [221].

Other Environmental Chemicals

Sakurai and Ogawa [209] reported the oxidation of aniline using thiosalicylic acid-Fe(II), or cysteine methyl ester-Fe(II). Optimal pH and reaction times were determined for maximum yields of o- and p-aminophenols. Ortho/para hydroxylation ratios could be obtained that were identical to those found with a given species in vivo. A $FeCl_3$/cysteine system has been shown to be a model of plant enzymes for the degradation of linoleic acid hydroperoxides to a epoxy-oxo product. All products in the model system were the same as in biological degradation experiments in plants [222, 223]. Recently Searle and Tomasi [224] have shown that ferrous sulphate and cysteine in a oxygenated aqueous solution produce $\cdot OH$ radical. The possible binding site(s) of ferric ion with various thiols has been described [153].

Hydroxylations occured in absence of haemoglobin in NADH/ascorbate systems with all substrates, therefore $\cdot OH$ was not involved in these systems. Hydroxylations by hydroxyl radical would be expected to give mostly the ortho isomer, whereas mainly para was found in these studies [153]. For aniline plus haemogloin or P-450, mainly the para isomer has formed, while without haemoglobin the ortho isomer was predominant. They showed that phenol and aniline but not acetanilide or p-toluidine are substrates in peroxidatic activity of haemoglobin. This was not so for P-450 which had very good activities for all four substances.

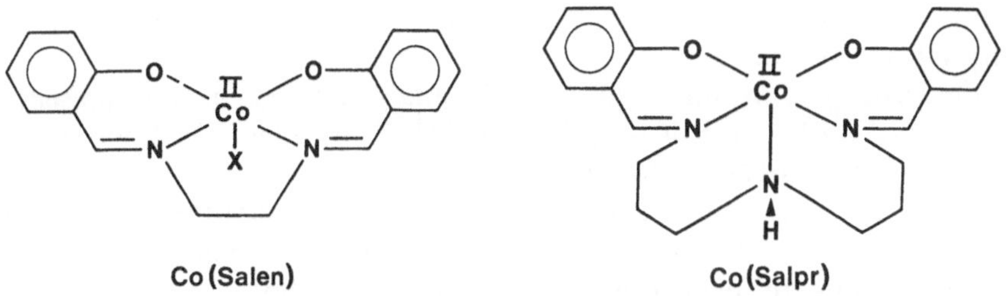

Co(Salen)   Co(Salpr)

Fig. 7. Structures of Co-containing complexes

Co(II)-Schiff base complexes such as Co(Salpr) and Co(Salen) (Salen = N,N'-disalicylideneethylenediamine) (Fig 7.) catalyze selective oxidation of organic substrates including environmental contaminants [225]. A cobalt dioxygen complex was described and its mechanism elucidated in a recent review [225]. Many biological oxidations are catalyzed by iron and copper complexes involved in the active site of relevant enzymes. The Catalysis by these metal complexes in enzymes is considered to involve the actions of dioxygen by forming a reactive metal-dioxygen complex of the types of Fe-O-O and Cu-O-O-Cu, which are formally equivalent to those of the natural oxygen carriers, hemoglobin and hemocyanin. Many Co(II)-complexes coordinated with N-bases are models for natural oxygen carriers and react with dioxygen to form CoO_2 or CoO_2Co (Fig 7.)

These cobalt-dioxygen complexes display reactivities to certain substrates leading to highly selective oxygenation analogous to those of dioxygenases. They were found to be good models for aromatic hydroxylating enzymes, exhibiting regio-specificity. Ravindranath et al. [226] studied a model for pyridine nucleotide-dependent flavoprotein hydroxylases. The enzyme model consisted of NADH, phenazine methosulphate and O_2. Hydroxylation products similar to those formed by hydroxylases were formed for the substrate m-hydroxybenzoate; maximum yields were obtained at acid pH. Inhibition by SOD was suggestive of superoxide anion formation. Thus the model was similar to flavoprotein hydroxylases and generates $\cdot O_2^-$. Hydroxylation of m-hydroxybenzoic acid by this model could be used as a probe to elucidate the mechanism of action of m-hydroxybenzoate-4-hydroxylase.

Co(Salpr) catalyzed the oxygenation of thyroxine analogs resulting in cleavage of the diphenylether linkage providing a model for peroxidase-catalyzed oxidation of thyroxine [227]; other examples for different substrates have been reported [228, 229]. Nakayama et al. [230, 231] reported the oxidation of 4-amino-1,2,4-triazine herbicides with t-butylhydroperoxide catalyzed by Co/Salen in CH_2Cl_2 at room temperature. This resulted in the specific deamination of the 4-amino group, providing a chemical model for the metabolism of the herbicides; as much as 80% yield in 7 hours was reported. Methylbenzenes were oxidized to aldehydes and acids with side chain acetoxylation using cobalt (II) or copper (II) acetates and NaBr in acetic acid at 150° [232].

Toluene was oxidized with a novel system consisting of catecholamine (3,4-dihydroxyphenylalanine)/H_2O_2/O_2 [233]. Products identified included: benzaldehyde, benzylalcohol, o-, m-, p-cresols; the mechanism was not similar to the

Udenfriend, Fenton or Hamilton reagents but was similar to an ascorbic acid/H_2O_2/O_2 system.

Anthracene was oxidized to phenols and quinones in the presence of iodosylbenzene/BF_3 or $RuCl_2$ $(PPh_3)_3$ [234]. Ruthenium has been shown to catalyze iodosylbenzene oxidation of alcohols, aldehydes, alkynes and sulfides [235-237]. Recently, Tolbert and Khanna [18] reported that the rat-liver *cytosol* oxidation products of 9,10-dialkylanthracenes were more similar to a biomimetic system containing tris (phenanthroline) tris (hexafluorophosphate) iron complex in aqueous CH_3CN under argon. Interestingly, rat-liver microsome oxidation products were *less* similar.

N-benzylamides react with potassium superoxide in benzene in the presence of a 18-crown-6 ether to give o- and p-hydroxylated products [238]. The superoxide anion is generated by this system and is a model for some peroxidase and xanthine oxidase catalysis. The oxidation of thioamides and thioureas to oxo compounds with potassium superoxide and crown ether as a model for oxygenases was also reported [239]. Oxidative demethylation of anisole [255] by S9 rat liver microsomes and 10 biomimetic systems were studied to elucidate mechanisms of O-demethylation. Only the tetraphenylporphinato iron(III)chloride-iodosylbenzene system was considered a suitable model for P-450 mediated oxidations based on kinetic isotope effect studies i.e. the ratio of the yield of phenol demethylation to that of 2-methoxyphenol hydroxylation from equivalent oxidation of the substrates.

When anisole was oxidized in the Hamilton system (H_2O_2 with catalytic amounts of Fe(III) and catechol) higher yields of hydroxylation products were obtained, as well as, all the hydroxylated isomers (o,m,p) of anisole compared with the Fenton and ascorbic acid oxidation systems [240]. Similar products to catalase, peroxidase, tyrosinase or phenolase were also found by Hamilton et al. [241] using the H_2O_2/Fe(III)/catechol system with anisole and other environmental contaminants (Table 1).

Aryl hydroxylation with peroxydisulphate in the presence of a transition metal was reported [242] for benzene, toluene and phenylacetic acid; products included phenols, alcohols and biphenyls.

Recently, a system was described in which a copper-ion mediated hydroxylation of an arene exhibited a mechanistic similarity to that known for the iron-dependent monooxygenases such as phenylalanine hydroxylase and P-450 [331]. The process of hydroxylation-induced migration was similar to the NIH proton shift and suggested that the reaction proceeds by the electrophilic attack of a Cu(I)/O_2 derived species upon the aromatic substrate. Similar studies by the same authors [331–336] used a dinucleating ligand in which two tridentate units of bis[2-(2-pyridyl)ethyl]amine are connected by a m-xylyl group. The three-coordinate dinuclear copper(I)complex reacts with dioxygen, resulting in the oxygenation of the ligand and concomitant formation of a phenoxo-and hydroxo-bridged dinuclear Cu(II) complex. The free phenol is formed and the hydroxylation reaction is analogous to the reactions mediated by the copper monooxygenases such as tyrosinase and dopamine β-hydroxylase.

The expoxidation of alkanes and hydroxylation of phenols by ozone may also be a model for flavin-containing monoxygenases [243]. For the CuCl-O_2 system [337] the attacking species on benzene (resulting in oxidation to phenol and dihydroxybenzenes and biphenyl) was not the hydroxyl radical; a percupryl ion (CuO_2^+) was proposed as a key step in the reaction.

Zinburg and Ballschmitter [83] reported the oxidation of hexachlorocyclohexanes (e.g. the insecticide lindane) to various chloro substituted phenols using NiO_2 or H_2O_2/H_2O oxidation systems.

An inorganic complex which mimics an oxidative enzyme activity and gives remarkable selectivity in partial oxidations of certain hydrocarbons under mild conditions, was recently described [245]. P-450 converts terminal methyl groups of linear hydrocarbons to $-CH_2OH$ with high selectivity due to steric constraints during the approach of a substrate hydrocarbon to a reactive ferryl group (FeO) inside the enzyme. The authors constructed oxidation catalyst systems involving iron-zeolite structures in which the silico-aluminate framework of the zeolite replaced the protein of the enzyme. Iron atoms are contained in a narrow zeolite channel of 5Å in diameter; molecular oxygen is the oxidant and molecular hydrogen a reducing cofactor. The catalyst also contains Pd(O) for the catalyst to function with oxygen. It is believed that Pd (palladium) catalyzes the reaction of molecular hydrogen oxygen to form hydrogen peroxide which reacts with the iron to form the reactive ferryl species. The catalyst Pd(O) was also effective in the metalloporphyrin biomimetic system discussed previously [190]. Octane/cyclohexane selectivity was greater than 150/1, with preferred hydroxylation of terminal methyl groups of n-octane taking place at 25° C and 50 psig. This demonstrates the possibility of applying an understanding of enzyme function to the design of a completely inorganic biomimetic system. Most recently, Brewer and Brewer [246] reported preparation of a biomimetic complex which proved useful toward modeling of cytochrome c oxidase. They synthesized a complex for a imidazolate-bridged model i.e. an imidazolate-bridged Fe(III)-Cu(II) complex. They formed this complex by reaction of an iron (III) porphyrin with a neutral metal Cu(II) chelate containing an imidazolate ring; axial ligation of the iron porphyrin with the imidazolate ring yielded a imidazolate-bridge with Cu(II)-Fe(III) centers.

The formation of hydroxyl radical (•OH) from the oxidation of glutathion, ascorbic acid, NADPH, hydroquinone, catechol and riboflavin reductants by hydrogen peroxide was studied [247] using several copper and iron complexes as possible catalysts. Copper-1,10-phenanthroline appears to catalyze the formation of OH• from H_2O_2 without superoxide radical being formed as an intermediate, and without the involvement of a catalyzed Haber-Weiss (Fenton) reaction. Superoxide radical was involved in the Cu(II)-catalyzed decomposition of H_2O_2 and in the oxidation of glutathione by atmospheric oxygen. The reaction studied for all these systems was 4-nitrophenol oxidation (hydroxylation) to 4-nitrocatechol. Copper-4,7-dimethyl-1,10-phenathroline was found to be a much more effective catalyst complex than the complex of 1,10-phenanthroline. Mechanisms for these reactions were proposed, and the toxicological significance of a variety of biological reductants to act as a source of OH • when oxidized by H_2O_2 is discussed

[247]. Rowley and Halliwell [248] also studied OH• formation from NADPH and NADH in N_2O_2 at physiological pH in the presence of copper salts; OH• concentration varied directly with the concentrations of H_2O_2, NADH/NADPH, and $CuCl_2$.

In contrast with Fenton's reagent which generates mainly the hydroxyl radical (•OH) [24–26, 76, 249, 250], the Udenfriend (ascorbic acid system) and the so-called *'oxene mechanism'* involve the perhydroxyl radical (•OOH) or the corresponding superoxide anion (•O_2^-) [249, 250]. As with the Fenton system, no epoxidation of aldrin was observed in the absence of BSA, and, even with its use, epoxidation was minimal with both the Udenfriend and mercaptobenzoic acid systems. The addition of either 6-nitro-1,3-benzodioxole or 6-nitro-1,2,3-benzothiadiazole caused no inhibition of epoxidation. Both synergists were completely recovered following the incubations indicating that they were not acted upon by these two systems.

The titanous chloride system can be formulated to produce either hydroxyl (•OH) or perhydroxy (•OOH) radicals in the presence of H_2O_2 and oxygen respectively [25]. In the presence of BSA, the titanous chloride H_2O_2 system produced dieldrin from aldrin and was inhibited by 6-nitro-1,3-benzodioxole. The titanous chloride O_2 system gave no epoxidation even in the presence of BSA.

Reversed micelles of the sodium salt of di(2-ethyl-hexyl) sulfosuccinic acid in octane catalyzes the oxidative demethylation of N,N-dimethylaniline in its reactions with molecular oxygen and/or cumene hydroperoxide [36]. Without surfactant no demethylation took place at 40° C.

Jonsson [251] studied the use of a transition metal, Cu(III) with trifluoroacetic acid for the hydrolysis of fluoro and chlorobenzenes and other environmental contaminants. Jonsson reported the formation of phenols, quinones and halophenyls, as well as, other hydroxylation products.

Pyridine-N-oxide photolysis was a useful model for MFO oxidations (e.g. aliphatic hydroxylations, dealkylations, S-oxidations, epoxidations of olefins and aromatic hydroxylations of several environmental contaminants [252].

30% aqueous H_2O_2 in trifluoroacetic and [82] oxidizes linear alkanes and cyclohexanes to an equilibrium mixture of secondary alcohols with 70% yields. Other novel systems reported include: $Fe(ClO_4)$-H_2O_2-CH_3CN, tris (acetylacetonato) iron(III)-BuOOH, KO_2-BuBr, and biacetyl-O_2-hv [104]. Systems were α or β stereoselective in the site of oxidation of cholesteryl acetate.

Factors Affecting Products Formed in Biomimetic Oxidation Systems

Several factors may affect the products formed and their yields in biomimetic systems; some of these are discussed below.

Solvent Effects

The solvating medium has been shown to have an influence on the model oxidation of α-tocopherol [253]. The inhibition of oxidation at very high concentrations of ethanol was due to solvation of the reactants; as in the inhibition of peroxide oxidation of phenols by high concentration of alcohols [254]. α-Tocopherol ($\pm$) was oxidized with t-butylhydroperoxide in $CHCl_3$ to simulate *in vivo* oxidation due to lipid hydroperoxides. No oxidation occurred after 3h at 60 °C but the addition of small amounts of ethanol resulted in immediate oxidation and formation of new product. After addition of 20% C_2H_5OH, further addition of C_2H_5OH brought a decrease in oxidation. Low concentrations of C_2H_5OH would promote decomposition of the hydroperoxide to alkoxy and hydroxy radicals which would then oxidize the α tocopherol.

Chemical Additive Effects

Controlled product formation and yields using a biomimetic system was demonstrated by Tsuchiya et al. [255]. Peracetic acid oxidation of phenol resulted in 1,2- and 1,4-benzenediols at 5–7% yields; with the addition of a chelating agent such as potassium diphosphate yields increased to 31–47%. This effect was attributed to the masking of traces of metals, especially iron and cobalt, in the reaction mixture which can further catalyze the oxidation of the benzene diols as they are formed. Twenty-two other additives were tested with various transition metals.

   For the activation of O_2, participation of an acylating agent is necessary for effective heterolytic cleavage of the O-O bond. The chemical system used for oxidation of alkanes (e.g. hexane, using Fe(III) TPP) in a solution of KO_2-18-crown-6 (in benzene and isopentane) plus an equal volume of acetic anhydride [105] is at present the best model of P-450. This system mimics P-450 oxidation of alkanes and initiates the process of O_2 activation (as does P-450) involving the formation of iron porphyrin peroxo complexes in the presence of acylating agents. The effect of various transition metals in the AAOS on the formation of parathion oxidation products has been reported [256].

   NADH, which normally has no effect upon the NADPH-dependent metabolism of aniline by MFOs, exerted a synergistic effect in the presence of various enhancing agents, such as acetone, maloxon or paraoxon. Several mechanisms may be involved, including enzyme induction, possibly by facilitation of transfer of the second electron to the P-450 substrate complex [105]. Such NADH effects and inhibition could be studied in biomimetic systems to further establish their "authenticity" or to study mechanisms of synergism, as well as, identification of synergistic xenobiotics.

   Rate enhancing effects of compounds added to a model system were reported recently for a metalloporphyrin system containing hypochlorite/MnTPP system [112, 113]. The order of rate enhancing effect for epoxidation of cyclohexene was as follows: 4-methylpyridine > pyridine > 4-cyanopyridine; imidazole blocked epoxidation completely. The rate of epoxidation in the presence of 4-methylpyridine increased with increasing concentration of 4-methylpyridine [112]. They proposed that pyridine and hypochlorite molecules compete for coordination to the manganese center of the MnTPP complex and in the presence of the stronger ligand imidazole, or at a high concentration of pyridine, coordination sites at manganese are blocked and epoxidation prevented.

Ligand Effects

The effects of various proteins (e.g. BSA) and other chemicals, the pH of the system, and the reaction time was studied for several biomimetic systems [31]. Proteins Triton B and BSA have an increasing effect on aldrin epoxidation [31].

BSA Effects

Marshall and Wilkinson [14] reported the epoxidation of aldrin by a modified Fenton system and its inhibition by substituted 1,3-benzodioxoles. Initial attempts to epoxidize aldrin to its 6,7-epoxide (dieldrin), by means of either the ascorbic acid of Fenton system were unsuccessful. It was established that the incorporation of bovine serum albumin (BSA) into the Fenton system resulted in a dramatic increase with epoxidation of aldrin to dieldrin. Although serum albumin and other sulfhydryl-containing proteins have been found to be necessary for some enzymatically catalyzed hydroxylation reactions [257], the role of BSA in Fenton's reagent is unclear. It is possible that the protein (BSA) supplies a surface upon which the highly lipophilic aldrin can be distributed in the aqueous Fenton's reagent. As a result of a series of experiments the Fenton system was modified for optimum conversion of aldrin to dieldrin.

P-450 is intimately associated with all membranes as a membrane bound hemoprotein of liver microsomes. The membrane was also suggested to take part in the transformation of chemical compounds ([258] and references therein). Stereoselective oxidation of aromatic sulphides and sulphoxides in the presence of BSA was reported by Sugimoto et. al. [259]. Oxidation of aromatic sulfides with achiral oxidizing agents (e.g. sodium metaperiodate ($NaIO_4$) and H_2O_2) bound to BSA resulted in a strong asymmetric bias to sulphoxide products in good yields (81%). Several optically active sulfoxides formed at > 90% optical purity in about 50% yield. They concluded that the method constitutes a successful biomimetic approach to achieving stereoselectivities as high as these obtained by sulfur-oxidizing microorganisms.

Previously it was shown that the binding domain of BSA serves as a chiral template for producing optically active aromatic alcohols of high optical purity (max 78%) in reduction model systems [259]. Many asymmetric reactions in which high stereoselectivity is sought cannot be readily obtained by methods using chiral agents. Optical purity is strongly dependent on BSA concentration and buffer pH in the oxidative [259]. The kind of oxidizing agent used greatly affected oxidative sensitivity to stereoselectivity and reactivity e.g. the reaction of phenylisopropylsulfide $\rightarrow$ sulfoxide $\rightarrow$ sulfone varies with the oxidizing agent [259]. Sugimoto et al. [259] studied several oxidants: $NaIO_4$, H_2O_2, MCPBA, t-butylhypochlorite, and t-butylhydroperoxide. H_2O_2 showed the highest stereoselectivity and reactivity; MCPBA always gave S enantiomer in excess while H_2O_2 gave the R enantiomer. Another characteristic of this oxidation was a remarkable change in optical purity and configuration of sulfoxide by the conformational changes of BSA occurring near pH > 8–9 and 5–6. BSA protein is present in a so called β-form at pH > 9 and N-form at pH 6–8 and an F-form at pH < 5. The respective form provides different chirality to the binding domain, where the observed stereoselectivity will be different. Of the three forms the binding domain in the β-form serves as a most desirable chiral environment for the highest stereoselectivity in the sulfide oxidation.

Besides the effects of BSA and Triton B previously discussed several other support-type chemicals can affect biomimetic systems. Razenburg et al. [112] found that for a hypochlorite/MnTPP model system the rate of epoxidation of cyclohexene was increased at least three times by attaching the manganese-porphyrin catalyst to a rigid support such as polymer of an isocyanide (RNC) < n. They proposed that this prevented the formation of inactive or less active dimers.

Cyclodextrin Effects

Cyclodextrins (CDs) consist of α-1,4 linked D-glucopyranose rings: α, β- and γ-CDs contain 6,7 and 8 glucose residues respectively in a molecule. X-ray crystallo-graphy shows CDs have a doughnut shape. CDs form inclusion complexes (as hosts) with many molecules and ions (guests) in solid phase or solution [260]. One of the main reasons for using CDs as models of enzymes is the formation of inclusion complexes between catalyst and substrate preceding the catalysis which is comparable to the forms of a Michaelis-Menton complex in enzymatic reactions.

In addition to complex formation the analogy between cyclodextrin catalyzed reactions and enzymatic reactions takes many other forms, they both:

1) are carried out in aqueous media;
2) show specificity with respect to both substrate and product and
3) show D, L optical isomer specificity.

Recent studies have shown that CDs are better models of enzymes than proposed previously. Some reactions accelerated by CDs are: cleavage of esters, amides, organophosphates, carbonates, and sulfates; intramolecular acylmigration; decar-boxylation; oxidation of phenyl esters, acetanilides, methyl phosphonates, arylcar-bonates, arylsulphates, α-keto acetate ions and α-OH ketones [260].

CDs exhibit catalysis of many organic reactions. A typical rate versus concentra-tion plot for CD catalysis is similar to enzyme saturation kinetics. A double reciprocal plot of the same data shows a straight line as enzymatic reactions do (i.e. Lineweaver–Burk plot in enzyme kinetics).

Catalysis by CDs is of two kinds:
1) Catalysis by the hydroxyl groups, in which they function as intra complex catalysts toward the substrates included in the CD cavity;
2) an effect of reaction field, in which the cavity of the CD serves as an apolar and sterically restricted reaction field. Both of these catalyses are important in enzy-matic reactions. Three kinds of catalysis by the hydroxyl groups of CD are known (a) nucleophilic catalysis (b) general base catalysis (c) general acid catalysis.

For the α-CD accelerated cleavage of the compound $H_3C\ P(O)FOCH(CH_3)_2$ the catalytic rate constant for the (R)(−) enantiomer is 35 times greater than for the (S)(+) enantiomer. This difference arises from the stereospecificity of the inclusion complexes [260]. CDs in their native forms show many features charac-teristic of enzymes such as: (1) specificity (2) formation of catalyst-substrate complexes prior to chemical transforms and (3) large accelerations in reaction rates. Modified CDs contain other functional groups that act as catalytic sites and modifications by introduction of certain noncatalytic groups on the CD can make it a better enzyme model through improvement in the binding process rather than in the catalytic process [260].

Several researchers [261–263] examined the effect of addition of cyclodextrin or surface active agents on their effect in several biomimetic systems. Predominant aliphatic hydroxylation with fairly good regioselectivity (γ-orientation) was observed in micellar systems; cetyltrimethyl-ammonium bromide (CTMABr) gave fairly good regioselectivity while with sodium lauryl sulphate, reactivity was greatly suppressed.

Other Effects

The increase of aliphated hydroxylation in micellar systems with CTMABr might be explained by the assumption of incorporation of the phenyl ring of bencyclane into micelles. Should this be the case, this system would be a good model for the binding of a substrate in a biological system. Also there was regioselective aliphatic hydroxylation in good yield, and this may be due to incorporation of the phenyl ring into the cavity of β-CD, however, stereoselectivity was unsatisfactory. When they [261] changed from 20% aqueous CH_3CN to pure water regio- and stereo-selective (γ-trans 7:1) aliphatic hydroxylation at the γ-position was obtianed in good yield (17.6%), leading to the target bencyclane metal. Only hemin/$PhIO/CH_2Cl_2$ or TPP $FeCl/PhOI/CH_2Cl_2$ systems gave some N-demethylated product. In the present study they found that the ferryl ion ligand complex, has an electrophilic nature leading to aromatic hydroxylation and the regio/stereo-selective aliphatic hydroxylation similar to that occurring in man. This complex is accepted as the most probable active species in enzymatic hydroxylation, and is induced by addition of β-CD in the model system [261-263]. These results suggest that the regio- and stereoselectivity of metabolic hydroxylation could be controlled by the binding of the substrate at the protein site rather than by the nature of the oxo-iron species.

Biomimetic Systems: Comparative Physicochemical Aspects

The current interest in mechanisms of oxygenation by P-450 dependent mono-oxygenation has led to the development of numerous chemical models. In addition, a knowledge of the mechanisms of the chemical systems will lead to a clearer understanding of the analogous biological processes. the relevance of a model depends upon how closely its reactions resemble those of the mono-oxygenases.

This has been assessed by such comparative criteria as the range of oxidation brought about by both systems and their product distribution, stereochemical studies, the origin of the oxygen and the magnitude of the NIH shift and kinetic isotope studies [264].

Kinetic Isotope Effects

Kinetic isotope effects are usually determined with deuterium or tritium as a replacement for hydrogen. Isotopic substitution may affect observed kinetics and thereby provide mechanistic information. The C-H bond is more reactive than the C-D bond and the rate of bond breaking for C-H would be greater than the rate of C-D reaction; kinetic isotope effect is defined on this basis as k_H/k_D. A primary isotope effect is defined as the effect of isotopic substitution at a bond that is broken during a reaction. A secondary isotope effect is defined as the effect of isotopic substitution at a bond that is *not* broken during a reaction [126].

The magnitude of deuterium kinetic isotope effects will depend on the extent of bond breaking in the transition state. if the CH or CD bond remains intact throughout the reaction there will be no primary kinetic isotope effect. The presence of a primary kinetic isotope effect in most electrophilic substitution reactions provides a clear indication that the hydrogen is not lost in the rate-determining step, thereby demonstrating that at least one intermediate must be involved in the reaction.

Kinetic isotope studies were determined for the oxidative demethylation of anisole (Me-^{2}H$_3$ and Me-H$_3$) by comparing the ratio of the yield of phenol (demethylation) to that of 2-methoxyphenol (hydroxlation) from equivalent oxidation of the two substrates [264]. Several model systems were studied (10 systems) including a rat liver microsome system. Biological systems showed a large kinetic isotope effect in agreement with the generally accepted mechanism for biological O-dealkylation of aromatic ethers involving an initial rate determining C-H bond cleavage i.e. $k_H/k_D = 7.3$–7.8. Model systems fall into 3 groups $k_H/k_D = 1.0$, 2–3 and the tetraphenylporphinato iron(III) chloride-iodosylbenzene system which shows an isotope effect comparable with that of the biological system ($k_H/k_D = 7$–8). Lindsay Smith et al. [195, 264] concluded that only the tetraphenyl-porphinatoiron(III)chloride-iodosylenzene system can be considered a sutiable model for P-450 mediated oxidations ($k_H/k_D = 9.0$).

Mori et al., [265] studied the aryl hydroxylation of H^2 and H^3-3-methylacetanilide by several biomimetic systems including: hv/pyridine-N-oxide, MCPBA, hv/H$_2$O$_2$, Fenton, and AAOS. The retention of H$_3$ (retention of isotopic hydrogen in phenolic product) in model systems was compared with retention in *in vivo* experiments using rats (for the 4-hydroxylation of acetanilide). The hv/pyridine-N-oxide system gave the same retention, 21%, as the *in vivo* experiments (enzymic oxidation) and MCPBA gave 12% retention similar to that for 3,4-benzpyrene treated rats; other sytems resulted in 9% retention.

NIH Shift

The NIH shift, the intramolecular migration of aromatic ring substituents such as deuterium, tritrum, halogens, and alkyl groups during metabolism of aromatic substrates to phenols, may be considered as a fundamental phenomenon associated with monooxygenases [266]. Such migrations are so characteristic of the enzyme reactions that oxidants which do not produce them may no longer be regarded as meaningful models for monooxygenases.

Detailed studies on model hydroxylating systems which exhibit the NIH shift should permit further elucidation of the mechanism of the more complex enzymic oxidations. Pyridine N-oxide/photolysis is of great value as a mechanistic model for enzymatic oxidation since it is capable of many oxidation reactions typical of MFOs and exhibits an NIH shift in which the retention values (of isotopic hydrogen) are similar to those observed during hydroxylation of several substrates with microsomes [267]. With Fenton or Udenfriend systems very high retention was reported [268] approximating half of that for a pyridine N-oxide system.

Ullrich et al., [269] studied three model systems and compared their oxidation products, for the substrates acetanilide and its halogenated analogues, to those formed by MFOs. The model systems studied included: $SnCl_2$ (oxene mechanism, no NIH shift), modified Fenton's (AAOS/H_2O_2, •OH, no NIH shift), and trifluoroperacetic acid/H_2O_2 (NIH shift); the last system most faithfully replicated the oxidation products as formed by MFOs, with similar ortho/meta/para hydroxylation product ratios. The NIH shift has been demonstrated in the following systems: pyridine N-oxide/hv, trifluoroperoxyacetic acid, MCPBA, t-butylhydroperoxide, and MFOs. The Fenton, Udenfriend, and Hamilton systems showed no 1,2-shifts [270].

Enzymatic Versus Nonenzymatic Oxidation Systems–A Mechanistic Comparison

For a discussion of specific mechanisms for enzymatic and nonenzymatic oxidative systems, the reader is referred to the following reviews [1, 107–111, 270, 271].

An electrophilic species of active oxygen with six electrons, which could be considered as a true model for the active oxygen species in enzymic reactions, is not certain. With the present knowledge on both MFO and model systems, it becomes obvious that the existing models may be insufficient in various respects. One essential difference is the environment of the active oxygen in enzyme and model reactions. Water is the most frequent reactant for the active oxygen in model systems, but is this also true for the enzymic active oxygen? An electrophilic species similar to an OH^+ ion must be unstable in the presence of hydroxyl ions, because the chemical equilibrium would be far on the side of the unreactive hydrogen peroxide [76]. In enzymatic reactions, an interaction with OH^- ions can be avoided either by a concerted mechanism where the oxygen molecule is reduced after formation of the enzyme-substrate-oxygen complex, or by reaction within lipid structures providing a water deficient environment.

A second difference is the fact that the mixed function oxygenases catalyze "vectorial" (i.e irreversible) reactions with respect to electron transfer and substrate oxidation. In model systems, however, we are dealing with a nonoriented reaction in homogeneous solution. The active oxygen is susceptible to an attack of both reducing agents and substrates so that a strong competition can occur. Some of the insufficiencies noted of model systems have become partially resolved by working in nonaqueous solutions and in heterogeneous systems (modified Fenton's system containing BSA) where electron transport and substrate oxidation can take place in different phases.

One conclusion to be derived from a survey of enzymic oxygen reactions is that there is no one species that can be labelled as "active oxygen". Perhaps this should not be surprising, since, we do not speak, for example, of "active water"; rather, water reactivity depends on its environment, whether it is in a polar or nonpolar solvent, in the vicinity of an acid or a base, complexed to a metal ion, etc. So it is with oxygen; it can have differing reactivities depending on its environment. As the discussion in the preceding sections indicates, it is becoming increasingly clear that oxygen itself frequently does not react with the most visible substrate of the oxygenase reaction; rather, the oxygen is converted first to a considerably more reactive product (frequently at the peroxide oxidation level) by reaction with a cofactor or cosubstrate by mechanisms often encountered in nonenzymic oxygen reactions.

Tanaka and Wein [272] reported the oxidation of the herbicide monuron (3-(4-chlorophenyl)-1,1-dimethylurea) in the Fenton and Udenfriend biomimetic systems. Both systems produced oxidation products similar to those formed by photodecomposition or plant metabolism. Since products formed in both model systems were not identical, they concluded that possibly combining the information obtained from several different biomimetic systems would be most useful to study mechanisms and products of biological oxidation reactions. This may be obvious since those systems which generate •OH radical are unable to hydroxylate aliphatic CH bonds and do not form epoxides or dihydrodiols at a double bond [273].

The presence of a "supporting surface" (BSA or cyclodextrin e.g.), a catalyst (hemin) or ligand (thiol) appear most promising as complete biomimetic systems. These systems appear to be among the best nonenzymatic models of enzymatic oxidation in regard to product and mechanistic similarities.

Future Studies

Pyridoxal with manganese ions has been shown to oxidize amino acids and may be a suitable biomimetic system for some amine oxidases [277, 278]; such a system could be studied for oxidation of environmental contaminants with amino functional groups. Other systems which have received cursory interest include a Cu(II) complex containing 3-n-nonylcatechol and a bidentate nitrogen counter ligand such a 1,10-phenathroline [280] and $CuCl/O_2$/pyridine/methanol [279, 280]; the former system has been shown to be a model for aromatic ring cleavage by dioxygenases. Cyclohexene was oxidized to its epoxide using a novel MnTPP/NaOCl biomimetic system [281]. The oxidation medium consisted of a two phase water/CH_2Cl_2 system with benzyltrimethylamonium chloride as a base transfer reagent; 4-methylpyridine enhanced the rate of reaction and imidazole blocked epoxidation. These are novel systems which maybe useful for studies with environmental chemicals or which maybe modified and their effectiveness as biomimetic systems extended.

The unique effect (enhancement) of acetone on the p-hydroxylation of aniline in microsomes [274] and P-450 [308] has been reported. Other enhancers of aniline hydroxylation included ethyl isocyanide, metyrapone, paraoxon, 2-pentanone, 2,2'-bipyridine, 1,10-phenanthroline, methoxyflurane, halothane and trichloroethane ([274] and rferences therein). Acetone also decreased N-demethylation of p-NO_2 anisole or p-NO_2 anisole blocked acetone enhancement of aniline metabolism [274]. Acetone inhibition was also observed for N- and O-demethylation of other xenobiotics, as well as, hydroxylation of benzo(a)pyrene. Model systems which mimic this effect would be most faithfully producing the oxidative products formed by multiple forms of aniline p-hydroxylase found in hepatic microsomal preparations. The oxidation of benzoate by the AAOS and Fenton's system was reported by Winston and Cederbaum [276]. They also studied the inhibition of oxidation by various chemical additives; this appeared to be a valid probe for detecting oxidizing species produced by chemical model systems generating the •OH species. Their studies with various •OH scavengers in model systems were compared to the effects of scavengers in S9 supernatant fractions. Studying these effects in biomimetic systems could further elucidate their mechanism and potentially open new approaches to studying synergism, inhibition and potentiation effects of chemical mixtures. This would also help to establish which biomimetic system is a faithful model of *in vivo* oxidative metabolism. The use of biomimetic systems to predict metabolic compounds responsible for toxicological profiles or the establishment thereof, has been recently elucidated [89, 91, 195, 275] and is most promising for further application.

DMSO inhibited the denaturation of oxytocin and insulin by the AAOS [142]. Preincubation of serum (mosquito fish) with endrin or aldrin significantly reduced the quantity of DDT subsequently bound to albumin while the reverse was not observed [282]. The reason for an inhibition in binding is unknown. The reduction in binding of a xenobiotic, due to the presence of a second, may have serious toxicological implications due to a potential increase in the bioavailability and hence metabolism of the second compound. This could be elucidated by studying reactive intermediates and final oxidative products formed in biomimetic systems for single xenobiotics compared with their mixtures. Such effects should be further studied with environmetnal contaminants to help elucidate the effects of admixtures on oxidative degradation and to help elucidate mechanisms of oxidation for complex-competitive systems. Biomimetic systems may be useful for studying the activation of environmental chemicals to reactive intermediates or products which bind to DNA. Such studies might help elucidate mechanisms of binding, as well as, screening candidate compounds for their capacity to bind. For example, the binding of estrogens to DNA via oxidative degradation was reported recently [283]. DNA reacted with estrogens (oestrone and oestradiol) in the presence of H_2O_2 and Fe(II)SO_4 or dihydrogen sodium citrate. Binding of the oestrogens to DNA was seven times greater when using sodium dihydrogen citrate in lieu of Fe(II)SO_4. The majority of chemical carcinogens, benzo(a)pyrene e.g., are believed to exert their carcinogenic activity via covalent interactions with the DNA of their target tissues; this has been especially well characterized for PAHs.

Benzidine (4,4'-diaminobiphenyl) is readily oxidized by horseradish peroxidase/H_2/O_2 [284]. Nucleophiles, e.g. endogenous thiols and DNA, can react with activated products of benzidine oxidation respectively inhibiting or promoting chemical carcinogenesis [285]. The oxidation of benzidine and other aromatic amines by the horseradish peroxidase/H_2O_2 system provides a useful model for metabolic activation of this carcinogen by mammalian systems. Similar studies using nonenzymatic biomimetic oxidation systems remain to be conducted and, indeed, may prove to be extremely fruitful.

The nitrogen of a chemotherapeutic imidazol was bound to the heme moiety of mouse liver microsomal P-450 resulting in inhibition of chloroethylnitrosourea hydroxylation by hemoprotein [19]. One could repeat this with the hemin/protoporphyrin system to study inhibition and for elucidation of mechanisms. For example, Mn(III) porphyrins with a substituted imidazole ligand have been shown to catalyze the transfer of an oxygen atom from hypochlorite to an olefin, [286].

The toxicity of the bipyridinium herbicide paraquat can be mediated by generation of reactive oxygen species *in vivo*, such as superoxideanion (O_2^-) and hydroxyl radical (•OH) [204]. Such effects might be predicted by studying the oxidative species formed in biomimetic systems with paraquat, e.g., since the above *in vivo* reactions have been shown to occur with microsomal P-450 [204].

Future work regarding mechanisms for modifying toxicity and herbicide synergists include the influence of monooxygenase inhibitors on the metabolism of the herbicides chlortoluron and metalachlor in cell suspension cultures. N-demethylation and benzyl alcohol formation were inhibited by 1-aminobenzotriazole and 3-(2,4-dichlorophenoxy)-1-propyne which are both enzyme-activated monooxygenase inhibitors [219]. The inhibition of biomimetic systems would be most interesting in paralleled studies with the above compounds. Fatty acids are also considered environmental contaminants, (e.g. fecal fatty acids such as stearic acid) and have had limited study in biomimetic oxidation systems [287, 288]. Biomimetic systems could also be used to study the activity and oxidation of food antioxidants. Such systems may also be useful to elucidate the effects of food components on antioxidant activity.

In conclusion, the iron porphyrin systems have received overwhelming attention in the last few years and appear to be the biomimetic systems of choice, especially in regard to mechanistic studies. However, the inorganic clay/palladium system, other metalloligand systems and electrochemical oxidation systems may be fruitfully applied to environmental chemicals in the future. The oxidation of environmental chemicals by nonenzymatic biomimetic systems have proven to be most useful for the study of enzymatic, chemical, and photochemical transformations.

List of Chemical Names and Abbreviations

Abbreviations

AAOS	ascorbic acid oxidation system
BSA	bovine serum albumin

CD (s)	cyclodextrin (s)
DMSO	dimethylsulfoxide
EDTA	(ethylenedinitrilo) tetraacetic acid
EPR	Electron proton resonance
FAD	flavin adenine dinucleotide
FMN	riboflavin mononucleotide
Fp	flavoprotein
MCPBA	m-chloroperoxybenzoic acid
MFO(s)	mixed function oxidase (s)
NADH	reduced nicotinamide adenine dinulecotide
NADP	nicotinamide adenine dinucleotide phosphate
NADPH	reduced nicotinamide adenine dinucleotide phosphate
NIH	National Institute of Health
P-450	cytochrome P-450
PAH	polycyclic aromatic hydrocarbons
PCB(s)	polychlorinated biphenyl (s)
PTPAA	peroxytrifluoroacetic acid
PVC	polyvinylchloride
SOD	superoxide dismutase
TFPAA	trifluoroperacetic acid
VIS	visible
α-MPG	α-mercaptopropionylglycine

Chemical Names

porphyrin derivatives defined in Table with Fig. 4.

acifluorofen	sodium5-[2-chloro-4-(trifluoromethyl) phenoxy]-2-nitrobenzoate
alachlor	2-chloro-N-(2,6-diethylphenyl)-N-(methoxymethyl)acetamide
allethrin	dl-2-allyl-4-hydroxy-3-methyl-2-cylcopenten-1-one ester of dl-cis, trans-chrysanthemum monocarboxylic acid
amitraz	N,N-di-(2,4-xylylimino methyl)-methylamine
Aroclors	mixtures of polychlorinated biphenyls
atrazine	2-chloro-4-ethylamino-6-isopropylamino-1,3,5-triazine
benzidine	4,4'-diaminobiphenyl
butachlor	N-(butoxymethyl)-2-chloro-N-(2,6-diethylphenyl)acetamide
chlordimeform	N-(2-methyl-4-chlorophenyl)-N',N'-dimethyl formamidine
chlortoluron	3-(3-chloro-4-methylphenyl)-1,1-dimethyl urea
DDT	1,1,1-trichloro-2,2-bis(p-chlorophenyl)ethane
diallate	diisopropylthiocarbamic acid S-2,3-dichloroallyl ester
EPTC	S-ethyl dipropylthioocarbanate
fonofos	O-ethyl S-phenyl ethylphosphonodithioate
glyphosate	N-(phosphonomethyl)glycine
halothane	2-bromo-2-chloro-1,1,1-trifluoroethane
lindane	γ-1,2,3,4,5,6-hexachlorocyclohexane
linuron	N-(3,4-dichlorophenyl)-N'-methoxy-N'-methylurea
metalochlor	2-chloro-N-(2-ethyl-6-methylethyl) acetamide
methoxyflurane	2,2-dichloro-1,1-difluoro-1-methoxyethane
metribuzin	4-amino-6-(1,1-dimethylethyl)-3-(methylthio)-1,2,4-triazin-5(4H) one
metyrapone	2-methyl-1,2-di-3-pyridyl-1-propanone
monuron	3-(4-Chlorophenyl)-1,1-dimethylurea
nitrofen	2,4-dichlorophenyl 4-nitrophenylether

oxyfluorofen	2-chloro-1-(3-ethoxy-4-nitrophenoxy)-4-(trifluoromethyl) benzene
paraoxon	phosphoric acid diethyl 4-nitrophenyl ester
paraquat	1,1'-dimethyl-4,4'-bipyridinium dication
profenofos	O-(4-bromo-2-chlorophenyl)O-ethyl S-propylphosphorothioate
pyridoxal	3-hydroxy-5-(hydroxymethyl)-2-methylisonicotinaldehyde
Salen	N,N'-disalicylidemethylenediamine
schradan	bis-N,N,N',N'-tetramethyl phosphorodiamidic anhydride
simazine	2-chloro-4,6-bis(isopropylamino)-S-triazine
sulprofos	O-ethyl-O-(4-methylthiophenyl)S-propyl phosphorodithioate
tetramethrin	3,4,5,6-tetrahydrophthalimidomethyl-2,2-dimethyl-3-(2-methyl propenyl) cyclopropane carboxylate
triallate	S-2,3,3-trichloroallyl diisopropyl thiocarbamate

References

1. Worobey B L, Webster G R B (1979) Toxicol. Environ. Chem. Rev. 3: 1
2. Kline E R, Mattson V R, Pickering Q H, Spehar D L, Stephan C E (1987) J. Water Poll. Control Fed. 59(6): 539
3. EPA Priority Pollutants Listed in Chemicals Service Inc. Catalogue, West Chester, PA, USA
4. Hites R A, Jungclaus G A, Lopez-Arila V, Sheldon L S (1979) in: Schuezle D (ed) Monitoring Toxic Substances, ACS Symposium Series No. 94, Chp 5
5. Williams R T (1959) in: Detoxification Mechanism, 2nd ed., John Wiley and Sons, New York, p 520
6. Parke D V, Williams R T (1969) Brit. Med. Bull. 25: 256
7. Matsumura F (1975) in: Toxicology of Insecticides, Plenum Press, New York
8. Gillette J R (1963) in: Tucker E. (ed) Progress in Drug Research, 6: 11
9. Snook M. E., Hamilton G. A. (1974) J. Am. Chem. Soc 93(3):860
10. Plapp Jr F W (1975) Environ. Qual. Saf. Suppl. Vol.3, iss: Pesticides 421
11. Gillette J R, Connery A H, Cosmides G J, Estabrook R W, Fouts J R, Mannering G J (eds) (1969) in: Microsomes and Drug Oxidations, Academic Press, New York
12. Casida J E (1970) J. Agric. Food Chem. 18: 753
13. Gram T E, Fouts J R (1968) in: Hodgson E (ed) The Enzymatic Oxidation of Toxicants, North Carolina State Univ., Raleigh, N. C. p 47
14. Jakoby W B, Bend J R, Caldwell J (eds) (1982) Academic Press, New York, N. Y.
15. Parke D R (1987) Arch. Toxicol. 60: 5
16. Halliwell B, Gutteridge JMC (1984) Biochem. J. 219: 1
17. Sligar S G, Kennedy K A, Pearson D C (1980) Proc. Natl. Acad. Sci. (USA) 77: 1240
18. Tolbert L M, Khanna R K (1987) J. Am. Chem. Soc 109: 3477
19. Lee F Y, Workman P, Cheeseman K H (1987) Biochem. Pharm. 36(8): 1349
20. Hamilton G A (1976) in: Jones J B, Sik C J, Perlman D, (eds) Techniques of Chemistry *10*, Applications of Biochemical Systems in Organic Chemistry, Part 2 John Wiley and Sons New York, New York, Chp. 6: 875
21. Estabrook R W, Hildebrandt A G, Baron J, Netter K J, Leibman K (1971) Biochem. Biophys. Res. Commun. 42: 132
22. Fenton H J H (1894) J. Chem. Soc., London 65: 899
23. Fenton H J H, Jones H O (1900) J. Chem. Soc., London 77: 69
24. Norman R O C, Lindsay Smith J R (1965) in: King T E, Mason H S, Morrison M (eds) Oxidases and Related Redox Systems, Wiley, New York, 1: 131

25. Staudinger H, Kerekjarto B, Ullrich V, Zubnzycki Z (1965) in: King T E, Mason H S, Morrison M (eds) Oxidases and Related Redox Systems, Wiley, New York, 2: 815
26. Ullrich V, Wolf J, Amadori E, Standinger H (1968) Hoppe-Seyler's Z. Physiol. Chem., 349:85
27. Bors W, Saran M, Lengelder E, Michael C, Firchs C, Frengel C (1978) Photochem. Photobiol. 28: 629
28. Walling C, Johnson R A (1975) J. Am. Chem. Soc 97: 363
29. Fridovich I (1975) Ann-Rev. Biochem. 44: 147
30. Hodgson E K, Fridovitch I (1975) Biochem. 14: 5294
31. Marshall R S, Wilkinson W F (1973) Pestic. Biochem. Physiol. 2: 425
32. Bors W, Saran M, Czapski G (1980) in: Bannister W H, Bannister J V (eds) Biological and Clinical Aspects of Superoxide and Superoxide Dismutase, Proc. Fed. Europ. Biochem. Soc. Symp. No. 62, Elsevier, p 1–31
33. Haber F, Weiss J (1934) Proc. Royal Soc. Serv. A 147: 332
34. Barnes A R, Sugden J K (1986) Pharm. Acta Helv. 61(8): 218
35. Walling C, El-Taliawi G M (1973) J. Am.Chem. Soc. 95: 844
36. Plimmer J R, Kearney P, Klingebiel V I (1971) J. Agric. Food Chem. 19(3): 572
37. Brown R F, Jamison S E, Pandit U K, Pinkers J, White G R, Braendlin H P (1964) J. Org. Chem. 29: 146
38. Boyland E, Sims P (1953) J. Chem. Soc. p 2966
39. Hewer A, Ribeiro O, Walsh C, Grover P L, Sims P (1979) Chem. Biol. Interac. 26: 147
40. Nagata C, Tagashira Y, Kodarua M, Ioki S, Osborne S (1973) Gann 64: 277
41. Walling C, Johnson R A (1975) J. Am. Chem. Soc. 97(2): 363
42. Kasai H, Nishimura S (1986) Environ. Heth. Perspec. 67: 111
43. Hamboeck H, Fischer R W, Di Iono E E, Winterhalter K H (1981) Molec. Pharmacol. 20: 579
44. Jarman M, Manson D (1986) Carcinogenesis 7(4): 559
45. Udenfriend S, Cooper J R (1952) J. Biol. Chem. 194: 203
46. Udenfriend S, Clark C T, Axelrod J, Brodie B B (1954) J. Biol. Chem. 208: 731
47. Brodie B B, Axelrod J, Shore P A, Udenfriend S (1954) J. Biol. Chem. 208: 741
48. Suzuki T, Casida J E (1981) J. Agric. Food Chem. 29: 1027
49. Hamilton G A, Hanfin J W, Friedman J P (1966) J. Am. Chem. Soc 88: 5269
50. Worobey B L (1986) Toxicol. Environ. Chem. 11: 117
51. Cohen G, Cederbaum A I, (1984) Arch. Biochem. Biophys. 199(2): 438
52. Hsieh S-T, Kraft P L, Archer M C, Tannenbaum S R (1976) Muta. Res. 35: 23
53. Rayman M P, Challis B C, Cox P J, Jarman M (1975) Biochem. Pharm. 24: 621
54. Preussmann R (1964) Arzneittel-Forsch 14: 769
55. Mayer V W (1971) Molec. Gen. Genet. 112: 289
56. Mayer V W (1966) Muta. Res. 3: 537
57. Archer M C, Eng V W-S (1981)Chem. Biol. Interactions 33: 207
58. Suzuki E, Mochizuki M, Shibuya K, Okada M (1983) Gann 74: 41
59. Preussmann R (1964) Arzneim.-Forsch. 14: 769
60. Rayman M P, Challis B C, Cox P J, Jarman M (1975) Biochem. Pharmacol. 24: 621
61. Hsich S T, Kraft P C, Archer M C, Tannenbaum S R (1976) Muta. Res. 35: 23
62. Mason D, Cox P J, Jarman M (1978) Chem. Biol. Interac. 20: 341
63. Mayer V W (1972) Mutation Res. 15: 147
64. Tierney B, Hewer A, MacNicoll A D, Gervasi P G, Rattle H, Walsh C, Grover P L, Sims P (1978) Chem. Biol. Interac. 23: 243
65. Boyland E, Kimura M, Sims P (1964) Biochem. J. 92: 639
66. Sullivan P D, Ellis L E, Calle L M, Ocasio I J (1982) in: Cooke M, Dennis A J, Fisher G L (eds) Polynuclear Aromatic Hydrocarbons: Physical Biological Chemistry, Inter. Symp. 6th Meeting p 779, CA99: 83502h
67. McNeil J M, Willis E D, Gower J D (1985) Biochem. Pharm. 34(22): 4066
68. Gower J D, Willis E D (1984) Carcinogenesis 5: 1183
69. Tierney B, Abercrombie B, Walsh C, Hewer A, Grover P L, Sims P(1978) Chem. Biol. Interac. 21: 289
70. Boyland E, Kimuia M, Sims P (1964) Biochem. J. 92: 631

71. Jones A R, Walsh D A (1979) Xenobiotica 9(12): 763
72. Jones Ar, Gibson J (1980) Xenobiotica 10(11): 835
73. Grinstead R R (1960) J. Am. Chem. Soc 82: 3472
74. Maissant J M, Bouchoule C, Blanchard M (1982) J. Molec. Catal. 14: 333
75. Casida J E, Ruzo LO (1986) Xenobiotica 16(10/11): 1003
76. Ullrich V, Standinger H (1969) in: Gillette J R, Conney A H, Cosmides G J, Estabrook R W, Fouts J R, Mannering G J (eds) Microsomes and Drug Oxidations, Academic Press, New York, p 199
77. Marshall R S, Wilkinson C F (1973) Pest. Biochem, and Physiol 2(4): 425
78. Jerina D, Daly J, Landes W, Witkop B, Udenfriend S (1967) J. Am. Chem. Soc. 89: 3347
79. Gniroff G, Daly J W, Jerina D M, Renson J, Witkop B, Udenfriend S (1967) Science 157: 1524
80. Jerina D M, Daly JW Witkop B (1971) Biochemistry 10: 366
81. Ikne-Rasa K M, Edwards J O (1962) J. Am. Chem. Soc 82: 763
82. Dens N C, Jedzwiak E J, Messer L A, Meyer M O, Strond S G, Tomezko E S (1977) Tetrahedron 33: 2503
83. Zimburg R, Ballschmiter K (1981) Chemosphere 10(8): 957
84. Aldrich Co. Technical Info. Sheet MCPBA Oct (1973)
85. Bellet E M, Casida J E (1974) J. Agric. Food Chem. 22(2): 207
86. Kuvatsuka S (1970) in: O'Brien RO, Yamamoto I (eds) Biochemical Toxicology of Insecticides, Academic Press, New York, p 131
87. McBain J B, Yamamoto I, Casida J E (1971a) Life Sci. 10: 947
88. McBain J B, Yamamoto I, Casida J E (1971b) Life Sci. 10: 1311
89. Gohre K, Casida J E, Ruzo L O (1987) J. Agric. Food Chem. 35: 388
90. Ruzo L O, Casida J E, Holden I (1985) J. Chem. Soc. Chem. Commun. 22: 1642
91. Kimmel E C, Casida J E, Ruzo L O (1986) J. Agric. Food Chem. 34: 157
92. Ruzo L O, Kimmel E C, Draper W M, Casida J E (1986) IUPAC Abstract 8A/7E-25 Ottawa, Ontario, Canada
93. Locke, R K (1972) J. Agric. Food Chem. 20(5): 1078
94. Hill H A O (1981) Philos. Ttans. R. Soc. Lond. Ser. B. 294: 119
95. Allahyani R, Lee P W, Lin G H Y, Wing R M, Fukuto T R (1977) J. Agric. Food Chem. 25(3): 471
96. Brodsky J, Andersson J T, Ballschmiter K (1986) Chemosphere 15(2): 139
97. Caughey W S, Alben J O, Beaudreau C A (1965) in: King T E, Mason H S, Morrison M (eds) Oxidases and Related Redox Systems, Proc. Symp. Amherst, Mass. July 15–19, 1964 Vol. I, p 97
98. Sano S (1982) in: Nozaki M, Yamamoto S, Ishimura Y, Coon M J, Ernster L, Estabrook R W (eds) Oxygenases and Oxygen Metabolism, p 551
99. Groves J T, Nemo T E, Meyers R S (1979) J. Am. Chem. Soc. 101: 1032 and later published work
100. Jones P (1982) in: Dunford H B, Dolphin D, Raymond K, Sieken L (eds) The Biological Chemistry of Iron, Publishing Co., p 427
101. Sakurai H (1986) Yakugaka Zasshi 106(8): 619
102. Guillochon D, Cambon B, Esclade L, Thomas D (1984) Enzyme Microb. Tech. 6: 161
103. Lindsay-Smith J R, Piggott R E, Sleath P R (1982) J. Chem. Soc. chem. Commun. p 55.
104. Muto T, Umehara J, Masumori H, Muira T, Kimura M (1985) Chem. Pharm. bull. 33(11): 4749.
105. Khenkin A M, Shteinman A A (1984) J. Chem. Commun. p 1219
106. Powis G, Lyon L, McKillop D (1977) Biochem. Pharmacol. 26: 137
107. Hamilton G A (1964) J. Am. Chem. Soc. 86: 3391.
108. Hayaishi O (1974) in: Hayaishi O(ed) Molecular Mechanisms of Oxygen Activation, Academic Press, New York, p 1.
109. Jerina D M (1973) Chem. Technol. 4: 120.
110. Ullrich V, Duppel W (1975) in: Boyer P D(ed) The Enzymes, 3rd edn. Academic Press, New York, 12 (Part B): 253
111. Bayer E, Krauss P, Roder A, Schretzmann P (1973) in: King T E, Mason H S, Morrison M (eds) Oxidases and Related Redox Systems, University Park press, 1: 227
112. Razenburg J A S J, Nolte R J M, Drenth W (1984) Tetra. Lett. 25(7): 789
113. Meunier B, Guilmet E, DeCarvalho M-E, Poilblanc R (1984) J. Am. Chem. Soc. 106: 6668
114. De Poorter B, Ricci M (1985) Tetrahedon Lett. 26: 4459

115. Renaud J P, Battioni P. Bartoli J , Mansuy, D J (1985) Chem. Soc. Chem. Commun. p 888
116. Mansuy D, Bartoli J F, Momenteau M (1982) Tetrahedon Lett. 23: 2781
117. Tabushi I, Koga N J, (1979) J. Am. Chem. Soc. 101: 6456
118. Perrée-Fauvet M, Gauderner A J (1981) Chem. Soc. Chem. Commun. p 874
119. Groves J T, Quinn R J (1985) J. Am. Chem. Soc. 107: 5790
120. Tabushi I, Marimitsu K (1984) J. Am. Chem. Soc. 106: 6871
121. Mansuy D, Fontecave M, Bartoli J F (1983) J. Chem. Soc. Chem. Commun. p 253
122. Falk J E (1964) Porphyrins and Metalloporphyrins, Elsevier New York, N. Y.
123 Chance B, Estabrook R W, Yonetani T (eds) (1966) Hemes and Hemoproteins, Academic Press, New York, N. Y.
124 Buchler J W, Prippe L, Rohbock K, Schneerage H H, Tsutsui M, Ostfield D, Hoffman L, Suzuki K, Cohen I, Kadish I A, Davis D G (1973) Annal. N. Y. Acad. Sci. vol. 206
125. Taylor T G, Xu F (1987) J. Am. Chem. Soc 109: 6201
126. Dolphin D, Muljiani Z, Rousseau K, Borg D C, Fajer J, Felton R H (1973) Annal. N. Y. Acad. Sci. 206: 177
127. Adams P A, Baldwin D B, Berman M C (1979) J. Chem. Soc. (Lond.) Chem. Commun. p 856
128. Adams P A, Berman M C (1982) Inorg. Biochem. 17: 1
129. Adams P A, Adams C, Baldwin D B, Bergman M C (1982) J. Inorg. Biochem. 17: 261
130. Haurowitz F, Groh M, Gansinger G (1973) J. Biol. Chem. 248(11): 3810
131. Mansuy D, Dansette P M, Peoquet F, Chattard J-C (1980) Biochem, Biophys. Res. Commun. 96(1): 433
132. Griffin B W, Ting P L (1978) Biochem. 17(11): 2206
133. Sakurai H, Mikito H (1975) Biochem. Pharmacol. 24: 1647
134. Sakarai H (1976) Chem. Pharm. Bull. 24(7): 1686
135. Sakurai H, Shimomura S, Ishizu K (1977) Chem. Pharm. Bull. 25(1): 199
136. Collman J P, Sorell T N, Dawson J H, Trudell J R, Bunnenberg E, Djerassi C (1976) Proc. Natl. Acad. Sci. (USA) 73(1): 6
137. Stern J O, Peisach J (1974) J. Biol. Chem. 249(23): 7495
138. Stern J O, Peisach J (1976) FEBS Lett. 62(3): 364
139. Tang S C, Koch S, Papaefthymiori G C, Foner S, Frankel R B, Ibers J A, Holm RH (1976) J. Am. Chem. Soc. 98(9): 2414
140. Roder A, Bayer E (1969) Europ. J. Biochem. 11: 89
141. Sakurai H, Shimomura S, Fukujawa K, Ishizu K (1978) Chem. Pharm. Bull. 26(5): 1348
142. Taylor P S, Dolphin D, Taylor T G (1980) J. Chem Soc. Commun. p 279
143. Lindsay-Smith J R, Sleath P R (1982) J. Chem. Soc. Perkin Trans. II p 1009
144. Sakurai H, Hatayama E, Nishida M (1983) Inorganica Chimica Acta. 80: 7
145. Adams P A, Berman M C, Baldwin D A (1979) JCS Chem. Commun. p 856
146. Adams P A, Adams C, Berman M C, Lawrence M C (1984) J. Inorg. Biochem. 20: 291
147. Ingleman M, Simdberg M, Ekstrom G (1982) Biochem. Biophys. Res. Commun. 106(2): 625
148. Sakurai H (1980) Chem. Pharm. Bull. 28(11): 3437
149. Sakurai H, Ogawa S (1975) Biochem. Pharm. 24: 1257
150. Sakurai H, Hatayama E, Fujitani K, Kato H (1982) Biochem. Biophys. Res. Commun. 108(4): p 1649
151. Sakurai H, Ogawa S (1975) Biochem. Pharmacol. 24: 1257
152. Sakurai H, Ogawa S (1979) Chem. Pharm. Bull. (Tokyo) 27: 2171
153. Hamed M Y, Silver J, Wilson M T (1983) Inorganica Chimica Acta. 78: 1
154. Sakurai H, Kito M (1975) Biochem. Pharmacol. 24: 1647
155. Norman R O C, Radda G K (1962) Chem. Soc. J. Proc. p 138
156. Langen H, Epprecht M, Linden T, Hehlgans T, Gutte B (1987) private communication
157. Moser R, Thomas R M, Gutte B (1983) FEBS Lett. 157: 247
158. Medeiros M H G, Wefers H, Sies H (1987) Free Radical Biology and Medicine 3: 107
159. Sakurai H, Hatayama E, Fryitani K (1984) Inorganica Chimica Acta. 91: 233
160. Sakurai H, Hatayama E, Fujitani K (1984) Inorganica Chimica Acta. 91: 233
161. Groves J T, Nemo T E, Myers R S (1979) J. Am. Chem. Soc 101: 1032
162. Groves J T, Kniper W J, Nemo T E, Myers R S (1980) J. Mol. Catal. 7: 169

163. Groves J T, Haushalter R C (1980) J. Am. Chem. Soc 102: 6375
164. Groves J T, Nemo T E (1983) J. Am. Chem. Soc 105: 6243
165. Groves J T, Myers R S (1983) Ibid 105: 5791
166. Groves J T, Nemo T E (1983) Ibid 105: 5786
167. Groves J T, Watanabe Y, McMurry T J (1983) Ibid 105: 4489
168. Hill C L, Schardt B C (1980) J. Am. Chem. Soc 102: 6374
169. Hill C L, Smegal, J A (1982) Nouv. J. Chim. 6: 287
170. Smegal J A, Schandt B C, Hill C L (1983) J. Am. Chem. Soc. 105: 3510
171. Smegal J A, Hill C L (1983) J. Am. Chem. Soc 105: 3515
172. Schardt B C, Hollander F J, Hill C L (1982) J. Am. Chem. Soc. 104: 3964
173. Groves J T, Nemo T E, Myers R S (1979) J. Am. Chem. Soc. 101(4): 1032
174. Nee M W, Bruice T C (1982) J. Am. Chem. Soc. 104: 6123
175. Fontecave M, Dansuy D (1984) J. Chem. Soc. Chem. Commun. 13: 879
176. Miyata N, Kirichi H, Hirobe M (1981) Chem. Pharm. Bull. 29(5): 1489
177. Groves J T, Nemo T E (1983) J. Am. Chem. Soc. 105: 5786
178. Groves J T, Meyers R S (1983) J. Am. Chem. Soc. 105: 5791
179. Santa T, Mori T, Hirobe M (1988) Chem. Pharm. Bull. 33(5)2175
180. Meunier B (1986) Bull. Chem. Fr. p 578
181. Mansuy D, Battioni J-P, Chottard J-C, Ullrich V (1979) J. Am. Chem. Soc. 101(4)3971
182. Mansuy D, Dansette P, Pecquet F (1980) Biochem. Biophys. Res. Commun. 96(1): 433
183. Ullrich V (1977) in: Jollow J D, Kocsis J J, Snyder R, Vainio H (eds) Biological Reactive Intermediates, Plenum Press, New York, N. Y. 65
184. Miyata N, Kiuchi H, Hirobe M (1981) Chem. Pharm. Bull. 29(5): 1489
185. White R E, Coon M J (1980) Ann. Rev. Biochem. 49: 315
186. Nordblom G D, White R E, Coon M J (1976) Arch. Biochem. Biophys. 175: 524
187. Hrcay E G, Gustafsson J, Ingelman-Sundburg M, Ernster L (1975) Biochem. Res. Commun. 66: 209
188. Ullrich V, Ruf H H, Wende P (1977) Croat. Chem. Acta. 49: 213 Lichtenberger F, Nastainczyk W, Ullrich V (1976) Biochem. Biophys. Res. Commun. 70: 939
189. Fontecave M, Mansuy D (1984) Tetra. 40(21): 4297
190. Backvall J E, Awasthi A K, Renko Z D (1987) J. Am. Chem. Soc. 109: 4750
191. Meunier B, Guilmet e, DeCarvalho M E, Poilblanc R (1984) J. Am. Chem. Soc 106: 6668
192. Fontecave M, Mansuy D (1984) Tetrahedron 40(21)4297
193. Frostin-Rio M, Pujol D, Bied-Charreton C, Perrée-Fauvet M, Graudemer A (1971) J. Chem. Soc. Perkin Trans. I, p 1984
194. Meunier B, Guilmet E, DeCarvalho M E, Poilbac R (1984) J. Am. Chem. Soc. 106: 6668
195. Lindsay-Smith J R, Mortimer D N (1985) J. Chem. Soc. Chem. Commun. p 64
196. Krman J (1801) Gilbert's Ann. 8: 206
197. Weinberg N L (ed) (1974) in: Technique of Electroorganic Synthesis, 5 (part I) of Techniques of Chemistry, John Wiley and Sons
198. Baizer M M. Lund H, (Eds) (1983) in: Organic Electrochemistry, 2nd edn., Marcel Dekker Publishers
199. Fry N L (ed) (1978) in; Synthetic Organic Electrochemistry, Harper and Row Publishers
200. Shono T (1984) Tetrahedron 40(5): 811
201. Fenn R J, Krantz K W, Stuart J D, (1976) J. Electrochem. Soc. 123(11): 1643
202. Stuart J ., Keenan R R, Fenn R J, Jensen R G, Pudelkiewicz W J (1972) ACS Natl. Meeting Extended Abstr. 12(2): 80
203. Fenn R J, Krantz K W, Stuart J D (1976) J. Electrochem. Soc. 123: 1643
204. Sokolovskii V D, Belyaer V D (1980) React. Kinet. Catal. Lett. 15: 357
205. Sokolovskii V D, Belyaer V D (1983) Syntnikova React. Kinet. Catal. Lett. 22: 127
206. Khenkin A M, Shilov A E (1987) React. Kinet. Catal. Lett. 33(1): 125
207. MacDonald T L, Gutheim W G, Guengrich F P (1987) Fed. Proceed. 46(6): 1956
208. Maissant J M, Bouchoule C, Canesson P, Blanchard M (1983) J. Molec. Catal. 18: 189
209. Sakurai H, Ogawa S (1975) Biochem. Pharmacol. 24: 1257
210. Akhiem A A, YuGerman S, Metelitsa D I (1978) React. Kinet. Catal. Lett. 8(2): 217

211. Miyamoto I, Yamamoto I (1981) Agric. Biol. Chem. 45(9): 1991
212. Barrett R, Pautet F, Daudon M (1987) Pharmazie 42(2): 132
213. Barrett R, Pautet F, Mathian B, Daudon M (1985) Pharmazie 40: 728
214. Ibid (1986) Xenobiotica 16(7): 615
215. McMahnon R E, Miller W M, Marshall F J (1978) in: Gorrod J W (ed) Biological Oxidation of Nitrogen, Elsevier, p 445
216. Panda A K, Mahapatro S N, Pangrahi O P (1981) J. Org. Chem. 46: 4000
217. Perrone R, Carbonara G, Tortorella V (1984) Arch. der Pharmazie 317: 635
218. Perrone R, Carbonara G, Tortorella V (1984) Arch. der Pharmazie 317: 21
219. Lee J-K, Foumier J-C, Catroux G (1977) J. Korean Agric. Chem. Soc. 20(1): 109
220. Worobey B L, Pilon J C, Sun W-F (1987) J. Agric. Food Chem. 35: 325
221. Terenter A P, Magilyansky Y D (1955) Dokl. Akad. Nauk SSSR 103: 91, CA50: 4807 (1956)
222. Gardner H W, Crawford C D (1981)Biochimica Biophysica Acta. 665: 126
223. Gardner H W, Jursinic P A (1981) Biochimica Biophysica Acta. 665: 100
224. Searle A, Tomasi A (1982) J. Inorg. Biochem 17: 161
225. Matsumura T, Nishinaga A (1982) in: Nozaki M, Yamamoto S, Ishimura V, Coon M J, Ernster L, Estabrook R W, Oxygenases and Oxygen Metabolism, Academic Press
226. Ravindranath S D, Kumar A, Kumar P, Vaidyanathan C S, Lao N A (1974) Arch. Biochem. Biohpys. 165: 478
227. Nagamachi T, Nishimaga A, Matsumura T (1972) Chem. Lett. p 111
228. Nishimaga A, Tomita H, Shimizu T, Matsumura T (1978) Fund. Res. in Homogeneous Catal. 2: 241
229. Nishinaga A, Tomita (1980) Molec. Catal. 7: 179
230. Nakayama Y, Tanemitsu Y, Yoshioka H (1982) Tetrahedon Lett. 23(24): 2499
231. Nakayama Y, Sanemitsu Y, Yoshioka H, Nishinaga A (1982) Tetrahedon Lett. 23(24): 2499
232. Okado T. Kamiyay (1981) Bull. Chem. Soc. Jpn. 54(9): 2727
233. Taniguchi M, Obata H, Takuyama T (1981) Technology Reports of the Kansai University 22: 119, CA 95(19)168659g
234. Mueller P, Gilabert D M (1986) Chimica 40(4): 127
235. Mueller P, Godoy J (1981) Tetrahedon Lett. 22: 2361
236. Ibid (1982) 23: 3661
237. Ibid (1983) Helv. Chim. Acta 66: 1790
238. Galliani G, Rindone B (1981) Tetrahedon 37: 2317
239. Ratori E, Nagano T, Kunieda T, Hirobe M (1981) Chem, Pharm. Bull. 29(10): 3075
240. Hamilton G A, Friedman J P (1963) J. Am. Chem. Soc. 88: 1008
241. Hamilton G A, Hamfin Jr J W, Friedman J P (1966) J. Am. Chem. Soc. 88: 5269
242. Walling C, Camaioni D M (1975) J. Am. Chem. Soc. 97(6): 1603
243. Keay R E, Hamilton G A, (1975) J. Am. Chem. Soc. 97: 6876
244. Guillochon D, Ludot J M, Esclade L, Cambon B, Thomas D (1982) Enzyme Microb. Tech. 4: 96
245. Hoggin J (1987) C & E News Sept 28, p 27
246. Brewer C T, Brewer G A (1987) Inorg. Chem. 26: 3420
247. Inone H, Hirobe M (1987) Biochem. Biophys. Res. Commun. 145(1): 596; Florence TM, (1984) Inorg. J. Biochem 22: 221
248. Rowley D A, Halliwell B (1985) Ibid 23: 103
249. Ullrich V, Standinger H (1966) in: Bloch K, Hayaishi O. (eds) Biological and Chemical Aspects of Oxygenases, Maruzen, Tokyo
250. Ullrich V (1969) Z. Naturforsch 246: 699
251. Jonsson L (1981) Acta Chemica Scandinavia B35 p 683 and related papers of this series
252. Jerina D M, Boyd D R, Daly J W (1970) Tetra. Lett. No. 6, p 457
253. Sumarno M, Atkinson E, Suarna C, Saunders J K, Cole E R, Southwell-Keely P T, (1987) Biochem. et. Biophys. Acta 920: 247
254. Walling C, Hodgdon R B (1958) J. Am. Chem. Soc. 80: 228
255. Tsuchiya T, Ikehira K, Imamura J (1982) J. Bull. Chem. Soc. Jpn. 55: 1926
256. Nakatsugawa T, Dahm P A (1965) J. Econ. Ent. 58(3): 500

257. Rhoads R E, Udenfriend S (1969) Biochem. J. 60: 1473
258. Eremin A E, Metelitsa D I (1985) React. Kinet. Catal. Lett. 27(1): 47
259. Sugimoto T, Kokubo T, Miyazaki J, Tanimoto S, Okano M, (1951) Bioorg. Chem. 10: 3111
260. Komiyana M, Bender M L, (1984) in: Page I (ed) The Chemistry of Enzyme Action, Elsevier, Chp 14, p 505
261. Kaiser E T, Kezdy F J (eds) (1971) Prog. in Bioorg. Chem. 2: 2
262. Vanetten R L, Sebastian J F, Clowers G A, Bender M L (1967) J. Am. Chem. Soc 89: 3242
263. Ono K, Katsube J (1983) Chem. Pharm. Bull. 31(4): 1267
264. Mori. Y, Toyoshi K, Baba S, (1976) Chem. Pharm. Bull. 24(6): 1387
265. Mansuy D, Battioni J-P, Chottard J-C, Ullrich V (1979) J. Am. Chem. Soc. 101(14): 3971
266. Jerina D M, Boyd D R, Daly J W (1970) Tetrahedon Lett. p 457
267. Mori Y, Toyoshi K, Baba S (1976) Chem. Pharm. Bull. (Tokyo) 24: 865
268. Ullrich V, Wolf J, Amadori E, Staudinger H (1968) Hoppe-Seyleer's Z. Physiol. Chem. 349: 85
269. Matsumura T (1977) Tetrahedon 33: 2869
270. Hamilton G A (1974) in: Hayaishi O (ed) Mechanisms of Oxygen Activation, Academic Press, New York, p 405
271. Tanaka F S, Wein R G (1979) J. Aric. Food Chem. 27(2): 311
272. Ullrich V (1969) Z. Naturforschg. 246: 699
273. Bidlack W R, Lowery G L (1982) Biochem. Pham. 31(3): 311
274. Ruzo L O, Ibid Abstr. 6C-15
275. Winston G W, Cederbaum A I (1982) Biochem. 21: 4265
276. Hamilton G A, Revesz A (1966) J. Am. Chem. Soc. 88: 2069
277. Riddle V M, Mazelis M (1965) Plant Physiology 40: 481
278. Rogié M M, Demmin T R (1978) J. Am. Chem. Soc. 100: 5472
279. Demmin T R, Rogié M M (1980) J. Org. Chem. 45: 4210
280. Razenberg J A S J, Nalte R J M, Drenth W (1984) Tetrahedron Lett. 25(7): 789
281. Denison M S, Yarborough J D (1985) Comp. Biochem. Physiol. 81C # 1 p 105
282. Blackburn G M, Kellard B, Orgee L, Thompson M H (1985) J. Chem. Soc. Perkin Trans. II p 287
283. Josephy P D, Ebing T E, Mason RP (1983) Molec. Pharmacol. 23: 766
284. Josephy P D, Iwaniw D C (1983) Carcinogenesis 6(1): 155
285. Collman J P, Brauman J I, Meunier B, Raybuck S A, Kodadek T (1984) Proc. Natl. Acad. Sci. (USA) 81: 3245 and references therein
286. Toyoda I, Terao J, Matsushita S (1982) Lipids 17(2): 84
287. Gardner H W, Weisleder D, Kleinman R (1978) Lipids 13: 246
288. Balba M H, Saha J G (1974) Bull. Environm. Contam. Toxicol. 11(2): 193
289. Knaak J B, Stahmann M A, Casida J. E. (1962) J. Agric. Food Chem. 10(2): 154
290. Ptashne K A, Neal R A (1972) Biochemistry 11(17): 3224
291. Marshall R S (1972) Ph.D. Thesis, Dis. Abstr. Int 32(9): 5233B
292. Metcalf R L (1968) in: Hodgson E (ed) Oxidation of Toxicants, North Carolina State Univ., Raleigh, N.C. p 151
293. Metcalf R L, Fukuto T R, Collins C, Brock K, Abd E.-Aziz S, Munoz R and Cassil C C (1968) J. Agric. Food Chem. 16(2): 300
294. Fenwick M L (1958) Biochem. J. 70: 373
295. Locke R K, Mayer V W (1974) Biochem. Pharmacol 23: 1979
296. Mead J A R, Smith J N, Williams R T (1958) Biochem. J. 68: 67
297. Buhler D R, Mason H S (1961) Arch. Biochem. Biophys. 92: 424
298. Fahmy M A H, Fukuto T R (1972) Tetrahedron Lett. 41: 4245
299. Tanaka F S, Wein R G (1979) J. Agric Food Chem. 27(2): 311
300. Singh J, Cochrane W P JAOAC 62 (4): 751
301. Brown M A, Casida J E (1987) 194th ACS Natl. Meeting, Abstr. AGRO # 171
302. Herriott A W (1971) J. Am. Chem. Soc. 93: 3304
303. Mayer V W (1972) Muta. Res. 15: 147
304. Booth J E (1955) Biochem. J. 60: 62
305. Boyland E, Mason D (1958) Biochem. J. 69: 601

306. Shannon P, Bruice TC (1981) J. Am. Chem. Soc. 103: 4580
307. Kitada M, Ando M, Ohmori S, Kabuto S, Kamataki S, Kitagawa T (1983) Biochem. Pharm. 32: 3151
308. Hino T, Yamaguchi H, Matsuki K, Nakano K, Sodeoka M, Nakagowa M (1983) J. Chem. Soc. Perkin Trans. I, p 141
309. Brodi B B, Axelrod J, Shore P A, Udenfriend S (1954) J. Biol. Chem. 208: 741
310. Smegal J A, Hill C L (1983) J. Am. Chem. Soc. 105(11): 3515
311. Cantoni L, Blezza D, Belvedere G (1982) Experimentia 38: 1192
312. Collman J P, Kodadek T, Raybuch S A, Brauman J, Papoyium L (1985) J. Am. Chem,. Soc. 107: 4343
313. Powell M F, Pai E F, Bruice T C (1984) J. Am. Chem. Soc 106: 3276
314. Collman J P, Brauman J, Meunier B, Hayashi T, Kodalek T (1985) J. Am. Chem. Soc. 107: 2000
315. Deno N C, Jedziniak E J, Messer L A, Meyer M D, Stroud S G, Tomezsko E S (1977) Tetrahedron 33: 2503
316. Al-Hayek N, Done M (1985) Environ. Tech. Lett. 6: 37
317. Razenberg J A S J, VanDerMade A W, Smelts J W H, Nolte R J M (1985) J. of Molec. Catal. 31: 271
318. Acheson R M, Hazelwood C M (1960) Biochem. Biophys. Acta 42: 49
319. Halliwell B (1977) Biochem. J. 167: 317
320. Richter H W, Fetrow M A, Lewis R E, Waddell W H (1982) J. Am. Chem. Soc. 104: 1666
321. Richter H W, Waddell W H (1982) J. Am. Chem. Soc. 104: 4630
322. Kumar P R, Ravindranath S D, Vaidyanathan C S, Rao NA (1972) Biochem. Biophys. Res. Commun. 48(5): 1049
323. Chambers R P, Gioggin P G, Musgrove W K R (1959) J. Chem. Soc. p 1804
324. Papouchado L, Bacon J, Adams RN (1970) J. Electroanal. Chem. 24: 1
325. Weinberg N L, Weinberg H R (1968) Chem. Rev. 68: 486
326. Walling C, Johnson R A (1975) J. Am. Chem. Soc 97: 363
327. Jerina D M, Daly J W, Witkop B (1971) Biochem. 10: 366
328. Wang X Y, Wong R, Li G Q, Li G N (1987) Acta. Chimica. Sinica 45(8): 780
329. Brook M A, Castle L, Lindsay-Smith J R, Higgins R, Morris K P (1982) J. Chem. Soc. Perkin Trans. II, p 687
330. Tsuji J, Mirato M Tetrahedron Lett. 28(33): 3683
331. Karlin K D, Cohen B I, Jacobson R R, Zubieta J (1987) J. Am. Chem. Soc. 109: 6194
332. Karlin K D, Gultneh Y, Hayes J C, Cruse R W, McKown J, Hutchison J P, Zubieta J (1984) J. Am. Chem. Soc. 106: 2121
333. Karlin K D, Gultneh Y (1985) J. Chem. Educ. 62(11): 983
334. Karlin K D, Gultneh Y (1987) Prog. Inorg. Chem. 35: 219
335. Karlin K D, Cruse R W, Gultneh Y, Hoyes J-C, McKown J W, Zubieta J (1986) in: Karlin K D, Zubieta J (eds) Biological and Inorganic Copper Chemistry, Adenine Press, Guilderland, N. Y., 2, p 101
336. Sasaki K, Ito S, Saheki Y, Kinoshita T, Yamasaki T, Harada J (1983) J. Chem. Lett. p 37

Quantitative Structure-Activity Relationships of Environmental Pollutants

Joop L. M. Hermens

Research Institute of Toxicology, University of Utrecht, P.O. Box 80176 NL-3508 TD Utrecht, The Netherlands

Summary

Techniques for the estimation of rates of environmental fate processes as well as effect concentrations of organic micropollutants are based on comparisons with descriptors related to the structure or to physical chemical properties by quantitative structure activity relationships (QSAR). The hydrophobic character of chemicals, estimated by the octanol water partition coefficient (K_{ow}), influences many processes. Simply due to the influence of hydrophobicity on the absorption in aquatic organisms, K_{ow} explains completely the variance in effect concentrations of relatively unreactive chemicals that act by narcosis [155, 156]; effect concentrations that cover seven orders of magnitude while the internal concentrations are almost constant [57, 83]. An expression of toxicity based on internal effect concentrations, instead of exposure concentrations as in LC50 data, is an interesting new development because it makes it more easier to compare data from different species, exposure regimes and exposure times [83]. QSARs established for these unreactive chemicals predict the minimal or base line effect concentrations of all organic micropollutants.

Effect concentrations of other classes of chemicals are usually lower than those from the first group with as examples: aromatic amines, phenols, some nitroaromatics, reactive chemicals and toxicants with specific modes of action. This higher activity can be the result of a specific interaction with a target or to a more or less unselective reactivity to nucleophiles. Because the affinity or reactivity to a target molecule can vary among chemicals from each of these classes, it is not surprising that effect concentrations are also influenced by descriptors other than K_{ow} and which are more related to electronic or steric properties. QSAR equations for aquatic toxicity data of these kind of chemicals are less well understood than the equations for unreactive chemicals because the processes, responsible for their effects, are more complex. Chemical reactivity seems an important feature, but most of the present QSAR equations for reactive chemicals are based on measured rate constants [205–207] and it will be a challenge to explore descriptors for chemical reactivity based on calculated parameters from e.g. molecular orbital calculations.

Outliers are usually recognized after a QSAR is established and explained in terms of differences in mode of action or toxicokinetic behaviour (e.g. metabolism). This points to the problem that the structural requirements, related to a particular QSAR, are usually not very well defined. More information on mode of action such as in studies to fish acute toxicity syndromes [249] and the structural requirements related to mode of action, are a necessity.

Predictions based on QSAR can be useful for a variety of purposes: regulating activities (evaluating new or existing chemicals); priority setting for existing chemicals; evaluation during the development of new chemicals. When predictions are used for practical purposes it is important to be familiar with the prediction models and their limitations. Besides the possibility to use QSAR for prediction, it is very helpful in the analysis of toxicity data, because it is a tool in the classification of large numbers of chemicals into a limited number of classes and in addition outliers can be recognized and studied in more detail. The distinction of classes of chemicals with similar modes of action is also valuable for the estimation of the effects of mixtures.

Introduction

Information on environmental fate and adverse effects of pollutants are basic needs in environmental hazard assessment (Fig. 1), [1]. Fate processes such as adsorption to sediment, chemical or microbial degradation and evaporation determine the concentration of a chemical that is accessible for uptake by biota from a specific compartment (bioavailability). When rates of these processes are known, environmental fate modeling [2, 3] can supply a rough estimate of the bioavailability of a chemical. Knowledge on toxicity is usually based on data for effect concentrations from studies with simple overall effects such as lethality (LC50s) or inhibition of reproduction with single species. Today, attention is also paid to more realistic effects on e.g. population and ecosystem level, although such data are very scarce. Another approach for setting 'safe levels' is by application of safety factors to results of e.g. chronic single species tests [4]. In general, there is a lack of experimental data for rates of fate processes as well as acute and chronic effect data from single species tests. It is not surprising, therefore, that attempts have been made to estimate fate as well as effects of pollutants. One of these estimation techniques, that are applied within environmental hazard assessment, are based on quantitative structure-activity relationships (QSAR).

As shown in Fig. 2, QSAR is based on comparisons of the toxicity or fate with chemical structure or physical chemical properties by several data analytical

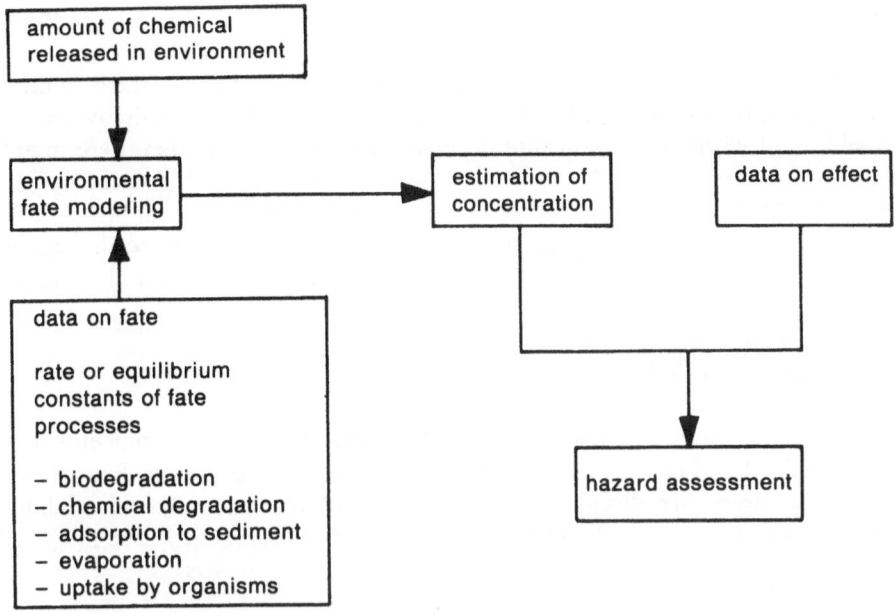

Fig. 1. A scheme for environmental hazard assessment in the aquatic environment

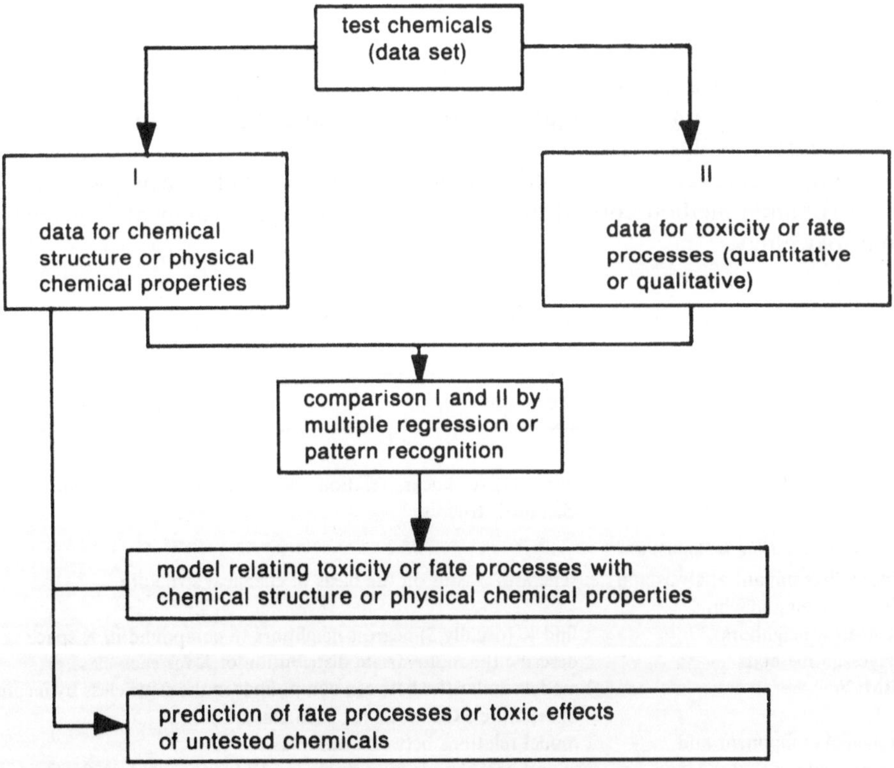

Fig. 2. Principle of QSAR analyses

models. Two broad classes of analytical models can be distinguished: multiple linear regression and pattern recognition. Before these models were discovered as a possible tool within environmental chemistry and toxicology, they were mainly applied and also developed within the pharmacological and pharmaceutical sciences [5–15]. Besides application of QSAR to toxicology in general [16–20], the use of QSAR in studying toxicity and fate of chemicals has been growing during the past decade [21–27]. Specifically, QSAR has been applied to the following fate processes:
- adsorption to sediment [28–32]
- biodegradation [33–42]
- uptake by aquatic organisms [43–68]
This paper will discuss QSAR models that are developed for environmental effect studies and will not include studies with environmental fate processes. Before some specific examples are discussed, attention is paid to the analytical methods, chemical descriptors and physical chemical properties, commonly used in QSAR analyses.

Data Analytical Methods in QSAR

Two broad classes of data analytical methods are applied within QSAR (see Table 1):

1. Multiple regression (MR)
2. Pattern recognition techniques (PARC), principal component analyses (PCA) and factor analyses (FA).

   Multiple regression (MR) will be discussed in more detail, because MR is the most common method applied to QSAR studies within environmental chemistry and toxicology.

Table 1. Multivartiate data analytical methods commonly used in QSAR (modified after Wold and Dunn [69]

method	scope
multiple regression	
Hansch and Free-Wilson analysis	: quantitative linear relation between biological activity and chemical structure
pattern recognition techniques	
linear discriminant analysis and linear learning machine	: seperates classes on the basis of chemical structure
K nearest neighbors	: find K (usually 3) nearest neighbors to compound in X space.
Bayesean methods	: describe the multivariate distribution of X for each class.
SIMCA	: model similarities between compounds in the same class by means of separate PC models for each class.
principal component and factor analysis	: model relations between variables x_i and compounds by a single model for the whole data set.

Multiple Regression Technique

The Hansch, as well as the Free Wilson approach, are based on multiple regression. Within the *Free Wilson approach*, biological activity (BA) is described as a linear function of structure indicator variables according to equation 1 [70].

$$BA = A + \Sigma^i \Sigma^j G_{ij} X_{ij} \tag{1}$$

in which: BA is the biological activity,

X_{ij} represents the presence ($X_{ij}=1$) or absence ($X_{ij}=0$) of substituent i at position j, and

G_{ij} is the group contribution to activity of substituent i at position j.

Equation 1 can be solved by multiple linear regression. The Free-Wilson approach assumes that the effect of the substituent at a specific position is constant and independent of the effect of substituents at other positions. Additivity will fail, however, if there are interactions between certain of the substituents [13]. This method is based on the structure of a chemical and yields no information on the relevancies of certain physico-chemical properties. The Free-Wilson method is only rarely applied in environmental toxicology and according to Martin [13], "the Free-Wilson approach would appear to be most useful in the design of a series of analogs of a lead and in preliminary QSAR analyses of a series of chemicals".

Most QSAR studies within the field that is discussed here, are based on the *Hansch approach*. Although the idea that physico-chemical properties are related to biological activities of chemicals is already known since the previous century from the work of Meyer [71] and Overton [72], Corwin Hansch [73, 74] is the pioneer and founder of modern QSAR analyses. The classical Hansch equation describes the influence of substituents on the biological activity of a parent molecule as follows:

$$\log 1/C = c\pi + c'\sigma + c''E_s + c''' \tag{2}$$

in which: C is the molar concentration of a chemical with a particular effect,

π is a substituent constant for hydrophobic effects,

σ is a substituent constant for electronic effects, and

E_s is a substituent constant for steric effects.

c constants that are obtained by fitting experimental data for 1/C with values for π, σ and E_s.

The substituent constants π, σ, and E_s will be discussed later in this paper. Equation 2 is solved by multiple linear regression for a dataset of a limited number of chemicals and this yields values for c, c', c" and c'''. This same equation can be used for predictions of C of related chemicals based on their substituent constants.

In several of his publications, Hansch has given a rationalization of this equation in terms of linear free energy relations [10–12, 73, 74]. The rationalization that is presented here is rather similar to the one given by Seydel and Schaper [75].

The activity of a drug or toxicant is dependent on:

- the probability that a chemical reaches its site of action (Pr_1),
- the probability that a chemical interacts with a receptor or target molecule (Pr_2), and
- the external concentration (C) or dose to which the organism is exposed.

It is assumed that, at a particular effect, the number of molecular events or the concentration of the target molecules (C_t) that has interacted with a toxicant is constant. C_t is a function of the external concentraion (C), the probability that the compound reaches its site of action (Pr_1) and the probability that it interacts with the target molecule (Pr_2). So, C_t can be written as:

$$C_t = cP_{r1}P_{r2}C = \text{constant} \tag{3}$$

Logarithmic transformation of equation 3 yields:

$$\log 1/C = c + \log Pr_1 + \log Pr_2 \tag{4}$$

In order to reach the site of action, a chemical must pass alternate aqueous and lipid compartments (membranes). Although there are examples of active transport through membranes, the movement of most organic chemicals is based on free diffusion. The coefficient of diffusion depends on molecular properties such as, hydrophobicity, polarity, degree of ionization, molecular shape and size [75]. The absorption, transport and distribution is often related to the octanol-water partition coefficient (K_{ow} or P_{oct}) of chemicals as a measure for its hydrophobic or lipophilic properties. Octanol was chosen mainly because it should resemble the apolar and polar characteristics of membrane lipids. It is assumed that the partition of a chemical between a biophase and water (K_{bio}) is related to the octanol-water partition coefficient (K_{ow}), according to a Collander type of equation [76]:

$$\log K_{bio} = c \log K_{ow} + c' \tag{5}$$

Numerous studies have shown that the uptake and transport of a chemical increases with its K_{ow} [75, 77–79]. The influence of hydrophobicity on toxico- or pharmacokinetic behaviour of chemicals is discussed in some excellent review articles as for instance from Seydel and Schaper [75], Lien [77] and Dearden [78]. These publications pay attention to different compartment models, non linear relationships, influence of ionization, thermodynamic aspects of partitioning and also discuss many examples of specific studies. Esser [79] have summarized many examples for bioconcentration by aquatic organisms. Non linear relationships are often observed in absorption and different models are developed to describe these non linear relationships such as the parabolic model from Hansch [10, 11] and the bilinear model from Kubinyi [80]. Based on a theoretical approach, Dearden and Townend [81] showed that the occurrence of an optimum in hydrophobicity depends strongly on non-equilibrium circumstances and hence on the duration of a test. In aquatic toxicity tests, equilibrium is usually established due to the test duration and the continuous exposure conditions. Only for very hydrophobic chemicals, the time of exposure might be too short to reach a steady state in internal concentrations. The influence of exposure time on the internal concentration and toxicity for aquatic organisms is discussed in more detail by McCarty et al. [82], Oliver and Niimi [61], Hawker and Connell [53] and Van Hoogen and Opperhuizen [83]. Uptake rate constants of the more hydrophobic chemicals usually decrease again; a phenomenon that is possibly related to the size of molecules [43, 52, 62]. Opperhuizen et al. [62] have suggested that the uptake is hindered when the diameter of the molecule exceeds a critical value of 9.5 Å. Bruggeman et al. [43] have shown that for such very hydrophobic chemicals uptake from food becomes more important than direct uptake from the aqueous phase. A decline in uptake is also observed for ionized chemicals because the ionized form of a chemical can hardly cross lipophilic barriers. The acid dissociation constant of a chemical (pKa) and the pH of a test solution are important features for ionizable compounds because they determine the degree of ionization. The subject of absorption of ionizable drugs is discussed in more detail by Schaper [84] and an example of the uptake of ionizable chemicals by fish is given by Saarikoski and Viluksela [64, 65].

The choice for the octanol-water system was also based on practical arguments and of course octanol and membranes are not completely similar. Based on thermodynamic arguments, Opperhuizen et al. [85] recently showed that octanol-water partitioning is different from fish-water partitioning. The high correlations that are usually observed between bioconcentration factors and K_{ow} may be partly due to the choice of chemicals. Most of such correlations are based on rather non-polar chlorinated hydrocarbons and the octanol-water partition coefficients of such compounds will be highly collinear with other parameters. It is indeed shown that correlations for bioconcentration based on chemicals with a wider range of structures are of lower statistical quality [46, 61], although in some cases this may also be related to metabolism of some of the chemicals in these data sets [61]. Regarding the use of octanol, I agree with Dearden [78] who has stated that: "as the octanol water system is the system of choice for practical purposes, we should use that system, at the same time seeking to unravel its complexities by fundamental investigations, including, if necessary, studies of hydrocarbon-water system".

In conclusion: absorption and transport of a chemical to the site of action by passive diffusion is strongly dependent on its hydrophobicity and the octanol-water partition coefficient is a rather successful descriptor for this process. So, Pr_1 from equations 3 and 4 can be written as:

$$\log Pr_1 = c \log K_{ow} + c' \tag{6}$$

It is obvious that equation 6 is not valid in those cases where other transport mechanisms are involved or when e.g. reactive intermediates are formed. Balaz et al. [86] has studied the consequences of the formation of reactive intermediates on QSAR equations.

In order to apply QSAR techniques for predicting biological activity, it is necessary to be able to predict the octanol-water partition coefficient from chemical structure. Optimization of the lipophilicity of a chemical, e.g. within drug design, is carried out preferably, not by testing all kind of analogues, but by estimating it from structure. Therefore, an important development for QSAR was the introduction by Fujita et al. [87] of a substituent constant π for hydrophobicity which is defined as:

$$\pi(X) = \log K_{ow}(RX) - \log K_{ow}(RH) \tag{7}$$

In equation 7, RX and RH are the substituted and unsubstituted parent compound, respectively. The basis of equation 7 is the assumption that the contribution of a substituent to the octanol-water partition coefficient is constant, regardless the structure. So, for a set of structure analogues, equation 6 can be written as:

$$\log Pr_1 = c\pi + c' \tag{8}$$

The second important aspect that affects the activity of a compound is the probability that a chemical interacts with the receptor site or target molecule (Pr_2). This probability is related to the affinity of a chemical to its target molecule according to equation 9, in which K represent the equilibrium constant of the interaction.

$$\log Pr_2 = c \log K + c' \tag{9}$$

Within pharmacology, different types of drug-receptor interactions are distinguished (see Table 2) and similar interactions can play a role between a chemical and a target molecule leading to a toxic effect. The interactions can be divided into three broad classes: hydrophobic interaction, electronic attraction and steric repulsion and classical chemical descriptors for these interactions are the substituent constants π, σ, (Hammett) and E_s (Taft).

The Hammett σ constant, developed around 1940 [88], represents the influence of substituents on the charge distribution within a molecule and is derived from the

Table 2. Physical properties correlated with the strength of noncovalent interactions (modified after Martin [13])

Interaction	parameter used for substituent effect
hydrophobic bond	octanol-water partition coefficient (P_{oct} or K_{ow}) aqueous solubility (S)
dispersion bond	molar refractivity × ionization potential (MR × IP) MR alone Hildebrand's molar attraction constant parachor
electrostatic bond	Hammett sigma (σ) values charges from molecular orbital calculation
hydrogen bond	same as for electrostatic bonds but only for the atom involved in the hydrogen bond
charge transfer bond	Hammett sigma (σ) values E(lumo) or E(homo) from molecular orbital calculations
steris repulsion	van der Waals radii Taft E_s values

influence of a substituent X on the acid dissociation constant (pKa) of benzoic acid (RH) according to:

$$\sigma(X) = pKa(RX) - pKa(RH) \tag{10}$$

The Taft substituent constant is, on the analogy of the Hammet constant, a well known descriptor for steric influences of a substituent X on reaction rates and this descriptor is defined as [89]:

$$E_s(X) = \log k\ X(CO)OCH_3 - \log k\ CH_3(CO)OCH_3 \tag{11}$$

In equation 11, k represents the rate constant of the hydrolysis of the ester methylacetate.

If the interaction between a drug and its receptor or between a toxicant and its target molecule has both electronic and steric characteristics, the equilibrium constant K in equation 7 can be written as:

$$\log K = c\sigma + c'E_s \tag{12}$$

Combination of equations 4, 8 and 12 yield the classical Hansch equation:

$$\log 1/C = c\pi + c'\sigma + c''E_s + c''' \tag{13}$$

The important contribution of Corwin Hansch is that he derived a substituent constant for hydrophobicity and that he applied substituent constants to biological processes; substituent constants that were originally developed for predictions of rates of organic reactions. Other chemical descriptors can replace the classical parameters in equation 13 and it is obvious that not all descriptors have to be relevant for a specific biological process.

Other QSAR Methods

Pattern recognition techniques are mainly applied to studies between biological activity and chemical structure. Most of these techniques are summarized in table 1. Pattern recognition is a qualitative technique because it separates active from inactive chemicals; e.g. carcinogens from non carcinogens. Such a separation is based on descriptors like chemical substructures or physical chemical properties. Recently, Wold et al. [90] have evaluated several of such techniques and point to several pitfalls. They conclude e.g. that the traditional pattern recognition techniques (the linear discriminant analysis and linear learning machine) are severely restricted in their application because they need the number of cases (compounds) to substantially exceed the number of variables (chemical descriptors) in the data set. Other methods, such as principal component and partial least squares analysis, do not have this disadvantage [90]. Because pattern recognition is not used very often in QSAR studies for environmental effects, the reader is refered to review papers for information on these techniques [13, 16, 17, 69, 90]. Examples of the application of these techniques, mainly in the field of carcinogenicity and mutagenicity, are discussed in many publications [16–20]

Very sophisticated QSAR techniques, such as molecular modeling, requires information on the structure of the target molecule and are therefore mainly applied for specific interactions of drugs [9, 14]

Some Remarks on Statistics

Wold [69, 90] has formulated statistical conditions for each of the data analytical methods commonly applied in QSAR analyses. Application of the traditional statistical methods (linear discriminant analysis, linear learning machine and multiple regression techniques) requires that the number of chemical descriptors is small in comparison with the number of chemicals in the dataset (at least 1:4) and the absence of collinearity between chemical descriptors [90]. The consequence of collinearity is discussed in more detail by Topliss and Edwards [91]. The SIMCA method and also partial least squares and principal component analysis can handle many variables and collinear data matrices [90].

The statistical quality of multiple regression correlations is expressed by the correlation coefficient (r) and standard error of estimate (s). The F test is suitable for testing the significancies of individual descriptors, as well as the significancy of an improvement of the correlation by adding more chemical descriptors to an equation [13].

Multiple regression correlations are generally solved by linear regression least squares procedures [13]. Halfon [92] and Cramer et al. [93] have suggested that other methods such as the geometric mean functional regression method, cross-validation, bootstrapping and partial least squares analyses are more appropriate.

Computer programs for multiple regression and some pattern recognition techniques, written in FORTRAN, are given by Jurs [94].

Chemical Descriptors

The different types of descriptors that are used in QSAR analyses can be divided into a few broad classes:

- descriptors for hydrophobicity
- descriptors for electronic effects
- descriptors for steric effects
- structural indices
- other descriptors

Descriptors for Hydrophobicity

Descriptors for hydrophobicity will be discussed in more detail, because hydrophobicity is one of the most dominant factors in QSAR studies.

The hydrophobicity of a chemical is usually expressed by its octanol-water partition coefficient (K_{ow}). The classical method for measuring K_{ow}, the shake flask method, is very well suited for chemicals with log $K_{ow} < 4$ to 5. The K_{ow} of more hydrophobic chemicals is more difficult to measure because emulsions of octanol in water yield inaccurate data for the real concentration of the chemical in the water phase. The most obvious expression of this problem is the high variance in K_{ow} data for rather hydrophobic chemicals. An alternative, especially for more hydrophobic chemicals, is the generator column method described by Miller et al. [95] and Woodburn et al. [96]. Brooke et al. [97] applied a slow stirring method that is very well suited for more hydrophobic chemicals. Using this same method, De Bruijn et al. [98] measured recently K_{ow} for about 60 chemicals, including many polychlorinated biphenyls, and they report a K_{ow} value of 8.27 for decachlorobiphenyl which is one of the highest values ever measured. These direct techniques are rather labour intensive because the concentration of each chemical must be analysed quantitatively.

A much faster and easier way for estimating K_{ow} is by RP-HPLC (reversed phase, high performance liquid chromatrography). Many studies [99–113] have shown that K_{ow} is related to capacity factors (k) on a reversed phase C18 column and hence K_{ow} can be estimated from its capacity factor and an established correlation between log K_{ow} and log k for a set of reference chemicals. The method is improved by measuring k at different volume fractions of methanol in the eluens and using k values extrapolated to 0% methanol [100, 103, 105, 106, 110], although this is not applicable to partially or completely ionized chemicals [103, 104]. Correlations between log K_{ow} and log k are reasonably good, but careful analyses of these data shows that those correlations are different for different chemical classes [98, 111]. So, the RP-HPLC method can yield estimates of K_{ow} but the precision depends on the specific method that is applied and ofcourse also on the accuracy of K_{ow} values for reference chemicals that are often measured by the shake flask method.

Another estimation technique for K_{ow} is simply by calculation. As mentioned before, the hydrophobic substituent constant π [87] is a good predictor of K_{ow}, but π is only applicable within congeneric data sets. The π substituent constants derived from K_{ow} values of benzene derivatives cannot predict very accurately K_{ow} of benzene derivatives that contain polar substituents such as NO_2, OH or NH_2 [114–116]. Therefore, different sets of π constants are available depending on the structure of the parent compound [114–116] and some illustrative examples are given in Table 3. A second method for calculation log K_{ow} uses fragment values. Starting from a large database of experimental octanol-water partition coefficients, Rekker [117, 118] calculated values for several fragments. The structure of a molecule is reduced to fragments and log K_{ow} can be simply calculated by summing values for these fragments. In some cases, e.g. when two electronegative groups are present within the same molecule, correction factors must be applied [117]. A rather similar method, also based on fragments values and correction factors, was introduced by Hansch and Leo [114]. Mayer et al. [119], who compared both methods, conclude that "the physicochemical nature of all correction factors is far from understood and that in most cases a posteriori manipulation of factors is necessary to approximate the observed values". Insight into both methods is necessary in order to be able to judge the probable precision of the prediction. The

Table 3. Some different hydrophobic constants for a few different aromatic substituents (from Norrington et al. [116])

	π			π^-		
	ortho	meta	para	ortho	meta	para
H	0.00	0.00	0.00	0.00	0.00	0.00
CH_3	0.84	0.52	0.60	0.49	0.50	0.48
OH	−0.41	−0.50	−0.61	−0.58	−0.66	−0.87
NH_2	−1.40	−1.29	−1.30	−0.84	−1.29	−1.42
NO_2	—	0.11	0.22	—	0.54	0.45
Cl	0.76	0.77	0.73	0.69	1.04	0.93
Br	0.84	0.96	1.19	0.89	1.17	1.13

method from Hansch and Leo is computerized (CLOGP) at Pomona College. A description of CLOGP is given by Chou and Jurs [120].

Aqueous solubility (S) is often used as descriptor for hydrophobic properties. It is not surprising that water solubilities are related to octanol-water partition coefficients [121–124]. Mackay et al. [124] calculated two correlations between ln K_{ow} and ln S, one for liquids (14a) and one for solids (14b). The equation for solids contains a melting point (T_M) and temperature correction term (T).

$$\ln K_{ow} = 7.494 - \ln s \tag{14a}$$

$$\ln K_{ow} = 7.494 - \ln s + 6.79(1 - T_M/T) \tag{14b}$$

The octanol-water partition coefficient or water solubility is also correlated with terms for molecular shape such as molecular volume (MV) and total surface area (TSA) [122, 125, 126]. Yang et al. [127] observed good correlations between water solubility and hydrophobic substituent constants (π) with molecular volume (or weight), dipole moments and hydrogen bonding (HB). Kamlet [128, 129] has shown that a large number of solubility and solvent dependent properties (XYZ) are well correlated by equations that include linear combinations of free energy or enthalpy contributions by three types of terms according to:

$$XYZ = XYZ_0 + \text{cavity term} + \text{dipolar term} + \text{hydrogen bonding term(s)} \tag{15}$$

Descriptors like total surface area (TSA) or molecular volume (MV) are related to the cavity term in equation 15. Therefore it is not surprising that within classes of non-polar chemicals, good correlations are observed between K_{ow} and e.g. TSA. When such correlations are based on data sets with more different structures, separate correlations between K_{ow} and TSA can be distinguished for each subclass of chemicals [130], very likely because of differences in the electronic character between classes of chemicals.

In conclusion: The most accurate determination of K_{ow} is still by experimental measurement by either the shake flask method, or for more hydrophobic chemicals, by, the generator column or slow stirring method. Estimation by calculation can yield rather accurate predictions for simple molecules if one is familiar with the techniques. Estimation from capacity factors on RP-HPLC or from total surface area is possible, especially for rather non polar chemicals and if one remains within a particular class of closely related chemicals.

Descriptors for Electronic Effects

The Hammett σ constant is the classical descriptor for the influence of substituents on charge distribution within a molecule. The standard σ constant, derived from pKa values of benzoic acid derivatives, is not applicable to substituents that can interact by resonance effects with polar substituents such as OH, NH_2 or NO_2 [114] and other σ constants, such as $\sigma(p)^+$ and $\sigma(p)^-$, can be applied in those cases [114]. Table 4 summarizes different types of σ values for a few substituents.

Table 4. Some different Hammett σ constants and Taft E_s constants for a few aromatic substituents (from Hansch and Leo, [114])

	σ_m	σ_p	σ_p^+	σ_p^-	E_s
H	0.00	0.00	0.00	0.00	0.00
CH_3	−0.07	−0.17	−0.31	−0.15	−1.24
OH	0.12	−0.37	−0.92	−0.16	−0.55
NH_2	−0.16	−0.66	−1.31	−0.15	−0.61
NO_2	0.71	0.78	0.79	1.24	−2.52
Cl	0.37	0.23	0.11	0.27	−0.97
Br	0.39	0.23	0.15	0.28	−1.16

Table 5. Some descriptors that are used in QSAR analyses to estimate electronic and steric effects

parameters for electronic effects:
Hammett σ constants
substituent constants for field and resonance effects (F, R)
dipole moments
molar refractivity (MR)
ionization potential
dielectric constant
hydrogen bonding
quantum mechanical variables:
– atomic charge densities
– net atomic and bond charges
– highest occupied molecular orbital (HOMO) electron density
– lowest unoccupied molecular orbital (LUMO) electron density
– nucleophilic and electrophilic superdelocalizabilities

Parameters for steric effects:
Taft constant E_s
van der Waals radii
Verloop Sterimol parameters
molar volume
molecular volume
total surface area
diameter of a molecule

Hammett constants for substituents in ortho positions are not available, because interactions with chemicals, substituted in ortho positions, have both electronic and steric aspects. This subject is discussed in detail by a.o. Fujita [131] and Bijlo and Rekker [132, 133]. Further details of different types of Hammett constants and their application in QSAR is given by Shorter [134].

Other chemical descriptors related to electronic interactions are summarized in Table 5. Each of these descriptors is characteristic for a particular type or types of interaction (Table 2). Molar Refractivity (MR), e.g., is a good descriptor for dispersion bonds (induced dipole-induced dipole interactions) and charge densities might be useful for electrostatic interactions [13]. In most cases, the type of interaction is unknown and the choice of a descriptor is usually based on trial and error. The aspect of electronic interactions is discussed in more detail by Schnaare [135]. Parameters derived from Molecular Orbital calculations are also applied in QSAR studies. The application of such descriptors is discussed in more detail by Kier [136] and examples of QSAR studies based on MO related parameters are given a.o. by Loew et al. [137].

Descriptors for Steric Effects

The Taft constant E_s is the classical descriptor for steric effects of substituents on chemical and biological processes. Value for E_s of a few substituents are given in Table 4. Other descriptors for steric effects are Verloop Sterimol parameters [138] and descriptors such as van der Waals radii, molecular volume, surface area and related parameters (see Table 5). The various steric descriptors are discussed in a review article from Fujita and Iwamura [139].

Structural or Topological Indices

Many QSAR studies, also within the field of environmental chemistry and toxicology, incorporate structural or topological indices such as the molecular connectivity (MC) indices (X) from Kier and Hall [140–142]. Molecular connectivity is an index that represents the relative branching of a molecule. The first order MC is obtained by adding bond contributions (δ) whose size is determined by the number of non hydrogen ligands attached to each atom i and j of the bond:

$$^1X = \Sigma(\delta_i \, \delta_j)^{-1/2} \tag{16}$$

Other structural indices are the mean information content (IC) from Shannon, the structural information content (SIC) [143–146] and the auto-correlation of a topological structure from Moreau [147]. Structural indices are also applied, e.g. by Basak et al. [146], in determining structural similarity of chemicals; an approach that is valuable for the selection of structurally related chemicals from large data bases.

The advantage of topological indices is that they can be calculated directly from the structure of a chemical, but QSARs based on such indices do not give direct information on the relevant physical-chemical properties. However, several studies have shown that the MC indices are related to particular physical-chemical properties (see e.g. examples given by Sabljic, [148]), although Dearden and Solanki [149] recently has shown that all different MC indices seem to reflect primarily steric properties.

Other Descriptors

Besides the descriptors given in the previous sections, many other parameters are used in QSAR studies such as: boiling and melting point, number of carbon atoms, chemical shift in NMR spectra. Also rate constants of reactions for a set of chemicals, e.g. alkylation rate constants with a nucleophile, are sometimes applied as a descriptor in QSAR equations.

Estimation of Physical-Chemical Properties

Estimation techniques for physical-chemical properties, including the octanol-water partition coefficient. acid dissociation constant, water solubility, boiling point and vapor pressure are described by Lymann et al. [150] and Yalkowsky et al. [151]. Although these models usually will yield rather good predictions, it is important to recognize that these estimated values remain predictions with a certain error and in some cases predictions will be completely wrong.

QSAR for Environmental Effects

Books or review papers on QSAR for pharmacological endpoints treat QSAR in the light of mode of action, pharmacological endpoint or target site. A book on QSAR of Drugs, edited by Topliss [15], is divided e.g. in several chapters entitled: synthetic antiinfective agents, semisynthetic antibiotics, antitumor agents, cardiovascular agents etc. The modes of action or target sites for most environmental pollutants are unknown and a discussion on QSAR for environmental effects cannot follow the same type of distinctions. To the opinion of the author, the following outline is a logical ordering of the literature and represents the present state of art of QSAR in environmental toxicology.

- QSAR for relatively unreactive organic chemicals (see p. 126)
- QSAR for less inert organic chemicals:
  * phenols, aromatic amines and nitroaromatics (see p. 132)
  * reactive organic chemicals (see p. 137)
  * other classes of chemicals (see p. 146)

The term "less inert organic chemicals" is a very broad one, but is used to distinguish QSAR studies for chemicals other than the unreactive ones. Still, some ordering is possible within the variety of "less inert chemicals". Some chemicals are able to react directly with nucleophiles by a chemical reaction with examples such as: some alkylhalides, epoxides, ethyleneimines, lactones and generally all alkylating agents. Other chemicals are metabolized to reactive intermediates (bioactivation) in the organism. Most examples of bioactivation are based on toxicity studies with mammals [152], with as a classical example the oxidation of the $P = S$ group in organophosphates to $P = O$ which is the active toxicant in vivo [153]. Also aromatic amines and nitroaromatics can be metabolized to their corresponding N-hydroxy or nitroso derivatives as reactive intermediates [154]. Phenols, aromatic amines and nitroaromatics will be discussed here separately, because they are extensively studied by QSAR.

In a certain way, the above classification is somewhat artificial, because to some extent there is overlap between these classes and for some chemicals it is not clear how to classify them.

Some general remarks on the QSAR equations presented are made on p. 146, and attention is paid to recent developments on subjects such as internal concentrations; mode of action (see p. 151); computer programs (see p. 153) and joint toxic effects of mixtures (see p. 154).

QSAR for Relatively Unreactive Organic Chemicals

The class of relatively unreactive organic chemicals is the most important group of chemicals that is studied in this area. QSAR is applied to toxicity data of such chemicals in different test systems; different regarding species and endpoints.

As shown in Fig. 3 and equations 17 and 18, Könemann [155] and Veith et al. [156] observed that LC50s to fish are strongly correlated with the octanol-water partition coefficient (K_{ow}) for a large number of chemicals. Without giving a complete survey of all QSAR studies for this particular group of chemicals, table 6 summarizes some important equations for different species and endpoint.

The equations in Table 6 have the general form:

$$\log 1/C = A \log K_{ow} + B \tag{36}$$

Values for 'A' in most of the equations are around 0.8 to 1.0. The influence of K_{ow} in these equations is related directly to the influence of hydrophobicity on the absorption and the internal concentration. The intercept B is a measure for the sensitivity of the test species or studied endpoint. The observation that values for B in the equations for fish LC50 data (equation 17 and 18) are rather close to each other justifies the conclusion that differences in species sensitivity are rather small; at least for this particular group of chemicals. A study from Slooff et al. [172] supports this argument because they observed that a set of QSAR equations based on LC50 data of 5 chemicals with 14 different aquatic species resulted in mean values and standard deviations for A and B of: 0.97 ± 0.07 and 0.89 ± 0.10. The lines drawn in Fig. 4 (a, b and c) that represent LC50 data also show that the differences in sensitivity are very small. More chronic studies and studies to sublethal effects are of course more sensitive and will result in higher values for B; e.g. B values for QSAR equations with LC50 and MATC to fathead minnow are 1.3

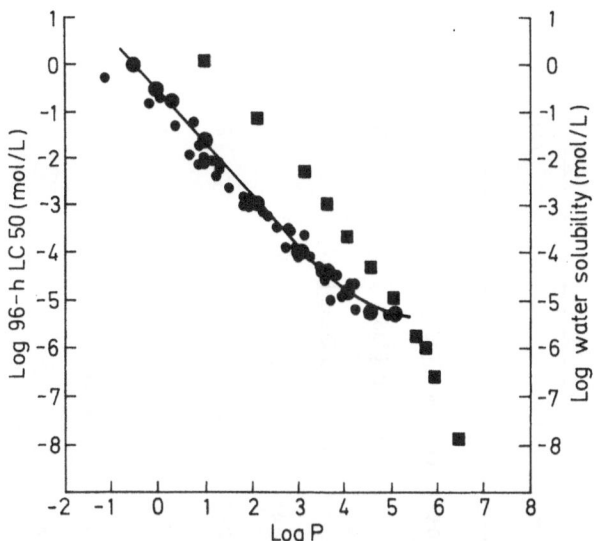

Fig. 3. Relationship between 96-h LC50 concentrations (small circles) for fathead minnows (*Pimephales promelas*) of some narcotic chemicals and log P superimposed on bilinear alcohol model. Solid squares represent water solubility data. Reprinted with permission from Veith et al. [156], copyright (1983), Canadian Journal of Fisheries and Aquatic Sciences.

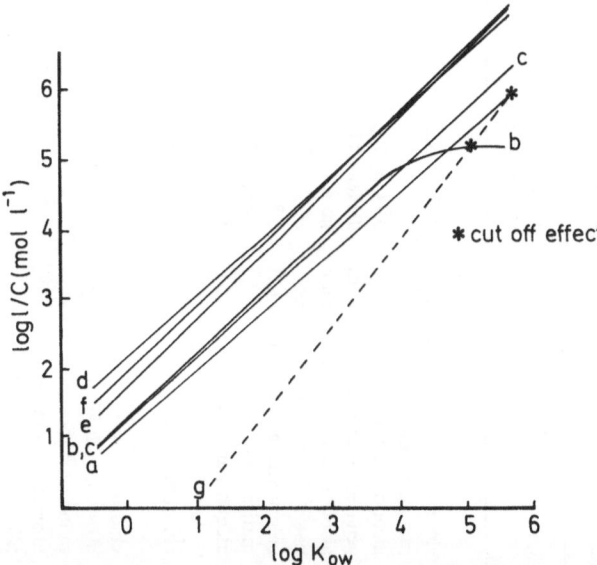

Fig. 4. Relationships between effect concentrations and octanol-water partition coefficients for relatively unreactive chemicals with a narcosis mechanism of action: line a, b and c: lethal effects. line d, e and f: sublethal effects. line a: LC50 to guppy/equation 17 from Könemann [155]. line b: LC50 to fathead minnow/equation 18 from Veith et al. [156]. line c: LC50 to Daphnia magna/equation 24 from Hermens et al. [163]. line d: MATC to fathead minnow/equation 34 from Call et al. [171]. line e: chronic toxicity to fish/equation 35 from McCarty et al. [82]. line f: NOEC (growth) to Daphnia magna/equation 27 from Hermens et al. [166]. line g: relation between K_{ow} and S (water solubility for a set of alcohols from Veith et al. (156)

Table 6. Some QSAR equations for unreactive organic chemicals

toxic endpoint[a]	type of chemicals[b]	equation[c] $\log 1/C$ (mol l^{-1}) =	n	r^2	reference	eqn.
lethal effects						
14-day LC50 to guppy	chlorinated alkanes and aromatics, alcohols and ethers	$0.87 \log K_{ow} + 1.1$	50	0.97	Könemann [155]	17
96-h LC50 to fathead minnow	alcohols, ketones, ethers, halogenated alkanes and aromatics	$0.94 \log K_{ow} - 0.94 \log (0.000068\, K_{ow} + 1) + 1.3$			Veith et al. [156]	18
48-h LC50 to golden orfe	alcohols, ketones, ethers, chlorinated alkanes and aromatics	no equation	47		Lipnick and Dunn [157]	
14-day LC50 to guppy	chlorinated alkanes and aromatics	$0.84 \log K_{ow} + 0.17 \log N - 0.10\,(^2X^v) + 1.8$	27	0.91	Koch [158]	19
96-h LC50 to goldfish	alcohols and ketones	$1.0 \log K_{ow} + 0.8$	8	0.97	Lipnick et al. [159]	20
96-h LC50 to fish	chlorobenzenes	$0.71 \log K_{ow} + 2.2$	13	0.70	McCarty et al. [82]	21
LC50 to fish	chlorobenzenes	$0.0086\, SC - 1.67^{d}$	8		LeBlanc [160]	22
48-h LC50 to Daphnia magna	alkanes, cycloalkanes, monoaromatics, polynuclear aromatics, chlorinated aromatics, chlorinated alkanes and aromatics	$-0.72 \log C_L + 0.61$	38	0.83	Abernethy et al. [161]	23a
		$-0.60 \log C_L + 0.09$	33	0.81	Bobra et al. [162]	23b
48-h LC50 to Daphnia magna	chlorinated alkanes and aromatics, alcohols and ethers	$0.91 \log K_{ow} + 1.3$	19	0.98	Hermens et al. [163]	24
24-h LC50 to Daphnia magna	chlorobenzenes	$0.78 \log K_{ow} + 2.3$	5	0.79	Calamari et al. [164]	25
18-h LC50 to Tetrahymea	aliphatic and aromatic hydrocarbons	$1.1 \log K_{ow} - 0.1$	10	0.81	Rogerson et al. [165]	26

sublethal effects

Endpoint	Class	Equation	n	r	Reference	
16-day NOEC (growth) to Daphnia magna	chlorinated alkanes and aromatics, alcohols	$0.95 \log K_{ow} + 2.0$	10	0.95	Hermens et al. [166]	27
14-day EC50 (reproduction) to Daphnia magna	chlorobenzenes	$0.73 \log K_{ow} + 3.0$	6	0.96	Calamari et al. [164]	28
4-h EC50 (cell growth) to green alga	chlorobenzenes	$0.99 \log K_{ow} + 0.4$	12	0.94	Wong et al. [167]	29
30 min. EC50 (bioluminescence) to photo bacteria	chlorobenzenes	$0.41 \log K_{ow} - 0.17 S + 3.15$	11	0.95	Ribo and Kaiser [168]	30
15-min EC50 (bioluminescence) to photo bacteria	alcohols, chlorinated alkanes and aromatics	$1.0 \log K_{ow} + 0.9$	22	0.91	Hermens et al. [169]	31
EC50 (bioluminescence) to photo bacteria	alcohols, esters, ethers, chlorinated alkanes and aromatics and some phenols	$4.1 V/100 + 1.5 \pi^* - 3.9 \beta + 1.5 \alpha_m - 1.6$	38	0.97	Kamlet et al. [170]	32
MATC to fish	chlorobenzenes	$0.012 SC - 1.43^{d)}$	3		LeBlanc [160]	33
32-d MATC to fathead minnow	chlorinated alkanes and aromatics, ketones and ethers	$0.89 \log K_{ow} + 2.2$	10	0.93	Call et al. [171]	34
chronic toxicity to fish	chlorobenzenes	$0.99 \log K_{ow} + 1.8$	12	0.79	McCarty et al. [82]	35

a) LC50: concentration with 50% mortality; EC50: concentration with 50% effect; NOEC: no-observed effect concentration; MATC: maximum acceptable toxicant concentration, derived from early life stage test (ELS).

b) The class of chlorinated alkanes and aromatics is restricted to unreactive representatives.

c) C: effect concentration (in mol l^{-1}); K_{ow}: octanol-water partition coefficient; X: MC indices; N: negentropy [158]; C_L: solubility of hypothetical subcooled liquid; S: number of symmetry planes; V: solute molar volume; π^*, β, α: solvatometric parameters for dipolarity/polarizability, hydrogen bond acceptor basicity and hydrogen bond donor acidity from Kamlet [170]; n: number of chemicals in dataset; r: correlation coefficient.

d) concentration units are not given.

and 2.2 (equations 18 and 34) and LC50 and NOEC (reproduction) with Daphnia magna are 1.3 and 2.0 (equations 24 and 27). The higher sensitivity of the tests with sublethal effects is also obvious from Fig. 4, because all three lines that represent sublethal effects (d, e and f) are situated above the lines for lethal effects (a, b and c).

Figure 3 and 4 and equation 18 show that the data from Veith are fitted by a bilinear equation. Especially the more hydrophobic chemicals are less toxic in the 96-h LC50 tests with fathead minnows than in the 14-day test with guppies. The time needed to establish equilibrium increases with increasing hydrophobicity [53]. Based on published relations between uptake, elimination rate constants and K_{ow}, Hawker and Connell [53] has shown that the time to reach 99% of the equilibrium concentration in fish (t_{eq}) is related to K_{ow} by a linear relationship (Fig. 5). For chemicals with log K_{ow} of 5 or 6, equilibrium is attained within 50 days and 1 year, respectively. The relationship depicted strictly hold only between log K_{ow} values of 2 and 6.5 (D. Hawker, personal communication). This study indicates that for the more hydrophobic chemicals 96 hours and 14 days is probably not long enough to establish an equilibrium and this very likely explains the differences in LC50s with guppies and fathead minnow. Figure 4, also shows the relation between K_{ow} and water solubility for a set of alcohols (line g). At a certain K_{ow}, line g will cross the lines that represent the QSAR equations: the so called "cut off effect", above which LC50s can no longer be determined because they exceed water solubility. From the above arguments, it will be clear that the K_{ow} value for the "cut off effect" will increase with increasing exposure times in toxicity tests.

Both Könemann [155] and Veith et al. [156] suggest that the effect of the test chemicals in their LC50 studies is related to narcosis. Many different classes of chemicals, in general rather unreactive and non-ionized, are known as general biological depressants. Albert [173] mentioned as examples: hydrocarbons (aliphatic and aromatic), chlorinated hydrocarbons, alcohols, ethers, ketones, sul-

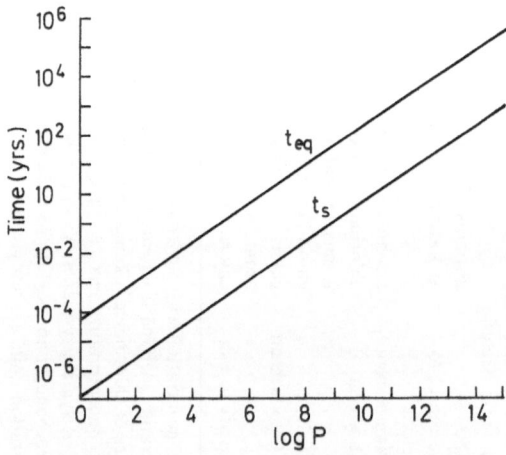

Fig. 5. Relationship between the octanol-water partition coefficient (log P) and the time taken by fish to establish effective 99% equilibrium (t_{eq}). Reprinted with permission from Hawker and Connell [53], copyright (1985), Pergamon Journals, Ltd.

phones, weak acids and bases and aliphatic nitro compounds. Studies on narcosis originate from the previous century. Classical studies are those from Meyer [71] and Overton [72], who demonstrated relationships between the olive-oil/water partition coefficient of certain non-electrolytes and their toxicity to tadpoles and other organisms. Lipnick [174] has recently discussed the important contribution of Overton and Meyer to the development of QSAR in environmental toxicology. Ferguson [175] provided in 1939 a thermodynamic explanation for correlations between depressant activity and lipophilicity. Ferguson indicated that the thermodynamic activity of narcosis type of chemicals lie within a relatively narrow range. In case of aqueous exposure at a concentration C, the thermodynamic activity A, needed to provide narcosis is related to the water solubility C_s, according to:

$$A = C/C_s \tag{37}$$

The exact mechanism of narcosis is still unknown. The lipid solubility theory suggests that changes in the lipid bilayer properties are responsible for the anaesthetic effect. Within this theory, two different classes are distinguished. The first and earliest theory, based on the work from Meyer and Overton, postulates that narcosis commences when any chemical, indifferent its structure, has attained a certain molar concentration in the lipids [176]. In the second class of theories, it is stated that anaesthesia occurs when the volume of a hydrophobic region is caused to expand beyond a certain critical volume [177]. More recently, it is suggested by a.o. Franks and Lieb [178] that the mechanism of anaesthetic action is related to direct binding on proteins. Because in the lipid solubility theory, narcosis is caused by the presence of a chemical in lipid bilayers and not by binding to a particular receptor, one often referes to narcosis as unspecific mode of action. Although at present, the exact mechanism of action is not unravelled, it is obvious that a wide diversity in effects, caused by rather inert chemicals is solely determined by the hydrophobic character. Before the first QSAR studies for fish LC50 data appeared, Hansch and Glave [179] published many correlations with the octanol-water partition coefficients based on different effects of several subclasses of unreactive organic chemicals. LD50 data to mammels and fish of several unreactive chemicals are analysed in QSAR studies by Lipnick et al. [180], while Roberts [181] calculated an equation for upper respiratory tract irritation.

Narcosis is a kind of minimal effect because in principle every organic chemical, even the inert gases, can cause narcosis [173]. So, the equations in table 6 represent, and at the same predict, the minimal or base line effect concentration of every organic chemical [155–157]. Ofcourse, as shown in Fig. 6, several organic chemicals will cause effects at much lower concentrations because they are more reactive or act by specific mechanisms of action [22]. It needs no surprise that LC50s of insecticides such as malathion, dieldrin, endosulfan and lindane are lower in comparison with unreactive organic chemicals, because all these insecticides act specific as neuro toxicants. LC50s of reactive chemicals will also be lower than those of unreactive ones. Reactivity of molecules can depend on very small changes in a molecule, such as e.g. the position of a double bond in chloroalkenes. Therefore the classification of a chemical into the class of unreactive chemicals must be based on a careful analysis of its structure.

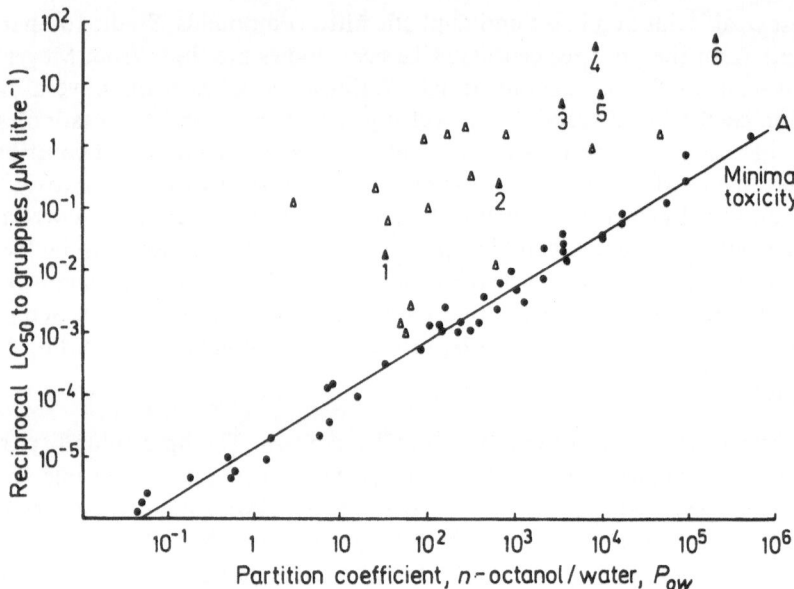

Fig. 6. Relationship between 7/14-day LC50 concentrations with guppies and octanol-water partition coefficient (P_{ow}) for: ● relatively unreactive chemicals with narcosis mechanisms of action (data from Könemann [155]. Δ reactive alkylhalides (data from Hermens et al. [205]. ▲ other chemicals: 1,3-dinitrobenzene (1); malathion (2); lindane (3); disulfiram (4); rotenone (5) and dieldrin (6) (data from Hermens et al. [266]. (modified from Hermens [22])

In conclusion, the major characteristics of QSAR equations for unreactive chemicals are the following:
- They can be applied to relatively unreactive organic chemicals such as: alcohols, ketones, ethers, unreactive chlorinated alkanes and aromatics. It is important to recognize that the presence of some substructures within a molecule from each of these chemical classes can result in much more reactive chemicals.
- They predict the minimal or base line effect concentration of each organic chemical.
- The octanol-water partition coefficient (K_{ow}) is the only predominant chemical descriptor.
- Differences in species sensitivities for this particular class of chemicals are very small.
- The K_{ow} for the "cut off effect", at which the LC50s exceed the water solubility, increases with the duration of toxicity tests.

QSAR for Phenols, Aromatic Amines and Nitroaromatics

Phenols

Toxicity data for substituted phenols are also frequently analysed in QSAR studies. Table 7 summarizes some of the derived equations for several endpoints.

In contrary to the equations in Table 6, most equations for phenols include a descriptor for electronic effects. Several types of electronic descriptors are applied but in many cases they will show a strong intercorrelation. The influence of electronic parameters is not surprising and can be explained in at least two different ways; by:
- the influence of ionization on uptake, or
- the influence of electronic effects on interactions with a target.

The ionized form of a chemical can hardly cross lipophilic barriers and this explains the decline in uptake rate constants of several substituted phenols with rising degree of ionization as observed by Saarikoski and Viluksela [64] in studies with guppies at different pH. Also LC50s to guppy depends on the degree of ionization and therefore on the pKa and pH. The data presented in Table 8 show a striking example. The LC50 of 4-chlorophenol is almost the same at the two different pH values of the test solutions while the LC50 of pentachlorophenol is lower at pH 6 than at a pH of 8, because at the lower pH a higher portion of pentachlorophenol is non-ionized. These studies indicate that it is important that LC50 tests with ionizable compounds are carried out at a constant and known pH.

Phenols are known as uncouplers of oxydative phosphorylation. Ravanel et al. [196] calculated a QSAR for the effects of chlorophenols on uncoupling activities in plant mitochondria and the equation included descriptors for steric as well as electronic properties (equation 57). So, electronic effects might also play a role in the interactions of phenols with their target site.

Lipnick et al. [189] compared fish toxicity screening data for 110 phenols with LC50s predicted by equation 17. Overall, the experimental toxicity data were consistent with the QSAR predictions and observed discrepancies were explained in terms of reactivity or metabolism.

Equation 38 to 42 are based on LC50 data to guppy and fathead minnow and, because guppy and fathead minnow are very similar in sensitivity, this is a good opportunity to compare these equations. Some of the equations contain K_{ow} and pKa as a descriptor but both descriptors can be highly intercorrelated for certain data sets as is the case in equation 39. In the discussion of these equations, it must be kept in mind that collinearity between descriptors will result in unreliable coefficients in a QSAR equation. Moreover, the influence of e.g. a descriptor for electronic effects may not show up because of collinearity with a descriptor for hydrophobic effects. The dataset of Saarikoski contains phenols with rather heterogeneous substituents, including chloro, alkyl, methoxy and nitro substituents. Equations 40a and 40b are based on all test chemicals, but it is obvious that the dinitrophenols are outliers (r^2 increased from 0.81 to 0.96 when these 2 chemicals were omitted). As shown in Fig. 7, Schultz et al. [185] observed two separate groups of phenols with different equations with K_{ow}. He suggested that one of these classes of phenols act as uncouplers and the other subclass as narcotics, represented by equation 42c and 42b, respectively. Veith and Broderius [184] also calculated an equation for phenols and some anilines, that act by narcosis (equation 41). The outliers in the study from Saarikoski might be considered as uncouplers, because dinitrophenols (especially 2,4-dinitrophenol) are classical uncouplers of the oxydative phosphorylation. So, in that case, equations 39a, 40c,

Table 7. Some QSAR equations for substituted phenols

toxic endpoint[a]	type of of chemicals	equation[b] $\log 1/C$ (mol l^{-1}) =	n	r^2	reference	eqn.
96-h LC50 to guppy	substituted phenols	$0.50\pi + 0.45 F + 0.64 R + 3.7$	14	0.96	Kopperman et al [182]	38
14-day LC50 to guppy (at pH of 7.3)	chlorophenols	$0.58 \log K_{ow} + 2.8$	11	0.91	Könemann and Musch [183]	39a
		$1.1 \log K_{ow} + 0.35\, pKa - 1.4$	11	0.96		39b
96-h LC50 to guppy (at pH of 7.0)	substituted phenols	$0.52 \log K_{ow} + 3.0$	21	0.64	Saarikoski and Viluksela [64]	40a
		$0.38 \log K_{ow} + 0.16\, DpKa + 3.1$	21	0.81		40b
	ibid without 2 dinitrophenols	$0.59 \log K_{ow} + 2.7$	19	0.95		40c
		$0.56 \log K_{ow} + 0.03\, DpKa + 2.7$	19	0.96		40d
96-h LC50 to fathead minnow	some phenols and anilines	$0.65 \log K_{ow} + 2.3$	39	0.90	Veith and Broderius [184]	41
96-h LC50 to fathead minnow	substituted phenols	$0.62 \log K_{ow} + 2.6$	27	0.82	Schultz et al. [185]	42a
		$0.60 \log K_{ow} + 2.4$	21	0.94		42b
		$0.59 \log K_{ow} + 3.2$	6	0.96		42c
96-h LC50 to Atlantic salmon	alkyl dinitro phenols	$0.31 \log K_{ow} + 5.3$	6	0.98	Zitko et al. [186]	43
96-h LC50 to rainbow trout	substituted phenols	$0.61 \log K_{ow} + 3.1$	5	0.96	McCarty et al. [82]	44
96-h LC50 to sheepshead minnow	chlorinated phenols and hydrocarbons	$0.50\, °X + 1.1$	19	0.84	Sabljic [187]	45
LC50 to fish	chlorophenols	$0.011\, SC - 1.33^{c)}$	8		LeBlanc [160]	46
24-h LC50 to guppy	chlorophenols	$-26.3\, \Sigma D + 5.7\, \Sigma D^2 + 35.2$	19	0.89	Benoit-Guyod et al. [188]	47

Biological endpoint	Chemical class	Equation	n	r	Reference	No.
fish toxicity screening data	phenols	comparison with QSAR predictions	110		Lipnick et al [189]	48
24-h LC50 to Daphnia magna	chlorophenols	$0.13 \pi^2 + 0.15 \pi + 1.1 F - 0.026 R - 0.028 MR + 3.5$	18	0.85	Durkin [190]	49
24-h LC50 to Daphnia magna	chlorophenols	$0.40 \log K_{ow} + 0.15 DpKa - 0.27 N + 3.21$	17	0.89	Devillers and Chambon [191]	50
96-h LC50 to shrimp	substituted phenols	$0.48 \log K_{ow} + 0.54 DpKa + 2.9$	23	0.92	McLeese et al. [192]	51a
14-day LC50 to two different earthworms	chlorophenols	$0.60 \log K_{ow} + 2.5$	5	0.89	Van Gestel and Ma [193]	51b
		$0.35 \log K_{ow} + 3.0$	5	0.71		
MATC to fish	chlorophenols	$0.014 SC - 1.09^{c)}$	3		LeBlanc [160]	52
chronic toxicity to fish	substituted phenols	$1.0 \log K_{ow} + 3.2$	5	0.95	McCarty et al. [82]	53
30-min. EC50 (bioluminescence) to photo bacteria	chlorophenols	$0.76 \log K_{ow} + 2.1$	18	0.91	Ribo and Kaiser [168]	54
48/60-h EC50 (population growth) to Tetrahymena	mono alkylated and halogenated phenols	$0.80 \log K_{ow} + 1.2 \sigma + 1.4$	27	0.90	Schultz and Cajina-Quezada [194]	55
MIC (cell growth) to E. coli	phenols	$0.59 \log k' - 0.05 pKa - 0.8^{c)}$	20	0.96	Nendza and Seydel [195]	56
EC50 (inhibition of uncoupling) in plant mitochondria	substituted phenols	$-12.1 \ \Sigma D^2 + 66.9 \ \Sigma D + 0.088 \ \sigma_1 - 0.32 A - 81.2$	19	0.91	Ravanel et al. [196]	57

a) LC50: concentration with 50% mortality; EC50: concentration with 50% effect; MIC: minimal inhibition concentration.

b) C: effect concentration (in mol l^{-1}); π: substituent constant for hydrophobicity; F: substituent constant for field effects; R: substituent constant for resonance effects; K_{ow}: octanol-water partition coefficient; DpKa: difference in pKa between substituted phenol and phenol; X: MC indices; SC: structure coefficient [160]; D: steric parameter [188] MR: Molar Refractivity; N: indicator variable for number of chlorine atoms in ortho position; k': capacity factor on RP-HPLC; D, A: steric parameters [196]; σ: Hammett substituent constant; n: number of chemicals in dataset; r: correlation coefficient.

c) concentration units are not given.

Table 8. LC50 to guppy of 4-chlorophenol and pentachlorophenol at pH 6 and 8 (data from Saarikoski and Viluksela [64])

	pH = 6		pH = 8	
	% non-ionized	LC50 (μmol 1^{-1})	% non-ionized	LC50 (μmol 1^{-1})
4-chlorophenol (pKa = 9.37)	>99	60	96	71
pentachlorophenol (pKa = 4.69)	5	0.44	<0.1	3.4

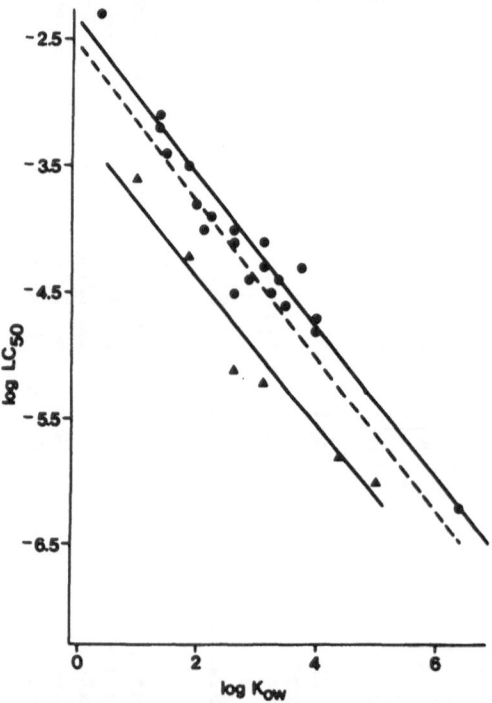

Fig. 7. Relationship between 96-h LC50 concentrations with fathead minnow for some phenols. LC50 data from Schultz et al. [185]. Reprinted with permission from Schultz et al. [185], copyright (1986), Academic Press

41 and 42b all represent the same group of chemicals or mechanism of action. Indeed, these equations are very similar with coefficients for the influence of log K_{ow} of 0.58, 0.59, 0.65 and 0.60 and intercepts of 2.8, 2.7, 2.3 and 2.4, respectively. The issue of QSAR and mode of action is discussed in more detail later on.

Almost all equations for phenols are based on toxicity data for aquatic species. An interesting exception is the study from Van Gestel and Ma [193] who described a relationship between effect concentrations of some chlorophenols to earthworms and K_{ow} (equation 51). Only the concentrations in the soil solution yielded a good correlation, which is an indirect proof that the uptake of chlorophenols from soil by earthworms proceeds mainly via the aqueous phase.

Aromatic Amines and Nitroaromatics

A few QSAR equations for aromatic amines and nitroaromatics are presented in Table 9. Effect concentrations are correlated with K_{ow} as well as descriptors for electronic effects. From mammalian studies it is well known that aromatic amines and nitro compounds are metabolized in vivo by reduction and oxidation to N-hydroxy or nitroso derivatives and that these reactive intermediates are responsible for their effects as ferrihemoglobin inducing agents [201]. A comparison of the intercepts of the equations for LC50 to guppy indicates that chloroanilines (equation 58) cause mortality at concentrations about 10 fold lower than unreactive organic chemicals (equation 17). The mechanism of action of aromatic amines, related to their higher activity, is unclear. Whether or not the formation of reactive intermediates play a role in studies with mortality as endpoint is unknown. Veith and Broderius [184] have suggested that some of the aromatic amines act simply by narcosis.

Baily and Spanggord [202] and Deneer et al. [47] have published QSAR studies for fish LC50 data of a set of nitroaromatics. Both studies indicate that metabolism is an important aspect in the correlation studies. Deneer observed that LC50s of dinitrobenzenes, and especially ortho and para dinitrobenzene, are much lower than LC50s of mononitroaromatics and that introduction of reduction potentials resulted in a much better equation (equations 62a,b). Based on these data, Deneer assumes that the dinitrobenzenes are reduced more easily to the corresponding reactive intermediates. Roberts [199], on the contrary, has suggested that the influence of electronic descriptors (equation 63) is related to differences in the reactivities of the reactive intermediates.

Newsome et al. [200] compared and evaluated literature data of 48 aromatic amines with predictions based on a QSAR equation and the observed deviations were explained in terms of reactivity.

QSAR for Reactive Organic Chemicals

It is well known that reactive chemicals are lethal at lower concentrations than unreactive compounds with equal K_{ow} values. Nucleophilic groups (NH_2, OH and SH) in biological macromolecules such as proteins and DNA are very likely the most important sites of action for such reactive chemicals. The effect of these chemicals can vary from e.g. enzyme inhibition to carcinogenicity.

Some classes of chemicals are reactive in a direct way, while other classes of chemicals are metabolized to reactive intermediates (bioactivation). Bioactivation processes are discussed in more detail by Anders [152]. Lipnick et al. [203] introduced the term "pro electrophile" to indicate those cases in which the actual reactive toxicant is produced metabolically.

Table 9. Some QSAR equations for aromatic amines and nitroaromatics

toxic endpoint[a]	type of chemicals	equation[b] $\log 1/C$ (mol l^{-1}) =	n	r^2	reference	Eqn.
14-day LC50 to guppy	chloroanilines	$0.92 \log K_{ow} + 2.3$	11	0.89	Hermens et al. [197]	58
30-min. EC50 (bioluminescence) to photo bacteria	chloroanilines	$0.57 \log K_{ow} + 3.0$	15	0.80	Ribo and Kaiser [198]	59
96-h LC50 to fathead minnow	some phenols and anilines	$0.65 \log K_{ow} + 2.3$	39	0.90	Veith and Broderius [184]	60
MIC (cell growth) to E. coli	anilines	$0.62 \log k' - 0.13 \, pKa - 0.9°$	19	0.89	Nendza and Seydel [195]	61
14-day LC50 to guppy	nitroaromatics	$-0.34 \log K_{ow} + 5.2$	20	0.11	Deneer et al. [47]	62a
		$0.96 \log K_{ow} + 8.8 \, E1/2 + 6.7$	20	0.93		62b
LC50 to guppy and fathead minnow	aromatic amines and nitroaromatics	$1.21 \, \Sigma\sigma + 3.6$	33	0.90	Roberts [199]	63
fish toxicity screening data	aromatic amines	comparison with QSAR predictions	48		Newsome et al. [200]	

a) LC50: concentration with 50% mortality; EC50: concentration with 50% effect; MIC: minimal inhibition concentration.
b) C: effect concentration (in mol/1); K_{ow}: octanol-water partition coefficient; k': capacity factor on RP-HPLC; pKa: acid dissociation coefficient; E1/2: half wave reduction potential; σ: Hammett substituent constant; n: number of chemicals in dataset; r: correlation coefficient. c: concentration units are not given.

Organic chemicals can react with nucleophiles according to different mechanisms of reaction. Reaction schemes for three of such mechanisms are given in Table 10: nucleophilic displacement reactions (scheme a), addition at a carbon-oxygen bond (scheme b) and addition at a carbon-carbon double bond (scheme c). Alkyl halides are an example of chemicals that react according to nucleophilic displacement reactions and other examples are given in Table 11. The reactivity of a compound within a particular class of reactive chemicals depends of course on its structure. Simple saturated alkyl halides are not very reactive, but introduction of a double bond, like in chloropropenes, generally yield more reactive chemicals.

LC50 data of some groups of reactive chemicals are analysed by QSAR and the derived equations are presented in Table 12. A study of LC50 data to guppy of several alkylating alkylhalides by Hermens et al. [205] indicates the necessity to include differences in reaction rates into QSAR equations. As shown in equation 64b, the LC50 data correlated much better with rate constants (k) of a reaction with 4-nitrobenzyl pyridine (4-NBP) than with K_{ow} (equation 64a). This reaction with 4-NBP is a kind of model reaction for nucleophilic displacement reactions and has been applied in correlations between mutagenicity and alkylating potency of several classes of organic chemicals [208–210].

Epoxides are also known as alkylating agents that react with nucleophiles in nucleophilic displacement reactions (Table 10, scheme a). Deneer et al. [206] calculated recently a QSAR equation for 14-day LC50s of a set of epoxides with K_{ow} and rate constants of reactions of the test chemicals with 4-NBP as chemical descriptors. It is obvious that neither of the equations using a single descriptor (equation 65a,b) lead to satisfactory correlations, but that only equation 65c employing both descriptors yields a highly significant correlation. From Fig. 8a it appears that most of the epoxides are lethal at much lower concentrations than the narcosis type of chemicals. Lipnick et al. [159], who compared LC50s of six epoxides with LC50s calculated with a QSAR for narcosis type chemicals, observed similar effects.

Table 10. Three different mechanisms of reactions of chemicals with nucleophiles

a. nucleophilic displacement reaction

Nu: $+$ $-C-Y \longrightarrow C-Nu + Y$:
Nu: nucleophile, e.g. $-NH_2$, $-OH$ or $-SH$ group in macro-molecules
Y   leaving group

b. addition to carbon-oxygen double bond (C=O)

$RNH_2 +$ C=O $\longrightarrow R-N=C$ $+H_2O$

with e.g. RNH_2 as nucleophile

c. addition to activated carbon-carbon double bond (C=C)

Nu: $+ A-CH=CH_2 \longrightarrow A-CH_2-CH_2-Nu$
A: e.g. NO_2, SO_2R, COR or COOR

Table 11. Some chemical structures that are known to act as alkylating agents and react with nucleophiles by nucleophilic displacement reactions (from Ross [204])

–C–Hal	$\begin{matrix} O \\ \| \\ -C-O-S-R \\ \| \\ O \end{matrix}$	$\begin{matrix} O \\ \| \\ -C-O-S-OR \\ \| \\ O \end{matrix}$
alkylhalides	sulphonic acid esters	sulphuric acid esters
$\begin{matrix} O \\ \| \\ -C-O-P-OR \\ \| \\ OR \end{matrix}$	$RO-CH_2-Cl$	$-C-N-R_3{}^+$ $-C-S-R_2{}^+$
phosphoric acid esters	chloromethyl ethers	ammonium/sulphonium compounds
$RS-CH_2-CH_2-Cl$	$R_2N-CH_2-CH_2-Cl$	$R-CH-CH_2$ $\backslash\,/$ O
2-chloroethyl sulphides	2-chloroethylamines	epoxides
$\begin{matrix} CH_2-CH_2 \\ \backslash\,/ \\ NHR \end{matrix}$	$\begin{matrix} (CH_2)_n \\ \|\ \| \\ C-O \\ \| \\ O \end{matrix}$	$-C-N_2$
ethyleneimines and -imides	lactones	diazoalkanes
$\begin{matrix} O \\ \| \\ RC-C-Hal \end{matrix}$		
halogenemethyl ketones and esters		

Aldehydes are known to interact with e.g. amino groups through addition at the carbon-oxygen bond (Table 10, scheme b). Deneer et al. [207] calculated a QSAR for LC50 data of 14 aldehydes using only octanol-water partition coefficients (equation 66 and Fig. 8b). The equation was not improved when rate constants of a reaction of the test chemicals with cysteine was added (equation 66b). This reaction was chosen because it quantifies differences in rates of addition of nucleophiles at the C=O bond in aldehydes. From the fact that K_{ow} alone is a good descriptor, Deneer concludes that "possibly, the rate of uptake of the compounds is the rate limiting process in the case of the compounds studied, but further research would be needed to establish this". They come to this conclusion because the predominant descriptors in a QSAR reflect the rate limiting processes involved in the development of the toxic effect [74]. Two compounds, 3-cyclohexene-car-boxaldehyde and 2-butenal (nr. 1 and 2 in Fig. 8b), are more toxic than predicted from equation 66a, possibly bcause of the presence of the double bonds which

Table 12. Some QSAR equations for some different classes of reactive chemicals

toxic endpoint[a]	type of chemicals	equation[b] $\log 1/C$ (mol l^{-1}) =	n	r^2	reference	eqn.
14-day LC50 to guppy	reactive alkylhalides	$0.47 \log K_{ow} + 4.0$	15	0.17	Hermens et al. [205]	64a
		$-1.3 \log (1604 + k_{nbp}^{-1}) + 10.4$	15	0.88		64b
14-day LC50 to guppy	epoxides	$0.18 \log K_{ow} + 4.0$	12	0.18	Deneer et al. [206]	65a
		$1.6 \log k_{nbp} + 4.3$	12	0.26		65b
		$0.39 \log K_{ow} + 3.0 \log k_{nbp} + 3.8$	12	0.89		65c
14-day LC50 to guppy	aldehydes	$0.36 \log K_{ow} + 3.5$	14	0.85	Deneer et al. [207]	66a
		$0.36 \log K_{ow} - 0.08 \log k_{cyst} + 3.7$	14	0.88		66b
fish toxicity screening data	alcohols	comparison with QSAR predictions	55		Lipnick et al. [203]	
24-hr LC50 to goldfish	epoxides	comparison with QSAR predictions	6		Lipnick et al. [159]	

a) LC50: concentration with 50% mortality.
b) C: effect concentration (in mol l^{-1}); K_{ow}: octanol-water partition coefficient; k_{nbp}: first order reaction rate constants of a reaction with 4-nitrobenzylpyridine (in day^{-1}) in equation 64b and in min^{-1}1 min^{-1}) in equation 66b); k_{cyst}: second order rate constants of a reaction with cysteine (mol^{-1}1 min^{-1}); n: number of chemicals in dataset; r: correlation coefficient.

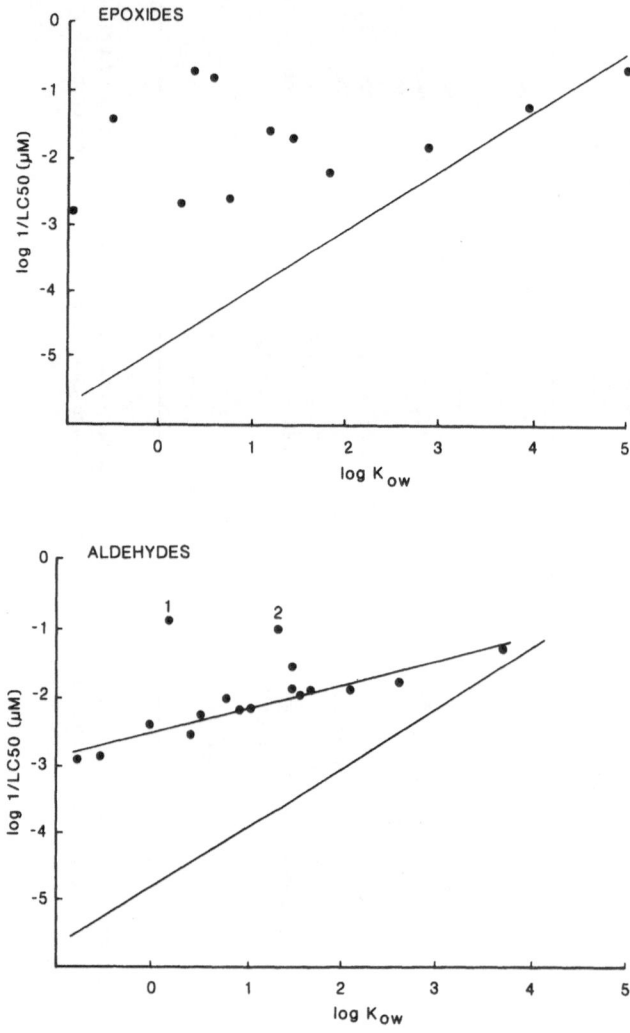

Fig. 8. Relationship between 14-day LC50 concentrations with guppies and octanol-water partition coefficient (K_{ow}) for epoxides (8a) and aldehydes (8b). LC50 data from Deneer et al. [206, 207]

enables them to react in a way different than the other aldehydes, e.g., by addition at the carbon–carbon double bond. (Table 10, scheme c).

An example of a QSAR study for chemicals that probably are activated to reactive intermediates is given by Lipnick et al. [203]. They observed that LC50s of allylic and propargylic alcohols are much lower than those calculated from a QSAR equation for narcosis type chemicals. It is proposed that the allylic and propargylic alcohols are metabolized to the corresponding aldehydes and that these aldehydes are the actual toxicants by interaction with nucleophiles through means of Michael addition (Table 10, scheme c). The proposed mechanism is depicted in Fig. 9.

$$CH_2 = CH-CH_2OH \longrightarrow CH_2 = CH-C = OH \qquad \text{ADH}$$

Fig. 9. Proelectrophile toxicity mechanism for primary and secondary allylic and propargylic alcohols as exemplified by allylalcohol that can undergo biochemical oxidation by the enzyme alcohol dehydrogenase (ADH) to generate an α,β-unsaturated aldehyde or ketone, which can undergo reaction and covalent bond formation with nucleophilic groups present in biological macromolecules. (after Lipnick et al. [203])

It is obvious that, in general, LC50s of reactive chemicals such as reactive alkylhalides, epoxides, aldehydes and unsaturated alcohols are lower than LC50s of unreactive chemicals. Figure 6 and 8 show LC50 data of such reactive chemicals and is a clear illustration of this general rule. Further, it is obvious that for obtaining significant correlations for these reactive chemicals, it is unnecessary to include rate constants as additional descriptors.

The chemical descriptors for reactivity that are used in the equations from Table 12 are all measured rate constants. The applicability of such equations will be strongly improved when differences in reactivity can be predicted through means of calculated descriptors. Within a set of structure analogues, Hammett σ constants and/or Taft Es constants can be useful and descriptors derived from Molecular Oribital calculations may be suitable for the prediction of reactivity of compounds with diverse chemical structures.

An interesting aspect of the QSAR equations for aldehydes and epoxides is the fact that ratios between observed LC50s and calculated LC50s (from a QSAR for narcosis type of chemicals) decrease with increasing values of K_{ow} (see Fig. 8). So, there appears to be a shift in mode of action towards narcosis for the more hydrophobic chemicals. Deneer [211] gives the following possible explanation: "As hydrophobicity increases, substances will show an increasing tendency to be deposited in the lipid phase. If the specific toxic effect is based on a reaction taking place in the aqueous phases, it will ultimately be superseded by the a-specific anaesthetic effect for hydrophobic chemicals". Similar effects are also observed by Veith et al. [212] for esters and by Lipnick et al. [159] for epoxides.

QSAR for other Classes of Organic Chemicals

Besides the classes of chemicals mentioned before, also other chemical structures are studied by QSAR. Table 13 gives a survey of these studies and a few will be discussed in more detail.

Moulton and Schultz [218] applied principal component analysis (PCA) to a dataset of EC50s (population growth) to Tetrahymena of selected pyridines. The following descriptors were applied: the hydrophobic substituent constant π, the ability to accept or donate hydrogen in a hydrogen bonding (Ha and Hd), molar refractivity (MR) and the electronic descriptors for field and resonance effects (F and R), Hammett σ constants and MC indices. The initial PCA reduced the original eight descriptors to four new variables (factors). The first factor (F1) was highly correlated with MR and the MC index and to a lesser extent with π. The

Table 13. QSAR studies for several classes of organic chemicals

type of chemicals[a]	toxic endpoint[b]	chemical descriptors[c]	reference
mono and 1,4-disubstituted benzenes	EC50 (bioluminescence) to photo bacteria	K_{ow}, MR, energy of UV absorption and EC50 of monosubstituted chemicals	Kaiser et al. [213, 214]
organotin compounds	LC50 for mud crab zoeae	K_{ow}, TSA	Laughlin [215]
organotin compounds	LC50 to Daphnia magna	K_{ow}, pKa and MC indices	Vighi and Calamari [216]
nitrogen heterocyclic compounds	EC50 (population growth) to Tetrahymena	K_{ow}, MC indices	Schultz [217]
pyridines	EC50 (population growth) to Tetrahymena	K_{ow}, MC indices, MR, Ha, Hd σ_p, F and R (principal component analysis)	Moulton and Schultz [218]
nitrogen containing aromatics	EC50 (population growth) to Tetrahymena	K_{ow}	Schultz and Moulton [219]
organophosphorus compounds	LC50 to guppy	K_{ow}, k_{nbp}, Hammett σ constants	Hermens et al. [220]
PAHs	LC50 to Daphnia pulex	K_{ow}, MC indices	Govers et al. [221]
alcohols	LC50 to bleak	K_{ow}	Bengtsson et al. [222]
esters	LC50 to fathead minnow	K_{ow}	Veith et al. [212]
esters	L50 to fathead minnow	K_{ow}, MC indices and other topological indices (IC, SIC and CIC)	Basak et al. [223]
ketones, esters, alcohols and nitriles	LC50 to fathead minnow	K_{ow}, Hammett σ constants	Purdy [224]
PAHs, halogenated and non halogenated alkanes and aromatics, nitrogen compounds and alkylether silanes	LC50 to Daphnia pulex	K_{ow}, S and molecular volume	Passino and Smith [225]

chlorobenzenes and chlorophenols	LC50 to guppy	groups accessible surface area	Gombar [226]
naphthalene derivatives	EC50 (immobilization) to Artemia	K_{ow}	Foster and Tullis [227]
naphthalene derivatives	EC50 (population growth) to Tetrahymena	K_{ow}, MC indices, MR, Ha, Hd σ_p, F and R (principal component analysis)	Schultz and Moulton [228]
substituted benzenes	LC50 fathead minnow	number of atoms/groups and group toxicity contributions (ΔT)	Hall et al. [229, 230]
dithiocarbamates	LC50, LRCT (population growth) to Daphnia magna	K_{ow}	Van Leeuwen et al. [231, 232]
thioureas	EC50 (bioluminescence) to photo bacteria	K_{ow}, MC indices	Govers et al. [233]
miscellaneous	LC50 to Daphnia magna	K_{ow}, MC indices, CIC	Vighi and Calamari [234]
miscellaneous	LC50 to fish	K_{ow}, S	Zaroogian et al. [235]
miscellaneous	LC50 to several species	K_{ow}, S	Yoshioka et al. [236]
miscellaneous	LC50 to Orizias latipes	K_{ow}, MW, i/o and MC indices	Yoshioka et al. [237]
miscellaneous	LC50 to fish	substructural fragments (discriminant analysis)	Komatsu et al. [238]

a) PAHS: polycyclic aromatic hydrocarbons.

b) LC50: concentration with 50% mortality; EC50: concentration with 50% effect.

c) K_{ow}: octanol-water partition coefficient; MR: molar refractivity; TSA: total surface area; pKa: acid dissociation constant; MC: molecular connectivity; MR: molar refracivity; Ha: indicator variable for hydrogen bonding acceptors; Hd: indicator variable for hydrogen donaters; σ: Hammett substituent constant; F, R: substituent constants for field and resonance effects; k_{nbp}: rate constants towards 4-nitrobenzylpyridine; IC: information content; CIC: complementary information content; SIC: structural information content; S: water solubility; MW: molecular weight; i/o: organic, inorganic characteristics [237].

second factor (F2) had high positive correlations with Ha and Hd and a high negative correlation with π. These two first factors also showed the highest correlation with the EC50 values according to:

$$\log 1/EC50 = 0.45\,F1 - 0.25\,F2 - 0.59 \quad n = 20 \; r^2 = 0.69 \tag{67}$$

The advantage of principal component analysis is that one can start with a larger number of descriptors and that collinear descriptors can be applied.

The second example has used discriminant function analysis as multivariate technique. Based on LC50 data for 280 compounds, obtained from the Registry of Toxic Effects of Chemical Substances of the US Department of Health and Human Services, Komatsu et al. [238] developed a model based on substructures and a few physico-chemical properties (K_{ow} and van der Waals volume). He performed a three-class discrimination with the following classes: weakly toxic (LC50 > 100 ppm), moderately toxic (10 < LC50 < 100 ppm) and strongly toxic (LC50 < 10 ppm). The analysis was started with 60 descriptors but the final subset of descriptors contained 2 numerical and 25 indicator variables for substructure fragments. The accuracy of the classification was about 85%.

Some General Remarks

The examples discussed in the previous sections show that several different descriptors are applied in QSAR studies: descriptors that are related to physico-chemical properties or to the structure such as in topological indices. The advantage of topological indices is that they can be calculated directly from structures, but such descriptors supply no information on mechanistic or kinetic aspects.

In correlations between biological activity and physical-chemical properties, only those descriptors will be important that describe differences in rates of the relevant processes. As already stated by Hansch et al. [74], the descriptors in a QSAR represent the rate limiting steps in the development of an effect. The relevancies of processes like uptake, elimination and metabolism (activation or deactivation) that yield to the toxic effect depends on the relative rates of these processes. A deactivation, e.g., will be of no importance when the rate of deactivation is much lower than the elimination rate of the unmetabolized compound. Moreover, the dominant descriptors in a QSAR equation can inform us on which processes are relevant and rate limiting. The influence of K_{ow} in the QSAR equations for relatively unreactive chemicals is simply related to the influence of hydrophobicity on the absorption of these chemicals. Also the influence of K_{ow} and pKa in the equations for phenols is, at least partly, related to kinetics. The equation for epoxides suggest that both uptake and reaction with the target are rate limiting steps. However, the fate of reactive chemicals and reactive intermediates are so complex, that it is difficult to explain the influence of a certain descriptor. It will also become more difficult to develop and explain a QSAR for chemicals with a specific mode of action, as e.g. for pesticides, because of possible very specific electronic and/or steric interactions with the target molecule.

With the help of the QSAR studies for phenols another interesting and more general aspect of QSAR can be discussed. As mentioned before, Saarikoski and

Viluksela [64], and also Schultz et al [185], found better correlations by leaving out a few data points. In both studies, dinitrophenols are outliers. Schultz et al. [185] recognized that also phenols with chloro substituents in the 4 or 5 position are outliers and based on this information two different equations were calculated. The authors have suggested that these two equations represent two different mechanisms that might be involved in the effect of phenols: polar narcosis and uncoupling of the oxydative phosphorylation in case of the dinitrophenols. Although dinitrophenols are known uncouplers, the outlying behaviour of these chemicals was recognized afterwards. The QSAR study with nitroaromatics from Deneer et al. [47] is another example. The lower LC50s of dinitrobenzenes, in comparison with mononitrobenzenes, were explained afterwards in terms of metabolism to reactive intermediates. The QSAR studies for aldehydes and epoxides showed that LC50s of the more hydrophobic representatives were well predicted by a QSAR for chemicals with a narcosis mechanism of action. This points to the interesting phenomenon that the mode of action within a homologuous set of chemicals changes, although one might expect similarity in mode of action for structure analogues. And again, these observations were made afterwards. Similar arguments are valid for several QSAR studies and point to a more general problem in QSAR. First of all, the structural requirements related to a particular QSAR are usually not well defined which makes it difficult to decide whether or not a QSAR can be applied for predictions of the effect concentration of a new or untested chemical. The QSARs for relatively unreactive chemicals are an exception because, from the many publications for this particular group, the structural requirements are more or less well defined. A second aspect is the general lack in knowledge on the mechanisms of action of chemicals in aquatic species and in particular in fish and it is generally recognized that QSAR should be applied to chemicals that has the same mode of action. Recently, more attention is paid to the distinction of different mechanisms of action in fish toxicity. These arguments point to the problems that arise when QSAR are applied for predictions. On the other hand, QSAR can also be helpful in the analysis of toxicity data. By studying correlations between effect concentrations and physico-chemical properties, outliers can be recognized and in some cases even can give some hints on mechanism of action. The same studies discussed above for phenols and nitroaromatics are good examples. It is beyond the scope of this paper to discuss the applicability of QSAR. In principle, predictions based QSAR can be useful for all kind of purposes: regulating activities (evaluating new or existing chemicals); priority setting for existing chemicals [239]; evaluation of new chemicals during product development. Evaluation of premanufactury notifications (PMNs) by the EPA is partly based on QSAR [240]; also the FDA uses QSAR in a more qualitative way to determine which tests will be relevant [241, 242]. A task force of the European Chemical Industry Ecology and Toxicology Centre (ECETOC) made a critical assessment of the applicability of QSAR for practical purposes in toxicology and ecotoxicology [20]. This task force came to the conclusion that, at present, QSARs should not be used in isolation for making decisions that affect the health of humans or other species, although QSARs may be useful within product development; priority setting and in the evaluation of joint toxic effects [20]. When predictions are used for practical purposes, it is important to keep a critical attitude towards every predicted endpoint: from a physico-chemical property such as pKa or K_{ow}, to bioconcentration factors, biodegradability and LC50s. Moreover, it is also important to be familiar with the prediction models and their limitations. Besides the possibility to use QSAR for prediction purposes, it is very helpful in analyzing data, because it is a tool in classifying large numbers of chemicals into a limited number of groups and outliers can be recognized and further studied.

QSAR and Internal Concentration

Recently, a few publications pay attention to internal concentration [47, 57, 83] with the idea that the internal concentration at a certain endpoint (e.g. lethality) is constant for chemicals from a particular group of chemicals. The advantages of this approach are that such data are easier to compare for different species, different exposure regimes and exposure times and even with field data [83]. The idea of a constant internal concentration will be valid in those cases where the intrinsic activity of chemicals is similar and when the target is located in the lipid phases such as membranes. The relatively unreactive chemicals that act by narcosis is a classical example for which it is assumed that the internal molar concentration is constant at a particular effect, although some theories state that the volume fraction and not molar concentration is constant.

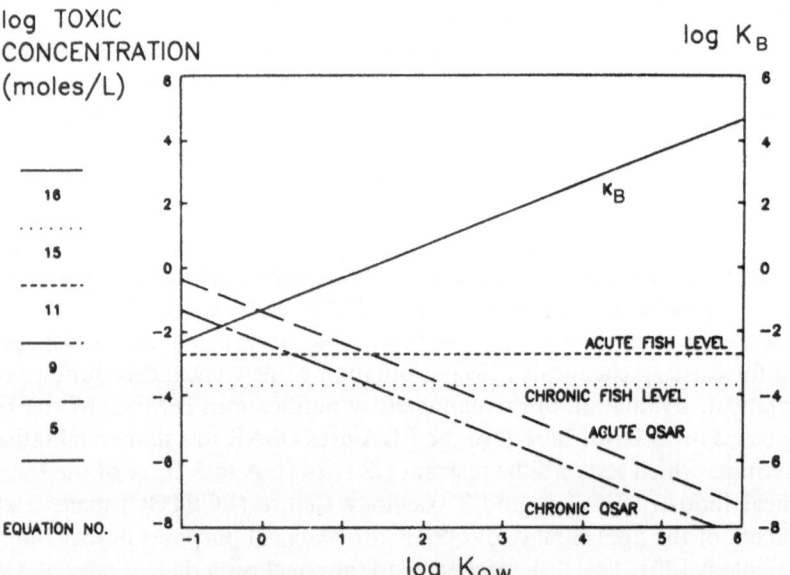

Fig. 10. Relationship between toxicity, bioaccumulation (log K_B) and octanol-water partition coefficient (log K_{ow}) for some narcotic organic chemicals. Reprinted with permission from McCarty [57], copyright (1987), Kluwer Academic Publishers

Table 14. Internal toxic concentration (ITC) for some classes of organic chemicals

type of chemicals	toxic endpoint	internal concentration (μmol g^{-1} fish)		reference
narcotic organics:				
chlorobenzenes	lethality in fish	1.0	(a)	McCarty [57]
various organics	lethality in fish	1.8–1.9	(a)	ibid
chlorobenzenes	lethality in fish	2.0–2.5	(b)	Van Hoogen and Opperhuizen [83]
chlorobenzenes	chronic toxicity in fish	0.40	(a)	McCarty [57]
various organics	chronic toxicity in fish	0.20	(a)	ibid
substituted phenols:				
substituted phenols	lethality in fish	0.31–0.55	(a)	McCarty [57]
substituted phenols	chronic toxicity in fish	0.03–0.22	(a)	ibid
others:				
mono nitroaromatics	lethality in fish	1.6	(a, c)	Deneer et al. [47]
1,2-dinitrobenzene	lethality in fish	0.004	(a, c)	ibid
1,4-dinitrobenzene	lethality in fish	0.0006	(a, c)	ibid
polychlorinated biphenyls	lethality in fish	2.0–2.5	(b)	Van Hoogen and Opperhuizen [83]

a) based on calculation from: BCF × LC50.
b) based on measured internal concentration of organisms after death.
c) original data are presented as μmol g^{-1} lipid; data are recalculated based on a lipid content of 5%.

Literature data of internal concentrations are given in Table 14. The data from McCarty [57] are also presented in Fig. 10. The internal concentration of chemicals that act by narcosis vary from 1.0 to 2.5 μ mol/g fish at lethality and are in the range of 0.2 to 0.4 at chronic effects. So, although the LC50s of such chemicals cover a range of seven orders of magnitude, the internal concentrations are quite constant and the differences in LC50s are completely due to differences in degrees of absorption. This idea is presented in a nice way in figure 10: K_B (the bio-concentration factor) increases with K_{ow}: LC50s decrease with K_{ow} but the toxic internal concentration is independent on K_{ow}. It is not surprising that the chronic fish level is lower than the acute fish level. The ratio between the acute and chronic toxic level is about 5 to 10 [57]. Based on data for LC50 and MATC (maximum acceptable toxicant concentration) from ELS tests with fathead minnows, Call et al. [171] observed a acute/chronic toxicity ratio in the same range (12.2 $\pm$ 8.5 / $n = 10$) for chemicals that act by narcosis.

The data for internal lethal concentrations indicate that phenols are slightly more toxic than narcotics and that PCBs, mononitrobenzenes and toluenes cause lethality at similar concentrations as chemicals that act by narcosis. As already mentioned in the section on "QSAR for Phenols, Aromatic Amines and Nitroaromatics", dinitrobenzenes are more toxic probably because of the formation of reactive intermediates.

In summary, evaluation of toxicity data based on internal toxic concentrations has the following advantages:

— ITC will be constant for classes of chemicals that act with unspecific mode of action; when the intrinsic activity does not vary among members of that class and when the site of action is located in a lipid phase.

— ITC values are independent of differences in exposure regimes and exposure times, which simplifies a comparison among different species and test protocols [83].

— Measured body burdens in the field can be evaluated based on literature data for ITC [83].

Instead of using internal concentrations, Abernethy et al. [243–244] have proposed that organisms will die at a constant volume fraction of a narcotic at the target site. This idea originates from Mullins [177] who has suggested that anaesthesia occurs when the volume of a hydrophobic region is caused to expand beyond a critical volume. According to Abernethy et al. [244], this volume fraction Y (in m^3 toxicant / m^3 target tissue) at the LC50 concentration (in mol m^{-3}) is related to the octanol-water partition coefficient (K_{ow}) and the molar volume V (in cm^3 mol^{-1}) as follows:

$$Y = 10^{-6} \text{ LC50 V } K_{ow} \tag{68}$$

Using equation 68, Abernethy et al. [244] analysed toxicity data to several aquatic species of 113 different chemicals, and observed that volume fractions of about 0.6% produces mortality in acute experiments and that a fraction of 0.06% produces chronic effects. It was also observed that log Y increased slightly, but significantly, with increasing log K_{ow} and Abernethy postulated that this is due to the fact that the organism-water phase partition coefficient is different, and generally lower, than the octanol-water partition coefficient [244].

A different approach for measuring internal lethal concentrations in fish was chosen by Hodson et al. [245-246], who determined median lethal doses (LD50s) to fish for various organics. LD50s were calculated from mortality data after intra peritoneal injection of different doses of a chemical dissolved in 5% ethanol in saline or in cod-liver oil. After removing 4 outliers, likely because of their specificity in mode of action, Hodson observed the following correlation between LC50 data and LD50s in combination with K_{ow} [246]:

$$\log LC50 = 0.33 - 0.77 \log K_{ow} + 0.57 \log LD50 \tag{69}$$

K_{ow} in equation 69 probably represents the influence of hydrophobicity on the internal concentration, while log LD50 is more related to the intrinsic toxicity of a chemical. This approach provides a mechanism for using mammalian data for estimating LC50s for aquatic species [246].

QSAR and Mode of Action

In the section on "Some General Remarks" it was recognized that the knowledge on mode of action of chemicals in aquatic species is rather limited.

McKim and coworkers [247–250] recently studied Fish Acute Toxicity Syndromes (FATS) in relation to QSAR. A FATS is defined as a specific combination of respiratory-cardiovascular and physiological responses in rainbow trout after an acute exposure to a chemical with a specific mode of action. Responses that are studies include: heart (HR), ventilation (VR) and cough rate (CR); ventilation

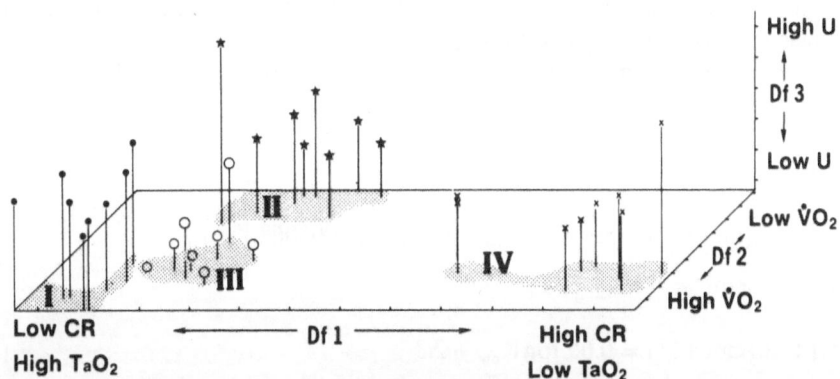

Fig. 11. Three dimensional plot of the first three discriminant functions that separate four FATS: respiratory uncoupler syndrome (I), narcosis sydrome (II), AChE inhibitor syndrome (III), respiratory irritant syndrome (IV). Each vertical line represents an individual fish exposed to one of the following eight chemicals: pentachlorophenol and 2,4-dinitrophenol, MS222 and 1-octanol, malathion, and carbaryl, acrolein and benzaldehyde. Reprinted with permission from McKim et al. [249], copyright (1987), Environmental Health Perspectives

volume (V_G), oxygen utilization (U) and consumption (VO_2) and some blood parameters. Based on this approach, McKim et al. [249] succeeded in distinguishing six different FATS. An example of this approach is given in Fig. 11, where four different FATS are successfully separated by using discriminant analyses. These four FATS include: respiratory uncoupling (pentachlorophenol and 2,4-dinitrophenol); narcosis (MS222 and 1-octanol), acetylcholinesterase (AChE) inhibition (malathion and carbaryl) and respiratory irritation (acrolein and benzaldehyde). The respiratory-cardiovascular responses to each of these FATS are as follows [249]: respiratory uncoupling is characterized by an increase in ventilation volume and oxygen consumption; narcosis by a lowering in heart rate and oxygen consumption; AChE inhibition by an increase in ventilation volume and respiratory irritation by a high increase in cough rate. The responses to chemicals that represent a particular mode of action correspond very well with the response that is expected. The increase in ventilation volume and oxygen consumption after exposure to pentachlorophenol and 2,4-dinitrophenol as classical uncouplers of oxydative phosphorylation, e.g., can be seen as an effort to generate ATP [249].

In a similar approach, Drummond et al. [251] succeeded in classifying similar groups based on behavioral parameters, although the classification was less successful because behaviour is less specific.

The idea behind this type of research is to develop a limited number of FATS and to classify a chemical into a FATS based on structural characteristics. The toxicity of a chemical can then be predicted by a QSAR equation, developed for the particular FATS and mode of action [249]. Although the measured responses may not be sufficient for classifying certain modes of action, the development of FATS is a powerful tool for getting more insight into major modes of action of the majority of the aquatic pollutants. Certainly if one recognizes that the diversity of chemicals is probably not so high as one would expect.

Based on this idea of mode of action, several QSAR equations are published for 96-h LC50 to fathead minnow representing at least 2 different modes of action (narcosis and respiratory uncoupling).

Two equations based on a rather large number of data are available for narcosis type of chemicals (equation 70 and 71).

Narco I chemicals [156]:

$$\log 1/LC50 (mol\ l^{-1}) = 0.94 \log K_{ow} - 0.94 \log (0.000068\ K_{ow}) + 1.3 \qquad (70)$$

Narco II chemicals [184]:

$$\log 1/LC50 (mol\ l^{-1}) = 0.65 \log K_{ow} + 2.3 \qquad (71)$$

Equation 70 represents the nonpolar narcotics such as the alcohols, ethers, hydrocarbons and chlorinated alkanes and aromatics, that are discussed in more detail in the section on "QSAR for Relatively Unreactive Organic Chemicals". Experimental LC50s of several aromatic amines and phenols are lower than calculated values using equation 70. Based on information in the literature on the mode of action of narcotics, on studies to behaviour and fish acute toxicity syndromes and on results from joint effects of mixtures, Veith and Broderius [184] developed a separate equation for polar narcotics (equation 71). Because both equations model different types of narcotics, Veith introduced the terms narcosis type I and II syndrome, related to equations 70 and 71, respectively. Esters are also more toxic than the narcosis I QSAR predicts, while the symptoms were similar to classic narcotics [212]. Veith et al. [212] calculated a separate equation for LC50s to fathead minnow of esters (equation 72).

$$\log 1/LC50(\text{mol } 1^{-1}) = 0.54 \log K_{ow} + 2.8 \tag{72}$$

It was recognized that equation 71 is valid for weakly acidic phenols and weakly basic aromatic amines. Phenols or anilines with two or more nitro substituents or four or more ring substituted halogens are more toxic and are likely oxydative uncouplers [184]. This same discrepancy was observed by Schultz et al. [185], who calculated a separate equation for chemicals that may act as uncouplers (equation 73). Because this equation is based on only 6 chemicals, this model may be refined when more data become available.

Uncouplers of oxydative phosphorylation [185]:

$$\log 1/LC50(\text{mol } 1^{-1}) = 0.59 \log K_{ow} + 3.2 \tag{73}$$

The idea of modeling chemicals according to their mode of action and to study the mode of action, e.g. by FATS, is an interesting development within QSAR and will certainly help us to understand the many QSAR equations that are published in the literature.

QSAR and Computer Systems

The above equations are the same that are implemented in the QSAR Information System that is developed at the EPA Environmental Research Laboratory in Duluth (ERLD)/Minnesota (Dr. G. D. Veith) in cooperation with the Center for Data System and Analysis at Montana State University (Dr. R. S. Hunter). This computer program contains several models to predict physico-chemical properties such as K_{ow}, water solubility, pKa, boiling point, vapor pressure etc. K_{ow} is calculated by CLOGP while calculation of the others descriptors is based on methods described by Lymann et al. [150]. Predictions are also made for bio-degradability, bioconcentration, adsorption and environmental partitioning by a fugacity model from Mackay et al. [3], LC50 for several species, chronic toxicity levels, phytotoxicity and genotoxicity. More information on the models that are used for predictions is given in the user manual from Hunter et al. [252] and by Crowley et al. [253]. Besides calculated values, the system also contains a lot of experimental data. The AQUIRE (Aquatic information retrieval toxicity data base) option contains measured toxicity data for approximately 3000 chemicals with many details on test protocols [254]. A chemical can be introduced by a simple notation technique, developed at ERL-D [255], called SMILES (simplified molecular identification and line entry system). Especially this notation technique is very useful in the implementation of QSAR models in large computer programs. LC50s, e.g., are predicted by an automatic classification of a chemical into a particular group (mode of action) according to its structure and the corresponding QSAR equation for that particular mode of action. The essential part in a prediction is the classification of a chemical into a particular group and the corresponding structural requirements. Therefore, our knowledge on mode of action and on the structural requirements must be extended. The SMILES notation technique is also used in the TOPKAT program from Health Design Inc. [256]; a computer program that is developed for predictions of mammalian toxic end-points.

QSAR and Joint Toxicity of Mixtures

Another aspect of QSAR and mode of action is reflected in the joint toxicity of mixtures. It is generally assumed that chemicals that belong to a particular QSAR equation act by a similar mode of action. Joint toxic effects of mixtures of chemicals with similar modes of action can be predicted by concentration addition [257–259] and this implicates that, e.g. in LC50 tests with a mixture of n chemicals, 50% mortality will arise at concentrations of the individual chemicals according to:

concentration addition (similar modes of action):

$$\Sigma \; c_i / LC50_i \; (i = 1 \; to \; n) = 1 \tag{74}$$

In equation 74, c_i and $LC50_i$ are the actual concentration and LC50 of component i. If we take a specific example of e.g. a mixture of 10 chemicals at equitoxic concentrations (concentrations in fractions of the LC50s are equal for all components), equation 74 means that 50% mortality will occur at a concentration of the individual chemicals of 0.10 of their corresponding LC50s. Such a low concentration of 0.10 LC50 will usually cause no mortality when the chemical is tested individually.

For mixtures of n toxicants with dissimilar and independent modes of action and when the susceptibilities of organisms to different toxicants are the same, 50% kill will arise at concentrations of the individual chemicals according to:

no addition (dissimilar and independent modes of action):

$$\Sigma \; c_i / LC50_i \; (i = 1 \; to \; n) = n \tag{75}$$

No addition corresponds in fact with the situation in which joint toxicity is disregarded, because the effect of a mixture does not exceed the effects of all of its components individually.

Several toxicity tests with aquatic organisms have shown that joint toxic effects of mixtures of chemicals that belong to a particular QSAR indeed correspond very well with concentration addition and are far away from no addition (see Table 15

Tables 15. Results of toxicity tests with some mixtures of organic chemicals

type of chemicals	toxic endpoint[a]	$\Sigma c_i/LC50_i$[b]	n[b]	reference
narcotic organics:				
various organics	14-day LC50 to guppy	0.9	50	Könemann [259]
various organics	96-h LC50 to fathead minnow	1.2	23	Broderius and Kahl [260]
various organics	48-h LC50 to Daphnia magna	1.2	50	Hermens et al. [163]
various organics	16-day NOEC (growth) to Daphnia magna	0.6	25	Hermens et al. [166]
various organics	16-day NOEC (reproduction) to Daphnia magna	1.1	10	De Wolf et al. [261]
other classes of chemicals:				
phenols	14-day LC50 to guppy	1.0	11	Könemann and Musch [183]
aromatic amines	14-day LC50 to guppy	1.1	17	Hermens et al. [197]
reactive organic halides	14-day LC50 to guppy	1.0	9	Hermens et al. [262]

a) LC50: concentration with 50% mortality; NOEC: no-observed effect concentration.

b) $\Sigma c_i/LC50_i = 1$: concentration-addition for mixtures of chemicals with similar modes of action.
$\Sigma c_i/LC50_i = n$: no-addition for mixtures of chemicals with dissimilar modes of action (see text for further explanation).

for a few representative examples). Recently, Deneer et al. [263] showed that chemicals that act by narcosis can give effects even at concentrations as low as 0.0025 of an LC50. A few other studies show that joint effects also occur for sublethal effects [166, 261, 264, 265] and for mixtures of chemicals with dissimilar modes of action [262, 264–266], although joint toxicity is probably lower at sublethal level for mixtures of chemicals with dissimilar modes of action [264, 265]. These examples show us that joint toxicity is a serious aspect that certainly has to be addressed in hazard assessment. The subject of joint toxic effects is discussed in more detail by a working party of the FAO [267].

References

1. Maki AW, Bishop WE (1985) In: Rand GM, Petrocelli SR (eds) Fundamentals of aquatic toxicology. Hemisphere, Washington, p 619
2. Burns LA, Baughmann GL (1985) In: Rand GM, Petrocelli SR (eds) Fundamentals of aquatic toxicology. Hemisphere, Washington, p 558
3. Mackay D, Paterson S, Cheung B, Neely WB (1985) Chemosphere 14: 335
4. Slooff W, Van Oers JAM, De Zwart D(1986) Environ. Toxicol. Chem. 5: 841
5. Ariëns EJ (ed) (1971) Drug Design, vol I. Academic, New York
6. Cavallito CJ (ed) (1973) Structure activity relationships, vol 1. Pergamon, Oxford
7. Chapman NB, Shorter J (eds) (1978) Correlation analyses in chemistry: Recent advances. Plenum, New York
8. Dearden JC (ed) (1983) Quantitative approaches to drug design. Elsevier, Amsterdam
9. Hadzi D, Jerman-Blazic B (eds) (1987) QSAR in drug design and toxicology. Elsevier, Amsterdam
10. Hansch C (1971) In: Ariëns EJ (ed) Drug design, vol 1. Academic, New York, p 271
11. Hansch C (1973) In: Cavallito CJ (ed) Structure-activity relationships, vol 1. Pergamon, Oxford, p 75
12. Hansch C (1978) In: Chapman NB, Shorter J (eds) Corelation analysis in chemistry: Recent advances. Plenum, New York, p 397
13. Martin YC (1978) Quantitative drug design. Dekker, New York
14. Seydel JK (ed) (1985) QSAR and Strategies in the design of bioactive compounds. Elsevier, Amsterdam
15. Topliss JG (ed) (1983) Quantitative structure-activity relationships of drugs. Academic, New York
16. Frierson MR, Klopman G, Rosenkranz HS (1986) Environ. Mutagen. 8: 283
17. Golberg L (ed) (1983) Structure-activity correlation as a predictive tool in toxicology. Hemisphere, Washington
18. McKinney JD (ed) (1985) Monograph on structure-activity correlation in mechanism studies and predictive toxicology. Environ. Health Persp. vol 61
19. Tichy M (ed) (1985) QSAR in Toxicology and Xenobiochemistry. Elsevier, Amsterdam
20. Turner L, Choplin F, Dugard P, Hermens J, Jaeckh R, Marsmann M, Roberts D (1987) Toxic. in Vitro 1: 143
21. Birge WJ, Cassidy RA (1983) Fund. Appl. Toxicol. 3: 359
22. Hermens J (1986) Pestic. Sci. 17: 287
23. Kaiser KLE (ed) (1984) QSAR in environmental toxicology. Reidel, Dordrecht
24. Kaiser KLE (ed) (1987) QSAR in environmental toxicology-II. Reidel, Dordrecht
25. Lipnick RL, Mackay D (eds) (1985) Symposium Structure-Activity Relationships. Environ. Toxicol. Chem. 4: 255
26. Nirmalakhandan N, Speece RE (1988) Environ. Sci. Technol. 22: 606
27. Veith GD, Konasewich DE (eds) (1975) structure-activity correlations in studies of toxicity and bioconcentration with aquatic organisms . Great Lakes Research Advisory Board, Windsor–Ontario

28. Briggs GG (1981) J. Agric. Food Chem. 29: 1050
29. Karickhoff SW, Brown DS, Scott TA (1979) Water Res, 13: 241
30. Schellenberg K, Leuenberger C. Schwarzenbach RP (1984) Environ. Sci. Technol. 18: 652
31. Sabljic A (1984) J. Agric. Food Chem. 32: 243
32. Oliver BG (1985) Chemosphere 14: 1087
33. Banerjee S (1987) In: Kaiser KLE (ed) QSAR in environmental toxicology-II. Reidel, Dordrecht, p 17
34. Beltrame P, Beltrame PL, Carniti P (1984) Chemosphere 13: 3
35. Calamari D, Da Gasso R, Galassi S, Provini A, Vighi M (1980) Chemosphere 9: 753
36. Dearden JC, Nicholson RM (1986) Pestic. Sci. 17: 305
37. Freed VH, Chiou CT, Schmedding DW (1979) J. Agric. Food Chem. 27: 706
38. Niemi GJ, Veith GD, Regal RR, Vaishnav DD (1987) Environ. Toxicol. Chem. 6: 515
39. Paris DF, Wolfe NL, Steen WC, Baughman, GL (1983) Appl. Environ. Microbiol. 45: 1153
40. Parsons JR, Opperhuizen A (1987) Chemosphere 16: 1361
41. Vaishnav DD, Boethling RS, Babeu L (1987) Chemosphere 16: 695
42. Wolfe NL, Paris DF, Steen WC, Baughman GL (1980) Environ. Sci. Technol. 14: 1143
43. Bruggeman WA, Opperhuizen A, Wijbinga A, Hutzinger O (1984) Toxicol. Environ. Chem. 7: 173
44. Butte W, Willig A, Zauke G-P (1987) In: Kaiser KLE (ed) QSAR in environmental toxicology-II. Reidel, Dordrecht, p 43
45. Chiou CT, Freed VH, Schmedding DW, Kohnert RL (1977) Environ. Sci. Technol. 11: 475
46. Davies RP, Dobbs AJ (1984) Water Res. 18: 1253
47. Deneer JW, Sinnige TL, Seinen W, Hermens JLM (1987) Aquat. Toxicol. 10: 115
48. Chiou CT (1985) Environ. Sci. Technol. 19: 57
49. Geyer H, Sheehan P, Kotzias D, Freitag D, Korte F (1982) Chemosphere 11: 1121
50. Gobas FAPC, Opperhuizen A, Hutzinger O (1986) Environ. Toxicol. Chem. 5: 637
51. Gobas FAPC, Shiu WY, Mackay D (1987) In: Kaiser KLE (ed) QSAR in environmental toxicology-II. Reidel, Dordrecht, p 107
52. Gobas FAPC, Mackay D (1987) Environ. Toxicol. Chem. 6: 495
53. Hawker DW, Connel DW (1985) Chemosphere 14: 1205
54. Kenega E.E. (1980) Environ. Sci. Technol. 14: 553
55. Könemann H, van Leeuwen K (1980) Chemosphere 9: 3
56. Mackay D (1982) Environ. Sci. Technol. 16: 274
57. McCarty LS (1987) In: Kaiser KLE (ed) QSAR in environmental toxicology-II. Reidel, Dordrecht, p 207
58. McCarty LS (1987) In: Kaiser KLE (ed) QSAR in environmental toxicology-II. Reidel, Dordrecht, p 221
59. McKim J, Schmieder P, Veith G (1985) Toxicol. Appl. Pharmacol. 77: 1–10
60. Neely WB, Branson DR, Blau GE (1974) Environ. Sci. Technol. 8: 1113
61. Oliver BG, Niimi AJ (1985) Environ. Sci. Technol. 19: 842
62. Opperhuizen A, Van Der Velde EW, Gobas FAPC, Liem AKD, Van Der Steen JMD, Hutzinger O (1985) Chemosphere 15: 1871
63. Opperhuizen A (1986) In: Poston TM, Purdy R (eds) Aquatic toxicology and environmental fate, Ninth Volume. American Society for Testing and Materials, Philadelphia, p 304
64. Saarikoski J, Viluksela M (1982) Ecotoxicol. Environ. Safety 6: 501
65. Saarikoski J, Lindström R, Tyynelä M, Viluksela M (1986) Ecotoxicol. Environ. Safety 11: 158
66. Sabljic A, Protic M (1982) Chem-Biol. Interactions 42: 301
67. Tulp MTh, Hutzinger O (1978) Chemosphere 7: 849
68. Veith GD, DeFoe DL, Bergstedt BV (1979) J. Fish. Res. Board. Can. 36: 1040
69. Wold S, Dunn WJ (1983) J. Chem. Inf. Comput. Sci. 23: 6
70. Free SM, Wilson JW (1964) J. Med. Chem. 7: 395
71. Meyer H (1899) Arch. exp. Pathol. Pharmakol. 42: 109
72. Overton E (1897) Z. Phys. Chem. 22: 189
73. Hansch C, Muir RM, Fujita T, Maloney PP, Geiger F, Streich M (1963) J. Am. Chem. Soc. 85: 2817
74. Hansch C, Fujita T (1964) J. Am. Chem. Soc. 86: 1616
75. Seydel JK, Schaper K-J (1982) Pharmac. Ther. 15: 131

76. Collander R (1951) Acta Chem. Scand. 5: 774
77. Lien EJ (1975) in: Ariëns EJ (ed) Drug Design, vol. V. Academic, New York, p. 81
78. Dearden JC (1985) Environ. Health Persp. 61: 203
79. Esser HO (1986) Pestic. Sci. 17: 265
80. Kubinyi H (1977) J. Med. Chem. 20: 625
81. Dearden JC, Townend MS (1977) In: McFarlane NR (ed) Herbicides and fungicides—factors affecting their activity. The Chemical Society, Burlington House, London, p 135
82. McCarty LS, Hodson PV, Craig GR, Kaiser KLE (1985) Environ. Toxicol. Chem. 4: 595
83. Van Hoogen G, Opperhuizen A (1988) Environ. Toxicol. Chem. 7: 213
84. Schaper K-J (1982) Quant. Struct-Act. Relat. 1: 13
85. Opperhuizen A, Serné P, Van Der Steen JMD (1988) Environ. Sci. Technol. 22: 286
86. Balaz S, Sturdic E, Augustin J (1986) Biophys. Chem. 24: 135
87. Fujita T, Iwasa J, Hansch C (1964) J. Am. Chem. Soc. 86: 5175
88. Hammett LP (1970) Physical Organic Chemistry. McGraw-Hill, New York
89. Taft RW (1956) In: Newman MS (ed) Steric effects in organic chemistry. Wiley, New York
90. Wold S, Dunn WJ, Hellberg S (1985) Environ. Health Persp. 61: 257
91 Topliss JG, Edwards RP (1979) J. Med. Chem. 22: 1238
92 Halfon E (1985) Environ. Sci. Technol. 19: 747
93. Cramer III RD, Bunce JD, Patterson DE (1988) Quant. Struct.-Act Relat. 7: 18
94. Jurs PC (1986) Computer software applications in chemistry. Wiley, New York
95. Miller MM, Ghodbane S, Wasik SP, Tewari YB, Martire DE (1983) J. Chem. Eng. Data 29: 184
96. Woodburn KB, Doucette WJ, Andren AW (1984) Environ. Sci. Technol. 18: 457
97. Brooke DN, Dobbs AJ, Williams N (1986) Ecotoxicol. Environ. Safety 11: 251
98. De Bruijn J, Busser F, Seinen W, Hermens J (1989) Environ. Toxicol. Chem. (in press)
99. Burkhard LP, Kuehl DW, Veith GD (1985) Chemosphere 14: 1551
100. Butte W, Fooken C, Klussman R, Schuller D (1981) J. Chromatogr. 214: 59
101. Carlson RM, Carlson RE, Kopperman HL (1975) J. Chromatogr. 107: 219
102. Eadsforth CV (1986) Pestic. Sci. 17: 311
103. El Tayar N, Van De Waterbeemd H, Testa B (1985) J. Chromatogr. 320: 293
104. El Tayar N, Van De Waterbeemd H, Testa B (1985) J. Chromatogr 320: 305
105. Garst JE, Wilson WC (1984) J. Pharm. Sci. 73: 1616
106. Hammers WE, Meurs GJ, De Ligny CL (1982) J. Chromatogr. 247:1
107. Klein W, Kördel W, Weiss M, Poremski HJ (1988) Chemosphere 17: 361
108. Könemann H, Zelle R, Busser F, Hammers WE (1979) J. Chromatogr. 178: 559
109. McCall JM (1975) J. Med. Chem. 18: 549
110. McDuffie B (1981) Chemosphere 10: 73
111. Opperhuizen A, Sinnige TL, Van Der Steen JMD, Hutzinger O (1987) J. Chromatogr. 388: 51
112. Sarna LP, Hodge PE, Webster GRB (1984) Chemosphere 13: 975
113. Veith GD, Austin NM, Morris RT (1979) Water Res. 13: 43
114. Hansch C, Leo AJ (1979) Substituent constants for correlation analysis in chemistry and biology. Wiley, New York
115. Leo A, Hansch C, Elkins D (1971) Chem. Rev. 71: 525
116. Norrington FE, Hyde RM, Williams SG, Wootton R (1975) J. Med. Chem. 18: 604
117. Rekker RF (1977) The Hydrophobic fragmental constant. Elsevier, Amsterdam
118. Rekker RF, De Kort HM (1979) Eur J. Med. Chem. - Chim. Ther. 14: 479
119. Mayer JM, Van De Waterbeemd H, Testa B (1982) Eur. J. Med. Chem.- Chim Ther. 17: 17
120. Chou JT, Jurs PC (1979) J. Chem. Inf. Comput. Sci. 19: 172
121. Chiou CT, Schmedding DW, Manes M (1982) Environ. Sci. Technol. 16: 4
122. Yalkowsky SH, Valvani SC (1979) J. Chem. Eng. Data 24: 127
123. Valvani SC, Yalkowsky SH (1980) In: Yalkowsky SH, Sinkula AA, Valvani SC (eds) Physical chemical properties of drugs. Dekker, New York, p 201
124. Mackay D, Bobra A, Shiu WY (1980) Chemosphere 9: 701
125. Doucette WJ, Andren AW (1987) Environ. Sci. Technol. 21: 821
126. Pearlman RS (1980) In: Yalkowsky SH, Sinkula AA, Valvani SC (eds) Physical chemical properties of drugs. Dekker, New York, p 321

127. Yang G-Z, Lien EJ, Guo Z-R (1986) Quant. Struct.-Act. Relat. 5: 12
128. Kamlet MJ, Doherty RM, Abboud J-L M, Abraham MH, Taft RW (1986) J. Pharm. Sci. 75: 338
129. Kamlet MJ, Doherty RM, Carr PW, Mackay D, Abraham MH, Taft RW (1988) Environ. Sci. Technol. 22: 503
130. De Bruijn J, Seinen W, Hermens J (1988) (Submitted)
131. Fujita T (1981) Analytica Chimica Acta 133: 667
132. Bijloo GJ, Rekker RF (1984) Quant. Struct.-Act. Relat. 3: 91
133. Bijloo GJ, Rekker RF (1984) Quant. Struct.-Act. Relat. 3: 111
134. Shorter J (1978) In: Chapman NB, Shorter J (eds) Correlation analysis in chemistry: Recent advances. Plenum, New York, p 120
135. Schnaare RL (1971) In: Ariëns EJ (ed) Drug Design, vol 1. Academic, New York, p 406
136. Kier LB (1971) Molecular orbital theory in drug research. Academic, New York
137. Loew GH, Poulsen M, Kirkjian E, Ferrell J, Sudhindra BS, Rebagliat M (1985) Environ. Health Persp. 61: 69
138. Verloop A, Tipker J (1977) In: Keverling Buisman JA (ed) Biological activity and chemical structure. Elsevier, Amsterdam, p 63
139. Fujita T, Iwamura H (1983) Topics in current chemistry 114: 119
140. Kier LB, Hall LH (1976) Molecular connectivity in chemistry and drug research. Academic, New York
141. Kier LB, (1980) in: Yalkowsky SH, Sinkula AA, Valvani SC (eds) Physical chemical properties of drugs. Dekker, New York, p 277
142. Kier LB, Hall LH, (1986) Molecular connectivity in structure—activity analysis. Research Studies, Letchworth
143. Basak SC, Harries DK, Magnuson VR (1984) J. Pharm. Sci. 73: 429
144. Basak SC,Magnuson VR, Niemi GJ, Regal RR, Veith GD (1987) Mathem. Model. 8: 300
145. Basak SC (1987) Med. Sci. Res. 15: 605
146. Basak SC, Magnuson VR, Niemi GJ, Regal RR (1988) Discr. Appl. Math. 19: 17
147. Devillers J, Chambon P, Zakarya D, Chastrette M (1986) Chemosphere 15: 993
148. Sabljic A (1981) Acta Pharm. Jugosl. 31: 189
149. Dearden JC, Solanki P (1988) In: Proceedings of 3rd International Workshop on Quantitative Structure—activity relationships in Environmental Toxicology, 22–26 May, Knoxville, TN
150. Lyman WJ, Reehl WF, Rosenblatt DH (eds)(1982) Handbook of chemical property estimation methods. McGraw-Hill, New York
151. Yalkowsky SH, Sinkula AA, Valvani SC (eds) (1980) Physical chemical properties of drugs. Dekker, New York
152. Anders MW (ed) (1985) Bioactivation of foreign compounds. Academic, Orlando
153. Eto M (1974) Organophosphorous pesticides: organic and biological chemistry, CRC, Cleveland
154. Nelson SD (1985) In: Anders MW (ed) Bioactivation of foreign compounds. Academic, Orlando, p 349
155. Könemann H (1981) Toxicology 19: 209
156. Veith GD, Call DJ, Brooke LT (1983) Can. J. Fish. Aquat. Sci. 40: 743
157. Lipnick RL, Dunn WJ (1983) In: Dearden, JC (ed) Quantitative approaches to drug design. Elsevier, Amsterdam, p 265
158. Koch R (1982) Chemosphere 11: 511
159. Lipnick RL, Watson KR, Strausz AK (1987) Xenobiotica 17: 1011
160. LeBlanc GA (1984) In: Kaiser KLE (ed) QSAR in environmental toxicology. Reidel, Dordrecht, p 235
161. Abernethy S, Bobra AM, Shiu WY, Wells PG, Mackay D (1986) Aquat. Toxicol. 8: 163
162. Bobra AM, Shiu WY, Mackay D (1983) Chemosphere 12: 1211
163. Hermens J, Canton H, Janssen P, de Jong R (1984) Aquat. Toxicol. 5: 143
164. Calamari D, Galassi S, Setti F, Vighi M (1983) Chemosphere 12: 253
165. Rogerson A, Shiu WY, Huang GL, Mackay D, Berger J (1983) Aquat. Toxicol. 3: 215
166. Hermens J, Broekhuyzen E, Canton H, Wegman R (1985) Aquat. Toxicol. 6: 209
167. Wong PTS, Chau YK, Rhamey JS, Docker M (1984) Chemosphere 13: 991
168. Ribo JM, Kaiser KLE (1983) Chemoshpere 12: 1421

169. Hermens J, Busser F, Leeuwangh P, Musch A (1985) Ecotoxicol. Environ. Safety 9: 17
170. Kamlet MJ, Doherty RM, Veith GD, Taft RW, Abraham MH (1986) Environ. Sci. Technol. 20: 690
171. Call DJ, Brooke LT, Knuth ML, Poirier SH, Hoglund MD (1985) Environ. Toxicol. Chem. 4: 335
172. Slooff W, Canton JH, Hermens JLM (1983) Aquat. Toxicol. 4: 113
173. Albert A (1979) Selective toxicity. Chapman and Hall, London
174. Lipnick RL (1986) Trends in Pharmacol. Sci. 7: 161
175. Ferguson J (1939) Proc. Roy. Soc. B 127: 387
176. Miller KW, Paton WDM, Smith RA, Smith EB (1973) Molec. Pharmacol. 9: 131
177. Mullins LJ (1954) Chem. Rev. 54: 289
178. Franks NP, Lieb WR (1986) Arch. Toxicol. Suppl. 9: 27
179. Hansch C, Glave WR (1971) Mol. Pharmacol. 7: 337
180. Lipnick RL, Pritzker CS, Bentley DL (1987) In: Hadzi D, Jerman-Blazic B (eds) QSAR in drug design and toxicology. Elsevier, Amsterdam, p 301
181. Roberts DW (1986) Chem-Biol. Interactions 57: 325
182. Kopperman HL, Carlson RM, Caple R (1974) Chem-Biol. Interactions 9: 245
183. Könemann H, Musch A (1981) Toxicology 19: 223
184. Veith GD, Broderius SJ (1987) In: Kaiser KLE (ed) QSAR in environmental toxicology-II. Kaiser KLE (ed) Reidel, Dordrecht, p 385
185. Schultz TW, Holcombe GW, Phipps GL (1986) Ecotoxicol. Environ. Safety 12: 146
186. Zitko V, McLeese DW, Carson WG, Welch HE (1976) Bull. Environ. Contam. Toxicol. 16: 508
187. Sabljic A (1987) In: Kaiser KLE (ed) QSAR in Environmental Toxicology-II Reidel, Dordrecht, p 309
188. Benoit-Guyod J-L, Andre C, Taillandier G, Rochat J, Boucherle A (1984) Ecotoxicol. Environ. Safety 8: 227
189. Lipnick RL, Bickings CK, Johnson DE, Eastmond DA (1985) In: Bahner RC, Hansen DJ (eds) Aquatic toxicology and hazard assessment (eighth symposium). American Society for Testing and Materials, Philadelphia, p 153
190. Durkin PR (1978) Tappy. Environ. Conf. Proc. p 165
191. Devillers J, Chambon P (1986) Bull. Environ. Contam. Toxicol. 37: 599
192. McLeese DW, Zitko V, Peterson MR (1979) Chemosphere 2: 53
193. Van Gestel CAM, Ma W-C (1988) Ecotoxicol. Environ. Safety 15: 289
194. Schultz TW, Cajina-Quezada M (1987) Toxicol. Letters 37: 121
195. Nendza M, Seydel JK (1987) In: Hadzi D, Jerman-Blazic B (eds) QSAR in drug design and toxicology. Elsevier, Amsterdam, p 361
196. Ravanel P, Taillandier G, Tissut M., Benoit-Guyod JL (1985) Ecotoxicol. Environ. Safety 9: 300
197. Hermens J, Leeuwangh P, Musch A (1984) Ecotoxicol. Environ. Safety 8: 388
198. Ribo JM, Kaiser KLE (1984) In: Kaiser KLE (ed) QSAR in environmental toxicology. Reidel, Dordrecht, p 319
199. Roberts DW (1987) In: Kaiser KLE (ed) QSAR in environmental toxicology-II. Reidel, Dordrecht, p 295
200. Newsome LD, Johnson DE, Cannon DJ, Lipnick RL (1987) In: Kaiser KLE (ed) QSAR in environmental toxicology-II. Reidel, Dordrecht, p 231.
201. Kiese M (1974) Methemoglobinemia: a comprehensive treatise. CRC, Cleveland
202. Bailey HC, Spanggord RJ (1983) In: Bishop WE, Cardwell RD, Heidolph BB (eds) Aquatic toxicology and hazard assessment, sixth symposium. American Society for Testing and Materials, Philadelphia, p 98
203. Lipnick RL, Johnson DE, Gilford JH, Bickings CK, Newsome LD (1985) Environ. Toxicol. Chem. 4: 281
204. Ross WCJ (1962) Biological alkylating agents. Butterworths, London
205. Hermens J, Busser F, Leeuwangh P, Musch A (1985) Toxicol. Environ. Chem. 9: 219
206. Deneer JW, Sinnige TL, Seinen W, Hermens JLM Aquat. Toxicol., submitted
207. Deneer JW, Seinen W, Hermens JLM (1988) Aquat. Toxicol. 12: 185
208. Eder E, Neudecker T, Lutz D, Henschler D (1980) Biochem. Pharmac. 29: 993
209. Eder E, Neudecker T, Lutz D, Henschler D (1982) Chem-Biol. Interactions 38: 303
210. Hemminki K, Falck K, Vainio H (1980) Arch. Toxicol. 46: 277

211. Deneer JW (1988) The toxicity of aquatic pollutants: QSARs and mixture toxicity studies. PhD Thesis, University of Utrecht, Utrecht, The Netherlands
212. Veith GD, De Foe D, Knuth M (1984-1985) Drug. Metab. Rev. 15: 1295
213. Kaiser KLE, Palabrica VS, Ribo JM (1987) In: Kaiser KLE (ed) QSAR in environmental toxicology-II. Reidel, Dordrecht, p 153
214. Kaiser KLE (1987) In: Kaiser KLE (ed) QSAR in environmental toxicology-II. Reidel, Dordrecht. p 169
215. Laughlin RB, Johannesen RB, French W, Guard H, Brinckman FE (1985) Environ. Toxicol. Chem. 4: 343
216. Vighi M, Calamari D (1985) Chemosphere 14: 1925
217. Schultz TW (1983) In: Nriagu JO (ed) Aquatic Toxicology. Wiley, New York, p 401
218. Moulton MP, Schultz TW (1986) Chemosphere 15: 59
219. Schultz TW, Moulton BA (1985) Environ. Toxicol. Chem. 4: 353
220. Hermens J, Bruijn J de, Pauly J, Seinen W (1987) In: Kaiser KLE (ed) QSAR in environmental toxicology-II. Reidel, Dordrecht, p 135
221. Govers H, Ruepert C, Aiking H (1984) Chemosphere 13: 227
222. Bengtsson B-E, Renberg L, Tarkpea M (1984) Chemosphere 13: 613
223. Basak SC, Gieschen DP, Magnuson VR (1984) Environ. Toxicol. Chem. 3: 191
224. Purdy R (1987) In: Kaiser KLE (ed) QSAR in environmental toxicology-II. Reidel, Dordrecht, p 271
225. Passino DRM, Smith SB (1987) In: Kaiser KLE (ed) QSAR in environmental toxicology-II. Reidel, Dordrecht, p 261
226. Gombar VK (1987) In: Kaiser KLE (ed) QSAR in environmental toxicology-II. Reidel, Dordrecht, p 125
227. Foster GF, Tullis RE (1984) Aquat. Toxicol. 5: 245
228. Schultz TW, Moulton MP (1985) Bull. Environ. Contam. Toxicol. 34: 1
229. Hall LH, Kier LB, Phipps G (1984) Environ. Toxicol. Chem. 3: 355
230. Hall LH, Kier LB (1986) Environ. Toxicol. Chem. 5: 333
231. Van Leeuwen CJ, Maas-Diepeveen JL, Niebeek G, Vergouw WHA, Griffioen PS, Luijken MW (1985) Aquat. Toxicol. 7: 145
232. Van Leeuwen CJ, Moberts F, Niebeek G (1985) Aquat. Toxicol. 7: 165
233. Govers H, Ruepert C, Stevens T, Van Leeuwen CJ (1986) Chemosphere 15: 383
234. Vighi M, Calamari D (1987) Chemosphere 16: 1043
235. Zaroogian G, Heltshe JF, Johnson M (1985) Aquat. Toxicol. 6: 251
236. Yoshioka Y, Ose Y, Sato T (1986) Ecotoxicol Environ. Safety 12: 15
237. Yoshioka Y, Mizuno T, Ose Y, Sato T (1986) Chemosphere 15: 195
238. Komatsu K, Hirono S, Moriguchi I (1985) Chem. Pharm. Bull, 33: 4081
239. Klein AW, Klein W, Kördel W, Weiss M (1988) Environ. Toxicol. Chem. 7: 455
240. Di Carlo F, Bickart P, Auer CM (1985) In: Tichy M (ed) QSAR in toxicology and xenobiochemistry. Elsevier, Amsterdam, p 433
241. Rulis AM, Hattan DG, Morgenroth VH (1984) Regul. Toxicol. Pharmacol. 4: 37
242. Rulis AM, Hattan DG, (1984) Regul. Toxicol. Pharmacol. 5: 152
243. Abernethy S, Mackay D, (1987) In: Kaiser KLE (ed) QSAR in environmental toxicology II. Reidel, Dordrecht, p 1
244. Abernethy SG, Mackay D, McCarty LS (1988) Environ. Toxicol. Chem. 7: 469
245. Hodson PV, Dixon DG, Kaiser KLE (1984) Environ. Toxicol. Chem. 3: 243
246. Hodson PV, Dixon DG, Kaiser KLE (1988) Environ. Toxicol. Chem. 7: 443
247. McKim JM, Schmieder PK, Carlson RW, Hunt EP, Niemi GJ (1987) Environ. Toxicol. Chem. 6: 295
248. McKim JM, Schmieder PK, Niemi GJ, Carlson RW, Henry TR (1987) Environ. Toxicol. Chem. 6: 313
249. McKim JM, Bradbury SP, Niemi GJ (1987) Environ. Health Persp. 71: 171
250. Bradbury SP, McKim JM, Coats JR (1987) Pest. Biochem. Physiol. 27: 275
251. Drummond RA, Russom CL, Geiger DL, DeFoe DL (1986) In: Poston TM, Purdy R (eds) Aquatic toxicology and environmental fate, Ninth Volume. American Society for Testing and Materials, Philadelphia, p 415

252. Hunter RS, Culver FD, Hill JR, Fitzgerald A (1986) QSAR system user manual. Institute for Biological and Chemical Process Analysis. Montana State University, Bozeman, Montana
253. Crowley M, Sassevile J-L, Couture P (1987) In: Kaiser KLE (ed) QSAR in environmental toxicology II. Reidel, Dordrecht, p 61
254. Russo RC, Pilli A (1984) AQUIRE: Aquatic information retrieval toxicity data base: Project description, guidelines, and procedures. Duluth, EPA report-600/8-84-021
255. Anderson E, Veith GD, Weininger D (1987) SMILES: A line notation and computerized interpreter for chemical structures. EPA Environmental Research Brief, Duluth, EPA/600/M-87/021
256. Health Design, Inc. (1987) Topkat: Technical Brochure. Available from Health Design, Inc., 183 East Main Street, Rochester, NY 14604, USA
257. Plackett RL, Hewlett PS (1952) J. Roy. Stat. Soc. B 14: 141
258. Anderson PD, Weber LJ (1975) In: Proceedings international conference on heavy metals in the environment, 27-31 October, Toronto, p 933
259. Könemann H (1981) Toxicology 19: 229
260. Broderius S, Kahl M (1985) Aquat. Toxicol. 6: 307
261. De Wolf W, Canton JH, Deneer JW, Wegman RCC, Hermens JLM (1988) Aquat. Toxicol. 12: 39
262. Hermens J, Leeuwangh P, Musch A (1985) Ecotoxicol. Environ. Safety 9: 321.
263. Deneer JW, Sinnige TL, Seinen W, Hermens JLM (1988) Aquat. Toxicol. 12: 33
264. Hermens J, Canton H, Steyger N, Wegman R (1984) Aquat. Toxicol. 5: 315
265. Deneer JW, Seinen W, Hermens JLM (1988) Ecotoxicol. Environ. Safety: 15: 72
266. Hermens J, Leeuwangh P (1982) Ecotoxicol. Environ. Safety 6: 302
267. EIFAC (1987) Revised report on combined effects on freshwater fish and other aquatic life of mixtures of toxicants in water. EIFAC technical paper no 37, FAO, Rome.

Biotransformation of Organic Chemicals by Fish: Enzyme Activities and Reactions

Dick T. H. M. Sijm[1] and Antoon Opperhuizen[2]

[1]Laboratory of Environmental and Toxicological Chemistry, University of Amsterdam, Nieuwe Achtegracht 166 NL-1018 WV Amsterdam, The Netherlands
[2]Department of Basic Veterinary Sciences, Environmental Toxicology Section, University of Utrecht, PO Box 80176, NL-3508 TD Utrecht, The Netherlands

---

*to whom correspondence may be addressed

Summary

Among biotransformation reactions the oxidation reactions and induction and inhibition of these reactions are the most studied. Many organic (aromatic) compounds can be oxidized in fish by the cytochrome P-450 dependent monooxygenases. Enzymes catalyzing these reactions can be induced under environmental conditions either by "natural" or by anthropogenic compounds.

Monooxygenase activities vary up to 600-fold within the fishes and average activities are about 3–36% of average mammalian activities. In general conjugating enzyme activities of fish are also lower than in mammals. Activities of these conjugating enzymes can vary enormously and are very dependent on the substrates which have been used to measure these activities. If compared to oxidation reactions, much less is known about the two other types of phase I reactions: reduction and hydrolytic reactions. Of the phase II reactions the most extensively studied reactions are glucuronide, glutathione and sulphate conjugation.

Biotransformation reactions are most abundantly studied in the liver. In general liver has highest activities per gram tissue and xenobiotics will be predominantly biotransformed there, since xenobiotics pass this organ after entering the organism. Other organs, however have high metabolic activities based on these whole organs, and thus can contribute significantly to metabolism of xenobiotic compounds.

Metabolism can be influenced by several factors, such as route of administration, diet, temperature and/or season, hormones, age, sex, species and strain. The importance of these factors on biotransformation however, is still unclear.

Compounds which are metabolized relatively fast are those which can undergo conjugation, e.g. compounds with functional groups such as –COOH, –OH and/or –NH$_2$.

The most extensively studied compounds in metabolism studies in fish are aldrin, benzo(a)pyrene, carbaryl, chlorophenols, DDT and structurally related compounds, naphthalene, parathion, phenols and some PCBs.

It may be clear from the data presented here, that fish are able to metabolize a number of xenobiotic compounds and that biotransformation can influence bioaccumulation or detoxifying or activating processes significantly. Hitherto, however only a limited number of degradation pathways have been elucidated, while very little is known about the bioaccumulation and toxicology of the metabolites which are formed. Hence a lot of research is required to increase the understanding of the role which metabolism plays in the bioaccumulation and toxicity of organic compounds and their biotransformation products.

Introduction

Until 1962 it was thought that fish did not metabolize xenobiotic compounds and that metabolism of these compounds was restricted to evolutionary higher organisms such as birds and mammals [1]. This was most likely due to insufficient methods or insensitive equipment.

More recently, however, many investigations have demonstrated that fish are able to metabolize many different classes of xenobiotics and some of the enzymes catalyzing these reactions have been identified and characterized. Amongst these the cytochrome P-450 dependent enzymes have been studied most often.

To distinguish the metabolism of xenobiotics from that of more endogenous compounds such as amino acids, carbohydrates and lipids, Lech and Vodicnik defined biotransformation as the biologically catalyzed conversion from one xenobiotic into another, which is catalyzed by enzymes [2]. A metabolite which is the product of a biotransformation reaction has other physical and chemical properties than its parent compound. Therefore its biotic fate will be different

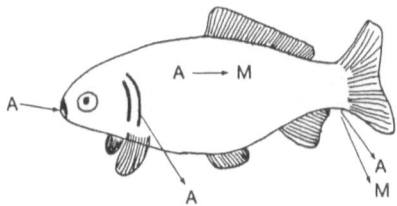

Fig. 1. Schematic representation of the accumulation of compounds in fish
(A: Parent compound, B: metabolite)

within an organism with respect to tissue distribution, persistence, and route and
rate of excretion. In Fig. 1 the uptake, elimination and biotransformation routes are
schematically represented. Furthermore, with respect to toxicity the metabolite
may be either less or more toxic than its parent compound due to detoxification or
activation reactions, respectively.

Therefore, biotransformation is of importance, both to bioaccumulation and to
detoxifying and activating processes. For instance, bioaccumulation can be decreased by altering a compound to a more excretable derivative. For instance,
pentachlorophenol is excreted from goldfish predominantly as a glucuronide
conjugate [3], which keeps pentachlorophenol from accumulating significantly in
goldfish.

Biotransformation which results in a compound which is more toxic than the
parent compound is exemplified by the transformation of fenitrothion by *Tilapia
mossambica*, to fenitrooxon, which is much more toxic than fenitrothion [4]. On the
other hand, an example of detoxification is transformation of 3-trifluoromethyl-4-
nitrophenol (TFM) by rainbow trout [5]. Salicylamide, an inhibitor of the
glucuronide formation, increased the acute toxicity of TFM to rainbow trout, which
indicates that TFM would be detoxified by glucuronidation by rainbow trout.

In the present paper biotransformation reactions in fish will be reviewed in
relation to the chemical structure of xenobiotics. The biotransformation reactions
of a few substances such as DDT and benzo(a)pyrene have been treated in many
articles. In the present paper biotransformation of these and other substances will
be reviewed, relative to the extent of biotransformation and products formed.
Emphasis will be on both the influence of chemical structure on the degradability of
the compound, and on the ability of various fish species to transform different
classes of xenobiotics.

Because routes of excretion (urine, bile, faeces, gill) and the induction and
inhibition of monooxygenases have been reviewed excellently by Lech and Vodicnik [2], Lech et al. [6] and Bend and James [7] these topics will not be discussed.

Types of Biotransformation Reactions in Fish

In general two types of biotransformation reactions are observed in fish, these have
been classified as phase I and phase II reactions. Phase I reactions are nonsynthetic

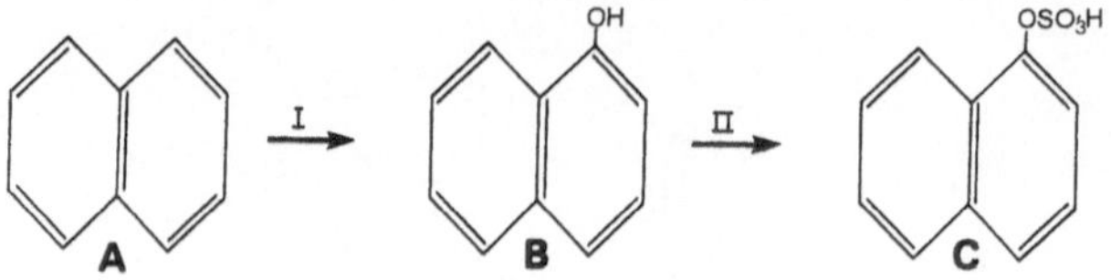

Fig. 2. Biotransformation of naphthalene in starry flounder (9). I: Phase I reaction, catalyzed by a monooxygenase; II: Phase II reaction, catalyzed by a sulphatase; A: naphthalene; B: 1-hydroxy-naphthalene; C: naphthalene-1-sulphate

Table 1. Relative enzyme activity in various tissues of rainbow trout [10]

Tissue	Relative tissue distribution[a]	Aniline hydroxylation activity*	Phenacetin O-dealkylation activity**
Liver	3%	100%	100%
Blood		19%	8%
Muscle	38%	5%	11%
Anterior kidney		40%	6%
Posterior kidney		45%	3%
Heart		7%	13%
Brain		0%	11%
Testes		0%	2%
Spleen		0%	14%
Gill		—	16%

*   = liver activity is 42 nmoles product formed per gram tissue per hour, which is
     arbitrarily set to 100%
** = liver activity is 318 nmoles product formed per gram tissue per hour, which is
     arbitrarily set to 100%
 a = tissue weight/body weight × 100 (data from [11])

reactions such as oxidation, reduction or hydrolysis. Phase II reactions are synthetic reactions, which are referred to as conjugation (Fig. 2, ref. [8, 9]). In these reactions xenobiotics are intermediate compounds and linked to other compounds such as sulphate or glucuronic·acid.

Enzyme activities are not equally distributed in the tissues of rainbow trout (Table 1). For example the hydroxylation of aniline and the O-dealkylation of phenacetin were measured in various tissues [10]. The relative tissue distribution data (tissue weight/body weight × 100%) from Kleeman et al. [11] multiplied by the tissue activity reveals that although activity in liver is the highest, there is an important contribution to hydroxylation or O-dealkylation by muscle which is 63% and 133%, of the total oxidation in liver, respectively.

Although the data of Kleeman et al. [11] may not be representative for all fish species, it is clear that enzyme activities in muscle and perhaps other organs should not be neglected.

For a range of biotransformation reactions, the predominant organs where the reactions take place, together with enzymes involved, are summarized in Table 2, which is constructed from the review of Lech and Vodicnik [2].

Table 2. Types and distribution of enzymatic reactions (constructed from Lech and Vodicnik [2])

Type of reaction	Place of reaction	Enzymes involved	Types of chemicals
PHASE I			
Hydrolysis	Liver, kidney, plasma	Esterase, amidase, epoxidehydrase	Esters, epoxides, amides
Reduction	Microsomal fraction of liver	Nitro, azo, keto, halo reductase	Halogenated organic compounds, ketones, nitro and azo compounds
Oxidation	Mitochondriae (liver)cytosol (liver)microsomes	Monoamineoxidase, alcoholdehydrogenase, monooxygenase	Many organic compounds
PHASE II			
Glucuronic acid conjugation	(Liver)cytosol	Dehydrogenase	Phenols, alcohols, arylamines
		Glucuronyltransferase	Carboxylic acids
Glutathione conjugation	(Liver)cytosol	Transferase, peptidase, acetylase	Aromatic ring or halide compounds
Sulphate conjugation	Cytosol	Sulfotransferase	Hydroxy or amino-compounds
Acetylconjugation		Ligase, transferase, transacylase	Aromatic or aliphatic carboxylic acids, aryl or alkylamine

Phase I Reactions

Oxidative Reactions

Oxidation comprises nine types of reactions (Fig. 3). The oxidative enzymes, located in the smooth endoplasmic reticulum of cells, are formed by a group of hemoproteins called cytochrome P-450 dependent enzymes. These proteins in reduced form, can combine with carbon monoxide to form products that have an absorption peak at 450 nm. These monooxygenases are also named mixed-function oxidases (MFO) because of their ability to oxidize a certain compound and nicotinamide adenine dinucleotide phosphate (NADPH) simultaneously. The monooxygenase activity depends on the reducing cofactor NADPH and molecular oxygen.

The amount of cytochrome P-450 can be determined for instance by the method of Omura and Sato [12] or the improved method of Johannesen and DePierre [13]. The concentrations of cytochrome P-450 varies among species of fish and of some other organisms (Table 3).

Fig. 3(a)

Fig. 3(b)

Fig. 3(c)

Fig. 3(d)

Fig. 3(e)

Fig. 3(f)

Fig. 3(g)

Fig. 3(h)

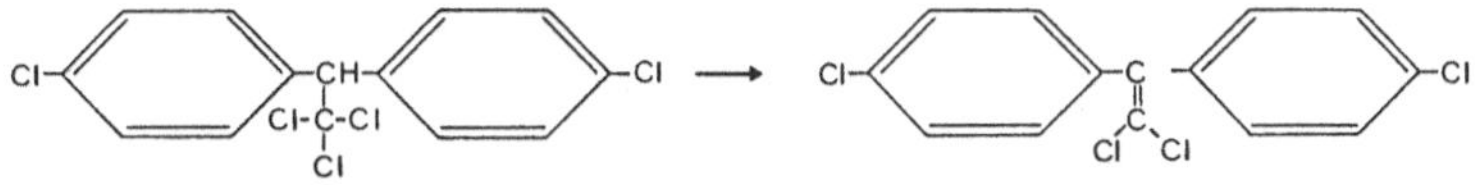

Fig. 3(i)

Fig. 3. Types of oxidation reactions in fish. a) epoxidation: aldrin to dieldrin; b) aromatic hydroxylation: aniline to aminophenol; c) oxidative O-dealkylation: anisole to phenol; d) oxidative N-dealkylation: aminopyrine to 4-aminoantipyrine; e) aliphatic hydroxylation: propoxur to hydroxypropoxur; f) oxidative decarboxylation: DDA to DBP; g) sulphuroxidation: methiochlor to methylsulphinylchlor; h) oxidative desulphonation: parathion to paraoxon; i) dehydrohalogenation: DDT to DDE

Concentrations of cytochrome P-450 range from 0.009 to 1.03 nmoles cytochrome P-450/mg, respectively in microsomal liver protein of killifish and kissing gourami [16]. Concentrations of cytochrome P-450 are generally about 0.30 nmoles cytochrome P-450/mg microsomal protein, which is approximately 30–60% of the concentration present in liver tissue from rats or 20% of the concentration present in rabbit liver. Fish and most crustacea have concentrations of cytochrome P-450 per mg microsomal protein which are higher than those in *Daphnia magna*.

Concentrations of cytochrome P-450 based on microsomal liver protein in fish are similar to these in mammals and birds, such as mouse, guinea pig, cat, dog,

Table 3. Cytochrome P-450 levels in fish* and some other organisms

Species	Cytochrome P-450 (nmoles/mg microsomal liver protein)	Ref.
Fish		
Atlantic stingray	0.43; 0.50	14–17
Black drum	0.14; 0.15	16, 17
Bluegill sunfish	0.28; 0.81	16, 18
Bluntnose stingray	0.30; 0.32	16, 17
Brook trout	0.15–0.56	16, 19, 20
Brown trout	0.20–0.40	6, 15, 16, 21, 22
California killifish	0.35	23
Carp	0.27; 0.38	16, 18
Catfish	0.16	15
Channel catfish	0.15–0.27	6, 24, 25
Cod	0.17	26
Dogfish shark	0.22–0.29	15, 16
Gibel	0.15	16
Golden orfe ide	0.18	18
Guppy	0.47; 0.58	18, 28

Table 3. (continued)

Species	Cytochrome P-450 (nmoles/mg microsomal liver protein)	Ref.
Japanese medaka	0.25	29
Killifish	0.009; 0.37	15, 16
Kissing gourami	1.03	16
Lake trout	0.17	30
Large skate	0.39; 0.41	6, 16
Little skate	0.22–0.32	14–16
Mangrove snapper	0.25	17
Mullet	0.28; 0.47	15, 17
Nurse shark	0.47	17
Perch	0.051	31
Pike	0.14	15
Rainbow trout	0.11–0.42	6, 15, 16, 18, 26, 31–40
Red killifish	0.80	18
Roach	0.089	31
Salmon	0.84	41
Sand dab	0.08	23
Scup	0.35–0.62	6, 15, 16
Sheepshead	0.18–0.44	6, 14, 15, 17, 42
Southern flounder	0.11	17
Southern stingray	0.31	17
Vendace	0.05; 0.074	27, 37
Winter flounder	0.12–0.85	15, 16, 43
Zebrafish	0.30	18
Mammals		
Cat	0.31	36
Dog	0.35	36
Guinea pig	0.53; 0.75	36, 44
Man	0.21–0.49	44, 45
Monkey	0.74	44
Mouse	0.48–0.99	16, 36, 44
Pig	0.38	44
Rabbit	0.67–1.55	7, 16, 36, 44
Rat	0.45–1.04	7, 16, 36, 37, 44
Birds		
Pigeon	0.22	46
Quail	0.23	36
Crustacea		
Barnacle	0.11	15
Blue crab	0.18; 0.29	7, 15, 17
Crayfish	0.31	27
Lobster	0.04–0.13	7
Spiney lobster	0.88; 0.91	7, 15, 17
Daphnia magna	0.03	47

* = from references with cytochrome P-450 induction studies, control values have been taken as the cytochrome-450 level.

pigeon, quail and man. The amounts of microsomal protein per gram of liver in these species, which is about 20 mg microsomal protein/g liver is also similar to that in fish [36]. Birds and mammals, however, have a greater ratio liver weight to body weight (25–70 g/kg) than fish. For instance rainbow trout has a liver to body weight ratio of approximately 10 g/kg [36]. Thus, fish possess approximately 60 nmoles cytochrome P-450 per kg body weight in the liver, whereas birds and mammals have 4 to 35 times this amount.

Several types of MFO (Fig. 4) exist in fish as well as in mammals. Therefore, several types of oxidation reactions can be catalyzed (Table 4). While MFO activity

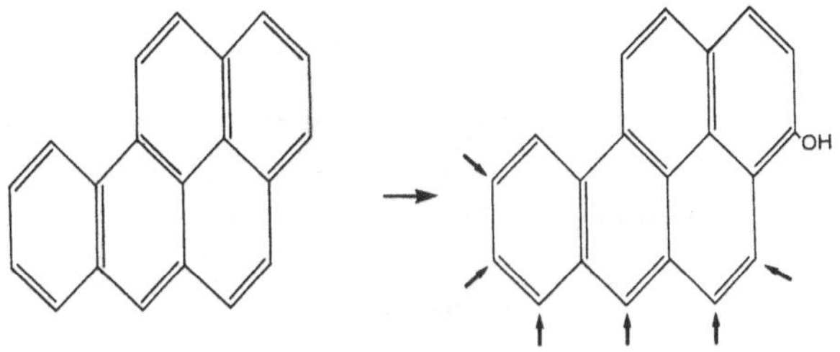

Fig. 4(a)

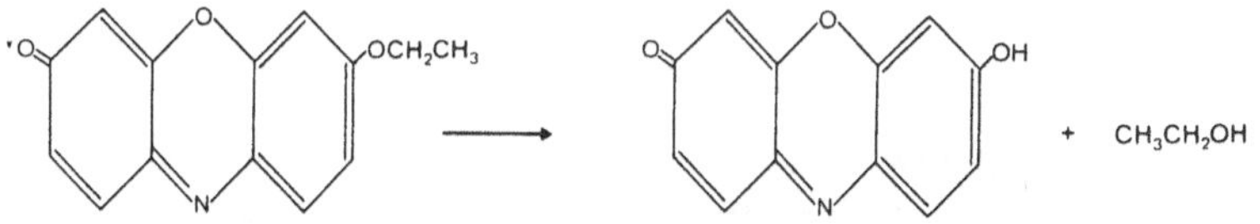

Fig. 4(b)

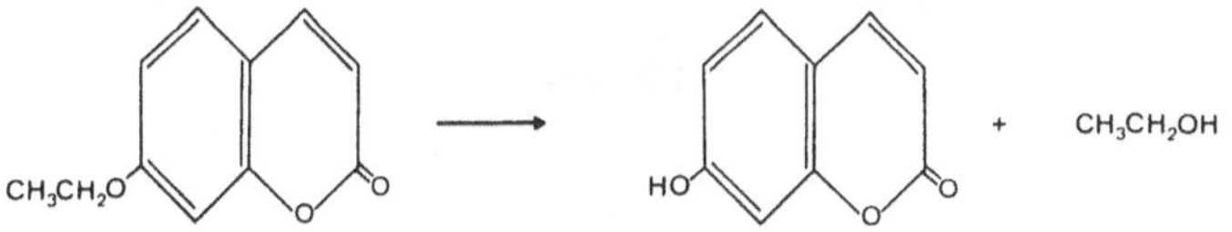

Fig. 4(c)

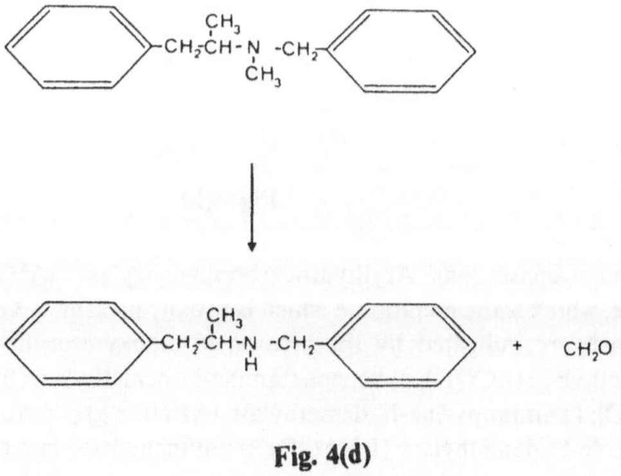

Fig. 4(d)

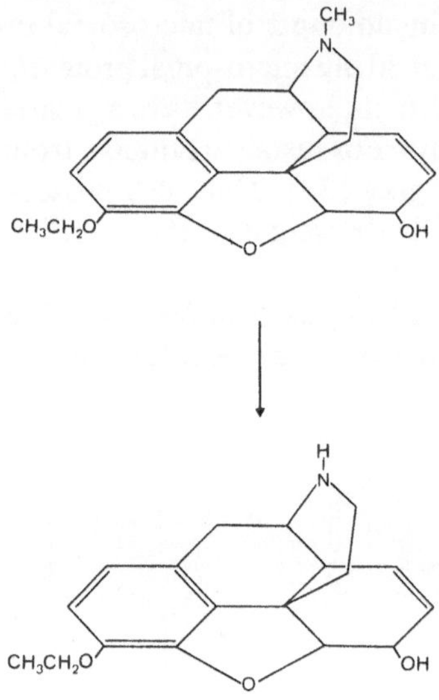

Fig. 4(e)

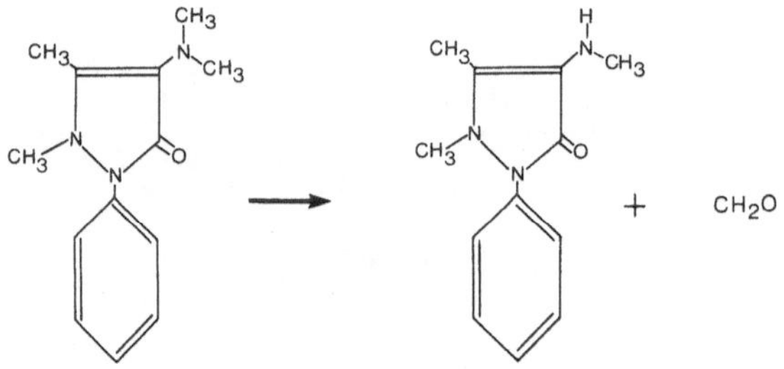

Fig. 4(f)

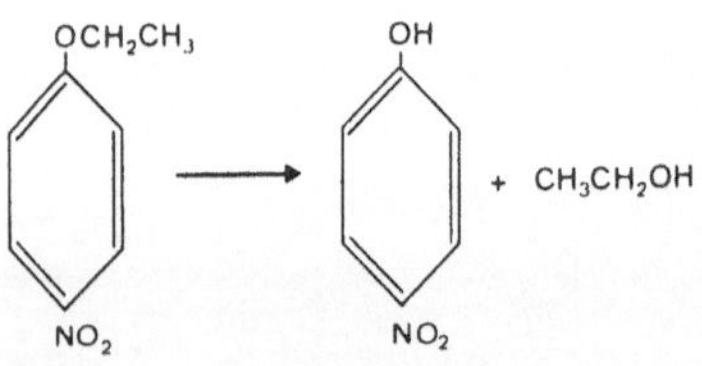

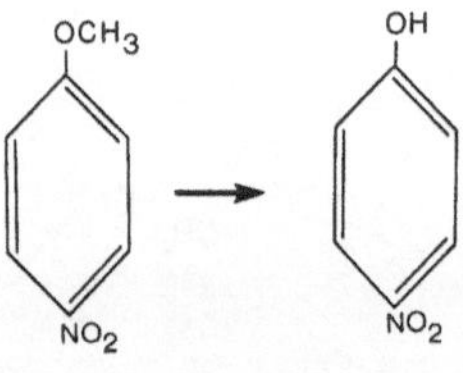

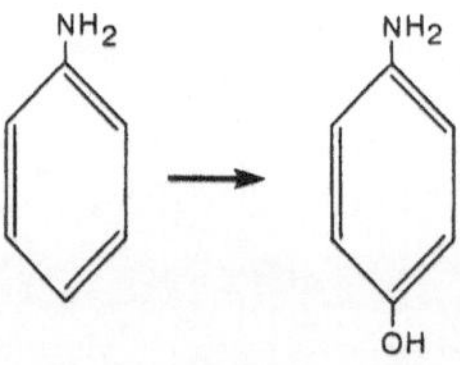

Fig. 4(g)   Fig. 4(h)   Fig. 4(i)

Fig. 4. Types of monooxygenases. a) Arylhydrocarbonhydroxylase (AHH): benzo(a)pyrene to 3-hydroxybenzo(a)pyrene, which is the metabolite which is usually measured. Other oxidation products can also be formed, which are indicated by the arrows; b) ethoxyresorufin-O-deethylase (EROD); c) ethoxycoumarin-O-deethylase (ECOD); d) benzphetamine-N-demethylase (BEND); e) ethylmorphin-e-N-demethylase (EMND); f) aminopyrine-N-demethylase (APND); g) p-nitrophenetole-O-deethylase (PNPOD); h) p-nitroanisole-O-demethylase (PNAOD); i) anilinehydroxylase (AH)

can be measured in other fish tissues they have generally been measured in microsomal liver proteins.

Arylhydrocarbonhydroxylase (AHH) is a monooxygenase which is usually measured by the rate of benzo(a)pyrene hydroxylation [64–69]; sometimes using radiolabelled [^{14}C–] or [^{3}H]-benzo(a)pyrene. The (hydroxylated) compounds formed are measured by fluorescence or by liquid scintillation counting (LSC) with either 3-hydroxybenzo(a)pyrene or a quinine sulphate solution as standard. AHH-activities cannot be compared among studies easily since different methods and different units have been used. Measurement of 3-hydroxybenzo(a)pyrene is probably not a succesful standard, since this metabolite constitutes only 30–40% of all metabolites formed from benzo(a)pyrene [68]. A better standard would be a quinine sulphate solution. Since AHH-activities are often determined by various methods only a selected number of studies can be used to compare these activities among species. In the interspecies study of Gregus et al. [36], rat and rabbit exhibited activities which were 34 and 7 times higher than the activity in rainbow trout.

Ethoxyresorufin-O-dealkylase (EROD) is usually measured according to the method of Burke and Mayer [70]. EROD-activities vary among fish species by about 200-fold (Table 4) and range from 0.0062 nmol/mg.min in winter flounder [63] to 1.18 nmol/mg.min in *Coryphaenoides armatus* from the Hudson Canyon (Table 4, [56]). An average activity for all fishes is about 0.18 nmol/mg.min, which is about 36% of the average mammalian level of 0.50 nmol/mg.min.

O-dealkylation of 7-ethoxycoumarin (ECOD) is also used to measure MFO activity [71]. In this method, the product formed, is umbelliferone, which is highly fluorescent. ECOD-activities vary about 600-fold among species of fish (Table 4) and range from 0.0014 nmol/mg. min in lavaret [58] to 0.89 nmol/mg.min in eel [7]. An average activity for all fishes is about 0.15 nmol/mg.min, which is about 11% of the average mammalian activity of 1.4 nmol/mg.min.

Other MFO-activities are not as frequently measured as AHH, EROD and ECOD. These are benzphetamine-N-demethylase (BEND), ethylmorphine-N-de-methylase (EMND), aminopyrine-N-demethylase (APND), p-nitrophenetole-O-deethylase (PNPOD), p-nitroanisole-O-demethylase (PNAOD) and aniline hydroxylase (AH).

BEND-activities vary about 20-fold among species of fishes (Table 4) and range from 0.12 nmol/mg. min in hagfish [7] to 2.5 nmol/mg. min in mangrove snapper [49]. An average activity for all fishes is about 0.75 nmol/mg. min, which is about 32% of the average mammalian activity of 2.4 nmol/mg. min.

EMND-activities vary about 10-fold among species of fishes (Table 4) and range from 0.07 nmol/mg.min in crayfish [27] to 0.80 nmol/mg.min in rainbow trout [34]. An average activity for all fishes is about 0.41 nmol/mg.min, which is about 21% of the average mammalian activity of 2.0 nmol/mg. min.

Comparison of APND-activities among different studies is difficult to make according to DeWaide [53]. APND-activity is measured as the amount of formaldehyde formed per minute per mg microsomal protein. When aminopyrine is completely demethylated two moles of formaldehyde are formed per mole of aminopyrine. At higher aminopyrine concentrations however this ratio was found

Table 4. Monooxygenase activities in fish (and some other animals) in liver microsomes

Species	Temperature (°C)	AHH* (.../mg. min)	EROD a	ECOD a	BEND a	EMND a	APND a	PNPOD a	PNAOD b	AH b	Ref
Fish											
Argentine	25	604, 702 AU									48
Atlantic stingray	35	0.86 FU	0.05		0.98						49
Atlantic stingray	20	0.77 FU	0.079		0.78						14, 50
Atlantic stingray	u	0.039 nmol	0.079		0.78					0.080	15
Atlantic stingray	35	0.77 FU	0.079		0.78					0.08	51
Black drum	35	1.62 FU	0.096		0.45						49
Black Sea bass	37	4.3 nmol									52
Bluegill sunfish	37, 30, 30	0.17 nmol		0.20		0.65					18
Blue whiting	25	22, 64 AU									48
Bluntnose stingray	35	0.17 FU		0.01	0.23						49
Bream	25	0.054 nmol						0.15		0.075	53
Brook trout	u	0.06 nmol									19
Brook trout	29	0.1 nmol									20
Brown trout	u						0.25	0.2			22
Carp	25–27		0.05								54
Carp	u	0.054								0.062	53
Carp	37, 30, 30	0.13 nmol		0.10		0.12	0.15				18
Channel catfish	30	0.0076 nmol	0.0067	0.0015			0.31				24
Channel catfish	23		0.63	0.69			0.34				25
Cod	25	6–185 AU									48
Cod	20	0.054	0.084	0.027		0.045					26
Coho salmon —female	u	0.0022 nmol									55
Coho salmon —male	u	0.00066 nmol									55
Common dab	25	188, 202 AU									48
Coryphaenoides armatus											
— Carson Canyon	20	0.045 nmol	0.18	0.097							56

— Hudson Canyon

Species	Temp	Activity							Ref
Dace	20	0.41 nmol	1.18	0.16			0.25	0.045	56
Dogfish	u								53
Dogfish shark	25	23.9 AU							48
Dogfish shark	20	0.11 FU		0.03				0.07	14
Dogfish shark	30	0.07 FU		0.08	0.15				7
Dogfish shark	30	0.014 nmol		0.08	0.15				49
Dogfish shark	u							0.07	15
Eel	25	0.21 FU		0.89	0.44		0.35	0.060	53
Eel	30							0.13	7
Golden orfe	25	0.18 nmol		0.10		0.14	0.10	0.17	53
Golden orfe	37, 30, 30			0.27					18
Gulf killifish	u		1.57	0.55					57
Guppy	37, 30, 30	0.58 nmol				1.65			18
Guppy	37, 35, 35	0.805 nmol			4.42	2.11			28
Gurnard	25	230–847 AU							48
Haddock	25	42–2229 AU							48
Hagfish	30	0.18 FU		0.20	0.12			0.12	7
Hake	25	17.7 AU							48
Horse mackerel	25	17–1141 AU							48
Houting	u						0.52	0.025	53
Ide	u						0.17	0.06	53
Jack crevalle	35	2.3 FU		0.49	1.1				51
Killifish	30	4.1 FU							7
Killifish	u	0.28 nmol						0.16	15
King of Norway	30	0.01 FU		0.06	0.16			0.01	7
Lake trout	u	0.31 nmol							15
Lake trout	30	0.8 nmol							30
Large skate	30	0.30 FU		0.47	1.07				49
Large skate	30							0.19	7
Lavaret	18	0.028 nmol		0.0016					58
Lavaret	25	0.026 nmol		0.0014					58
Lemon sole	25	840–4529 AU							48
Ling	25	7.3–89 AU							48
Little skate	20	0.18 FU		0.64	1.07				14, 50

Table 4. (continued)

Species	Temperature (°C)	AHH* (.../mg. min)	EROD a	ECOD a	BEND a	EMND a	APND a	PNPOD* a	PNAOD b	AH b	Ref
Little skate	20	0.17 FU		0.32	1.07						49
Little skate	30									0.56	7
Little skate	u	0.009 nmol		0.32	1.07					0.56	15
Loach	u						0.07			0.035	53
Long rough dab	25	49, 236 AU									48
Mackerel	25	249 AU									48
Mangrove snapper	35	6.64 FU		0.16	0.88–2.5						49
Mangrove snapper	35	6.3 FU		0.17	0.90; 2.5						51
Megrim	25	41–546 AU									48
Mullet	35	2.9; 3.5 FU		0.147	2.78					0.34	51
Mummichog	u			0.49	1.13					0.17	15
Nurse shark	35	1.4 FU		0.107	0.31; 0.41						51
Perch	25			0.029			0.07			0.018	53
Perch	18	0.020 nmol									31
Pigfish	35	5.1 FU		0.06							51
Pike	25						0.25			0.023	53
Pike	u	0.12 nmol									15
Poor cod	25	146 AU									48
Rainbow trout											32
— juvenile	u			0.05							32
— mature male	u			0.09							32
— mature female	u			0.04							32
— 0.5 yr juvenile	u	240 cpm						3.5			59
— 1.5 yr juvenile	u	380 cpm						3.7			59
— mature female	u	450 cpm						2.2			59
— mature male	u	800 cpm						5.9			59
Rainbow trout	25	0.056 nmol	0.020	0.075	0.29 (29°C)						33
Rainbow trout	30		0.054	0.080							54
Rainbow trout	u	0.025 nmol	0.030	0.025	1.10	0.80					34

Rainbow trout	u		0.029				0.25			35
Rainbow trout	20		0.020							60
Rainbow trout	20, 30	0.020 nmol		0.082			0.15			37
Rainbow trout	25			0.0376	0.34				0.047	53
Rainbow trout	18	0.020 nmol		0.044						31
Rainbow trout	28	0.068 nmol				0.178		0.082(a)		38
Rainbow trout	u	0.002–2.28 nmol								15
Rainbow trout	30, 25, 25	0.02 nmol		0.05		0.22				18
Rainbow trout	28.5	0.0012 nmol							0.028	61
Rainbow trout	25	0.022 nmol			0.768					39
Rainbow trout	30		0.136	0.101						39
Rainbow trout	18	0.005–0.060 nmol		0.005–0.090						62
Rainbow trout	20	0.052	0.019	0.064		0.472				26
Red fish	25	146–456 AU								48
Red killifish	37, 30	0.90 nmol		0.60			0.70		0.16	18
Roach	25									53
Roach	18	0.0041 nmol		0.19						31
Roach	30	0.014 nmol								30
Rudd	25						0.63		0.16	53
Saithe	25	24–346 AU								48
Salmon	u						0.018		0.008	53
Scup	30	0.13 FU								7
Scup	u	0.69 nmol								15
Sea bass	35	3.6 FU							0.007	51
Sea lamprey	25						0.17			53
Sheepshead	u	3.5 FU		0.175	0.92					42
Sheepshead	20	3.1 FU		0.19	1.01					14, 50
Sheepshead	35	1.38 FU		0.05	1.10					49
Sheepshead	u	0.16 nmol		0.19	1.01				0.25	15
Sheepshead	35	3.1 FU		0.19	1.01				0.25	51
Sheepshead minnow	u		1.56	0.22						57
Southern flounder	20	0.25 FU		0.005	0.18, 0.32					14, 50
Southern flounder	u	0.012 nmol		0.005	0.18, 0.32				0.27	15
Southern flounder	35	0.25 FU		0.005	0.18; 0.32				0.27	51

Table 4. (continued)

Species	Temperature (°C)	AHH* (../mg. min)	EROD a	ECOD a	BEND a	EMND a	APND a	PNPOD a	PNAOD b	AH b	Ref
Southern stingray	35			0.28						0.014	51
Splake	18	0.091 nmol		0.0101							58
Splake	25	0.114 nmol		0.0106							58
Spur dog	25	72 AU									48
Tench	25						0.10			0.033	53
Thornback ray	25	13–60 AU									53
Thorny skate	30	0.12 FU		0.12	0.45					0.16	7
Torsk	25	138 AU									48
Trout	25	0.01 nmol	0.42		0.2						36
Twait	u					0.1				0.033	53
Vendace											
— female	20, 30	0.0060 nmol		0.0047							37
— male	20, 30	0.0075 nmol		0.0068							37
Vendace	18	0.084 nmol		0.029							31
White bream	25						0.28			0.056	53
Whiting	25	6–878 AU									48
Winter flounder	30	1.3–2.0 FU	0.062–0.237		0.36–0.78						43, 63
Winter flounder	30	2.54 FU			0.59					0.19	49
Winter flounder	u	0.21 nmol		0.32	0.59					0.19	7
Winter flounder	25			0.32							15
Witch	25	7.5–1316 AU									48
Zebra fish	37, 30, 30	0.10 nmol		0.35		0.14					18
Other organisms											
Blue crab	35	0 nmol	0	0.02	0.015					0.016	51
Crayfish	18			0.1		0.07					27
Frog	25						0.20			0.14	53
Hen	42						0.93			1.00	53
Leopard frog	u	0.2 nmol					0.8	0.6			22

Species	Temp									Ref.
Lizard	25					0.55			0.25	53
Mool-handed crab	u					0.01			0.087	53
Pigeon	42					2.63			0.46	53
Quail	37	0.15 nmol	0.05	1.1	0.5					36
Snake	u	0.05 nmol				1.4	0.6			22
Spiny lobster	35	0.03 FU		0.11					0.08	51
Cat	37	0.06 nmol	0.75	0.5	1.0					36
Dog	37	0.04 nmol	0.44	1.0	1.1					36
Guinea pig	37	0.08 nmol	0.30	1.3	1.2					36
Hamster	37					2.33			0.27	53
Man	u	0.18 nmol	0.61	1.5				2.5(a)	0.49	45
Mouse	u					1.83			0.57	53
Mouse	37	0.15 nmol	0.28	1.9	3.4					36
Rabbit	37	5.0 FU	3.5	6.5					0.9	7
Rabbit	37	0.07 nmol	1.1	4.0	1.3					36
Rat	37	0.4 nmol				1.5			0.13	53
Rat	u					10	3.0			22
Rat	20, 30	0.160 nmol	0.049							37
Rat	37	0.34 nmol	0.10	2.2	3.8					36

a   = activity expressed as nmol/mg (microsomal protein).min
b   = activity expressed as nmol/g (liver).min
*   = activity expressed per mg microsomal protein per minute; unities are presented in the table:
AU = arbitrary units, based on the fluorescence of a reference compound
FU = fluorescence units, based on the fluorescence of a reference compound
u   = unknown.

to be higher. No linear relationship is thus expected between aminopyrine concentration and demethylation activity [53]. The APND-activities mentioned in Table 4 vary about 40-fold and range from 0.018 nmol/mg.min in salmon [53] to 0.70 nmol/mg.min in roach [53]. An average activity for all fishes is about 0.25 nmol/mg.min, which is about 6% of the average mammalian activity of 3.9 nmol/mg.min.

PNPOD-activities have only been measured in bream [53] and brown trout [22] with activities of 0.15 and 0.2 nmol/mg.min, respectively. These activities are approximately 6% of the activity in rat, which is 3.0 nmol/mg.min [22].

Several additional PNAOD-activities have been measured. These values, however, are difficult to compare since their units are not the same. The PNAOD-activity in rainbow trout [38] is about 3% of that of man [45].

AH-activities vary about 70-fold among species of fishes (Table 4) and range from 0.008 nmol/g liver.min in scup [53] to 0.56 nmol/g liver.min in little skate [7]. An average activity for all fishes is about 0.13 nmol/g liver.min, which is about 27% of the average mammalian activity of 0.47 nmol/g liver.min.

In summary average MFO-activities for all fishes studied (Table 4), ranged from 0.15–0.75 nmol/mg.min, which are 3–36% of the average mammalian activities. Average EROD-, ECOD-, PNPOD- and AH-activities are all about 0.2 nmol/mg. min for fish. In addition, MFO-activities vary up to 600-fold among fishes. Therefore it is difficult to point out one fish, which is representative for all fishes. Rainbow trout, which is the most extensively studied fish has MFO-activities lower than the average activities of all fishes, while guppy and red killifish have cytochrome P-450 concentrations and MFO-activities higher than the average concentration and activities, respectively. The choice of fish may thus influence the results of metabolism, toxicity or bioaccumulation tests [18].

Although highest MFO-activities (based on mg microsomal protein) are usually measured in the liver, other organs may play a significant role in metabolizing xenobiotics. This since some organs such as muscle have a relatively large size, compared to liver (Table 1). Some extrahepatic MFO-activities have been summarized (Tables 5A and 5B). Especially kidney has MFO-activities comparable with the liver. Microsomal cytochrome P-450 content in the kidney of rainbow trout was approximately 20% of that in the liver. EROD- and ECOD- activities calculated on a per cytochrome P-450 basis in the kidney, however were 10 times higher than in the liver [40].

The influence of the possible entero-hepatic circulation of xenobiotics in fish has not been well studied. The role of gut microflora in biotransformation of propachlor has been studied with rats [72]. The products formed in liver and bile reentered the liver after being transformed by intestinal microflora and/or intestinal enzymes. In goldfish it was observed that phenylglucuronide was hydrolyzed by intestinal mucus and/or flora [73]. After addition of an inhibitor of β-glucuronidase, D-saccharid acid-1,4-lactone, more phenol was released as phenylglucuronide than when this inhibitor was not added. Intestinal mucus and/or flora thus prevented phenylglucuronide, which was synthesized in liver, to be excreted in the water.

Induction and Inhibition of Oxidative Reactions. The study of induction and inhibition of monooxygenases in fish has been excellently reviewed by Lech et al. [6]

and Netter [74] and will thus be discussed here only briefly. More attention will be addressed to the inducers and inhibitors of monooxygenases in fish.

Induction of MFOs is an increase in the amount and/or an increase in the activity of these enzymes. A two phase induction of cytochrome P-450 and PNAOD-activity was found in rainbow trout after pretreatment with Clophen A50 [75]. The first phase includes activation of existing enzymes, the second phase includes *de novo* synthesis of enzymes.

Induction is usually the result of the administration of an inducer, commonly a xenobiotic, to the fish. Inhibition is the opposite of induction. In this case enzymes are blocked in activity possibly due to a strong binding or complex formation of the enzyme with the inhibitors.

Induction and inhibition are very important to biotransformation, since it increases or decreases the rate of transformation of a compound. Sometimes, also other biotransformation products are found after induction or inhibition [76]. If there exists a balance between the production of highly reactive intermediates by the cytochrome P-450 dependent enzymes and the deactivation of these metabolites by hydrolytic and conjugating enzymes, this balance can be disturbed by induction of the MFOs. In this case the hydrolytic and conjugating enzymes cannot cope with the increased production of highly reactive intermediates [30]. Benzo(a)pyrene for instance is metabolized to several products. Absolute higher amounts of metabolites and relatively higher amounts of certain metabolites are formed in mullet [77] and in small estuarine and freshwater fish [57] when pretreated with 3-MC.

Monooxygenases are cytochrome P-450 dependent enzymes. To investigate whether the concentrations of these MFOs are increased or decreased after induction or inhibition one has attempted to isolate and identify these different MFOs with the aid of sodium dodecyl sulphate-polyacrylamide gel electrophorese (SDS-PAGE). Three types of cytochrome P-450 isozymes were isolated from rainbow trout by Williams et al. [78]. One isozyme had an absorption maximum of 449.5 nm and a molecular weight of 54,000 and was able to oxidize aflatoxin B1. Two other isozymes had absorption maxima of 447.0 nm and molecular weights of 58,000 [78]. The former cytochrome P-450 isozyme was believed to be responsible for the oxidation of lauric acid in rainbow trout [79]. Klotz et al. isolated three P-450 isozymes with molecular weights of 52,700, 45,900 and 54,300, respectively from scup [80]. These three MFOs oxidized testosterone at different positions (the 6β-, 2α-and 15α-positions). The monooxygenase with the molecular weight of 54,300 oxidized benzo(a)pyrene, ethoxycoumarin, acetanilide and BNF [80].

Elevated AHH-activities, caused by induction, concomitantly occurred with elevated amounts of a polypeptide with a molecular weight of 57,000 [63, 81]. This induction was shown with 1,2,3,4-dibenzanthracene and BNF in winter flounder [43, 81].

EROD-activity in the liver increased concomitantly with cytochrome P-450 content in channel catfish after BNF treatment [25]. MFO-activities in the kidney however were not elevated after BNF treatment [25].

In rainbow trout a novel protein with a molecular weight of 57,000 appeared on SDS-PAGE in liver microsomes after pretreatment of the fish with BNF [39], isosafrole or TCDD [82]. In cod, Japanese medaka and rainbow trout a novel

Table 5A. ECOD- and AHH-activities in (extra-)hepatic organs

Species	Temperature (°C)	Activity (./mg. min)	Liver	Kidney	Heart	Gills	Intestine	Ref.
Channel catfish	23	ECOD (pmol)	630	130	0.0	0.0	2.2	25
Perch	18	ECOD (pmol)	28.5	—	1.3	2.8	1.0	31
Perch	18	AHH (pmol)	20.1	—	2.0	1.2	4.4	31
Rainbow trout	18	ECOD (pmol)	44.4	27.5	1.7	0.4	2.6	31
Rainbow trout	18	AHH (pmol)	20.3	3.0	0.0	1.4	4.0	31
Roach	18	ECOD (pmol)	192	1.4	0.3	2.3	1.2	31
Roach	18	AHH (pmol)	4.1	2.2	0.1	1.7	0.2	31
Vendace	18	ECOD (pmol)	11.2	6.4	0.3	1.0	0.4	31
Vendace	18	AHH (pmol)	8.4	1.8				31

Table 5B. Monooxygenase activities in (extra-)hepatic organs

Species	Crayfish	Little skate	Little skate	Little skate	Scup	Vendace	Vendace
(Ref)	(27)	(7)	(7)	(7)	(15)	(37)	(37)
Activity (../mg. min)	ECOD[a] pmol	AHH[b] FU	BEND[b] nmol	AH[b] nmol	AHH[c] nmol	ECOD[b] pmol	AHH[d] pmol
Liver	100	0.17	1.25	0.97	0.69	6.8	7.5
Kidney	—	0.03	0.35	0.52	0.22	5.1	3.16
Heart	1	0.02	—	—	0.01	0.1	0.18
Gills	2.6	0.02	0.12	0.10	0.012	0.87	1.42
Intestine	22.9	—	—	—	—	0.43	0.26
Hepatopancreas	2.3	0.03	—	—	—	—	—
Green glands	2.3	—	—	—	—	—	—
Skin	—	—	—	—	—	0.1	0.12
Testis	—	—	—	—	0.059	—	—
Hindgut	—	—	—	—	0.006	—	—
Spleen	—	0.03	<0.01	<0.01	—	—	—
Spiral valve	—	<0.003	<0.01	<0.01	—	—	—
Stomach lining	—	0.02	0.12	0.04	—	—	—

— = not measured
FU = fluorescence units, based on the fluorescence of a reference compound
[a] = assay temperature is 18 °C
[b] = assay temperature is 30 °C
[c] = assay temperature is unknown
[d] = assay temperature is 20 °C

protein appeared with a molecular weight of 58,000 after pretreatment with BNF [29, 40]. Other proteins, which were induced in rainbow trout and were found with the aid of SDS-PAGE, had molecular weights of 45,000, 48,000, 51,000 and 59,500 [39].

Since the cytochrome P-450 concentration in the liver of fish is about 0.3 nmol/mg protein and the molecular weight of cytochrome P-450 is about 50,000, microsomal protein contains about 1.5% (w/w) cytochrome P-450. Microsomal proteins are not entirely functional in the oxidation of organic chemicals. It is therefore unlikely that cytochrome P-450 can be induced infinitely.

In mammals two major types of inducers are known [6]: 3-methylcholanthrene (3-MC) and phenobarbital (PB) type inducers. 3-MC type inducers increase the concentration of cytochrome P-448, which is an isozyme of cytochrome P-450, and increases AHH-activity. PB type inducers increase the concentration of cytochrome P-450. In fish however, PB does not induce any MFO-activity [29, 30, 35, 53, 71, 77, 84], except EMND- and BEND- activities in guppy, however, only after 2 weeks of exposure to PB [28]. 3-MC does induce certain MFO-activities such as AHH- and EROD-activity (Table 6). Some inducers and inhibitors are summarized in Table 7.

An increased ethylmorphine-N-demethylation activity is the result of a PB-type induction [6]. In rainbow trout EMND-activity was not or only slightly induced after treatment of Aroclor 1242 (150 mg/kg), Aroclor 1254 (150 mg/kg), BP-6 (150 mg/kg) or BNF (100 mg/kg) [34]. This is in agreement with the observation that there is no PB type induction in fish. Benzphetamine-N-demethylation activity however, is affected by Aroclor 1254 [34, 42], Aroclor 1242, BP-6 [34], BNF [32] and 3-MC [49]. This enzyme was supposed to be induced by a PB type inducer, but was found to be induced by a 3-MC type inducer in fish.

AHH- and EROD-activities were both induced by BNF and DBA, and were both inhibited by ANF in winter flounder [43]. These activities are not manifestations of the same enzyme, since AHH- and EROD-activities decreased 26% and 74%, respectively in winter flounder in one year [63]. They obviously represent activities of distinct monooxygenases.

Benzo(a)pyrene showed a nonlinear dose-response relationship in induction of AHH- and AH-activities [61]. Doses of 0.3 mg/kg or more increased enzyme activities in rainbow trout [61]. Induction of MFO-activities seems to occur under

Table 6. Ability of some common inducers to increase MO-activities in fish [15]

Active inducers	Inactive inducers
3-MC	PB
Benzanthracene	Dichlorodiphenyltrichloromethane
2,3,7,8 TCDD	Phenylbutazone
Dibenzanthracene	2,2',4,4' tetrachlorobiphenyl
Dimethylbenzanthracene	2,2',4,4',5,5' hexachlorobiphenyl
BNF	2,2',4,4',5,5' hexabromobiphenyl
Benzo(a)pyrene	
3,3',4,4' tetrachlorobiphenyl	
3,3',4,4',5,5' hexachlorobiphenyl	

environmental conditions. Induction of AHH-activity was shown in winter flounder by "natural" products, such as flavonoids, safroles and indoles [43].

Kurelec et al. [90] found induced AHH-activity in Blennius pavo in contaminated
areas. AHH-induction was also observed in these fish after treatment with 170 ppb
Diesel 2 oil. Consequently induction may be caused by oil spills. In Microcosmos
sulcatus however, no induced AHH-activity was found in these contaminated areas,
indicating that there may be differences in sensitivity between different fish [90].

Codfish and sculpin flounder also had elevated MFO-activities when they were
exposed to 0.15–0.60 ppm petroleum in water. Such petroleum levels may be found
in some environmental areas [85]. PCB-concentrations, as low as can be measured
in environmental areas, were sufficient to cause induction of MFO-activities in
rainbow trout and carp (Table 7, [54]), although route of administration in the
experiments done by Melancon and Lech is different than in environmental
exposure.

Spies et al. [91] observed higher MFO-activities in flatfish in more polluted areas,
near a sewage outfall and near a petroleum seep, than in a relatively unpolluted
area.

Carp and bullhead had higher EROD-activities when they were transferred from
dechlorinated and filtered water to Kinnikinnic river water. When returned to the
dechlorinated water the EROD-activity decreased again [92]. DeWaide obtained
similar results with roach which were held in river water or in tap water [53].
AHH-activity in Black Sea bass was induced with benzo(a)pyrene (Table 7).
Benzo(a)pyrene administered together with cadmium (CdCl$_2$), however did not
increase AHH-activity as much as benzo(a)pyrene alone [52].

The route of administration of the chemicals is important in induction or
inhibition studies which was shown in sheepshead (Table 7). When 3-MC was
administered intramuscularily (i.m.) ECOD- and EROD-activities were more
induced than when 3-MC was administered intraperitoneally (i.p.). The reverse,
however was found for AHH-induction: less induction was achieved by 3-MC when
it was administered i.m. than when it was administered i.p. [14].

Very interesting is the effect of tricaine and quinaldine sulphate on AHH-activity.
AHH-activities were decreased by these compounds in brook trout [19]. Since
these compounds are used as narcotic agents, care must be taken in interpreting
AHH-activities from these anaesthesized fish. Other MFO-activities, such as
ECOD-, BEND- and APND-activities seem not to be influenced by tricaine in
rainbow trout [35].

In addition to some potent inducers, some other familiar MFO-inducers in
mammalian studies, which are not active inducers in fish, are listed in Table 6.

Hydrolytic Reactions

Little is known about hydrolytic reactions in fish compared with oxidative
reactions. Hydrolytic reactions are catalyzed by esterases (Fig. 5), amidases and
epoxidehydrases (Fig. 6) and occur in several organs (Table 2 [2, 7, 93]). These
enzymes catalyze hydrolysis of amides, phosphates, carbamate esters and alkene or

Table 7. Inducers and inhibitors of fish monooxygenase activities

Species	Inducers, Inhibitors	Treatment*	Liver enzymes	Activity (ratio treated/control)	Ref.
Atlantic stingray	3-MC	20 mg/kg, day18	AHH, ECOD	1.1; 0.9	14
Black Sea bass	BP	0.075 mg/kg, ip, 3 days	AHH	0.8	52
	BP	0.75 mg/kg, ip, 3days	AHH	2.0	52
	BP	7.5 mg/kg, ip, 3days	AHH	1.6	52
Bluegill sunfish	Metyrapone	0.5 mM	AHH	1.25	18
	ANF	0.5 mM	AHH	1.22	18
	Metyrapone	0.1 mM	EMND	0.40	18
	ANF	0.1 mM	EMND	0.54	18
Brook trout	Tricaine	112 mg/L, 5 min	AHH	0.22	19
	Quinaldine sulphate	30 mg/L, 5 min	AHH	0.54	19
Carp	Aroclor 1254	0.3 mg/kg, ip, 5days	EROD	2	54
	3,3',4,4'PCB	0.2 mg/kg, ip, 5days	EROD	20	54
Channel catfish	Aroclor 1254	1 mg/kg, day4	EROD, ECOD, AHH	1.8; 1.0; 1.2	24
	Aroclor 1254	10 mg/kg, day4	EROD, ECOD, AHH	5.3; 2.2; 3.3	24
	Aroclor 1254	100 mg/kg, day4	EROD, ECOD, AHH	15; 3.3; 8.7	24
Channel catfish	BNF	50 mg/kg, ip, day2	EROD	2.8	25
	BNF	50 mg/kg, ip, day5	EROD	2.3	
	BNF	50 mg/kg, ip, day12	EROD	1.2	
Channel catfish	BP	0.1 mg/kg, 15.5°C	AHH	2.2–8.4	84
	BP	0.1 mg/kg, 19°C	AHH	1.8–5.5	
	BP	0.1 mg/kg, 26.5°C	AHH	2.4–4.6	
Cod	BNF	100 mg/kg	ECOD, EROD, EMND, AHH	4.9; 13; 1.6; 12	26
Codfish	petroleum	0.3–0.6 ppm in water	AHH	3.8	85
Coryphaenoides armatus	ANF	0.0001 M	AHH	0.04–0.38	56
Dogfish shark	DBA	10 mg/kg, day7	AHH, ECOD	5.6; 1.3	14
Fathead minnow	3-MC	38 mg/kg, ip, day5	EROD, ECOD	3; 2	57
Gulf killifish	3-MC	20 mg/kg, ip, day5	EROD, ECOD, AHH	10; 9; 8	57
Guppy	metyrapone	0.5 mM	AHH	0.79	18
	ANF	0.5 mM	AHH	1.40	18
	metyrapone	0.1 mM	EMND	0.35	18
	ANF	0.1 mM	EMND	0.65	18

Species	Inducer	Dose/regimen	Enzymes	Activity	Ref.
Guppy	PB	50 mg/1, 2 weeks	AHH, EMND, BEND	0.68; 1.6; 1.4	28
Japanese medaka	BNF	10% in diet, ip, day3	EROD	1.9	29
Japanese medaka	3-MC	12 mg/kg, ip, day3	EROD, ECOD, AHH	4; 1.3; 2	57
Killifish eggs	Aroclor 1254	20 ppb, 12days	AHH	nd/nd	86
	Aroclor 1254	100 ppb, 12days	AHH	30/nd	86
	Aroclor 1254	200 ppb, 12days	AHH	62/nd	86
	No 2 Fuel Oil	1000 ppb, 12days	AHH	39/nd	86
King cobra guppy	3-MC	9 mg/kg, ip, day3	EROD, AHH	8; 2.5	57
Little skate	DBA	10 mg/kg, day10	AHH, ECOD	44; 3.6	14
Rainbow trout	Aroclor 1242	150 mg/kg, ip	BEND, AHH, ECOD, EROD	1.3; 11; 8.1; 14	34
	Aroclor 1254	150 mg/kg, ip	BEND, AHH, ECOD, EROD	0.5; 13; 5.1; 15	34
	BP-6	150 mg/kg, ip,	BEND, AHH, ECOD, EROD	1.1; 7.0; 5.5; 16	34
	BNF	100 mg/kg, ip,	AHH, ECOD, EROD	41; 12; 45	34
Rainbow trout	Tricaine	50 mg/L, 24h	EROD, ECOD, BEND, APND	0.4; 0.7; 1.0; 0.9	35
	Tricaine+BNF	50 mg/L + 100 mg/kg	EROD, ECOD	21; 6.8	35
	BNF	100 mg/kg, ip, 24h	EROD, ECOD	23; 8.3	35
Rainbow trout	Clophen A50	500 mg/kg, ip, day7	EROD	44	60
	BNF	100 mg/kg, ip, day3	EROD	189	60
Rainbow trout	Clophen A50	1000 mg/kg, ip, 2weeks	AHH, ECOD	4.4; 22	87
	Estradiol 17β	325 mg/kg, ip, 2weeks	AHH, ECOD	0.6; 0.6	87
	Clophen A50+ Estradiol 17β	1000 mg/kg, ip + 325 mg/kg, 2weeks	AHH, ECOD	2.4; 1.1	87
	Testosterone	130 mg/kg, ip, 2weeks	ECOD	0.85	87
	Clophen A50+ Testosterone	1000 mg/kg, ip + 130 mg/kg, 2weeks	ECOD	2.5	87
Rainbow trout, juvenile (40–100 g)	Estradiol 17β	25 mg, 2weeks	AHH, PNAOD, EMND	1.2; 0.76; 0.60	38
	Estradiol 17β	50 mg, 2weeks	AHH, PNAOD, EMND	0.56; 0.56; 0.61	38
	Estradiol 17β	75 mg, 2weeks	AHH, PNAOD, EMND	0.59; 0.58; 0.66	38
Rainbow trout	BNF	500 ppm in diet, 3weeks	ECOD	6.3	33
	BNF	100 mg/kg, ip, 3weeks	ECOD	7.1	33
	BNF	500 ppm in diet, 1week	AHH	16	33
	BNF	100 mg/kg, 1week	AHH	11	33
	BNF	500 ppm in diet, 3weeks	EROD	83	33
	BNF	500 ppm in diet, 3weeks	BEND	0.65	33

Table 7. (continued)

Species	Inducers, Inhibitors	Treatment*	Liver enzymes	Activity (ratio treated/control)	Ref.
Rainbow trout	Aroclor 1254	0.2 mg/kg, ip, 5days	EROD, ECOD	2; 2	54
	Aroclor 1254	0.5 mg/kg, ip, 5days	EROD, ECOD	7; 3	54
	3,3',4,4'PCB	0.2 mg/kg, ip, 5days	EROD, ECOD	52; 7	54
	3,3',4,4',5 PCB	0.6 mg/kg, ip, 5days	EROD, ECOD	25; 7	54
Rainbow trout	BP	30 mg/kg, ip, day3	AHH, AH	17; 2.3	61
	Aroclor 1254	30 mg/kg, ip, day3	AHH, AH	5; 2.9	61
	Phenanthrene	30 mg/kg, ip, day3	AHH, AH	1; 2.0	61
	Chrysene	30 mg/kg, ip, day3	AHH, AH	3.2; 2.0	61
Rainbow trout	Aroclor 1242	150 mg/kg, ip, day3	EMND, AHH, EROD, ECOD	1.0; 10; 14; 2.8	39
	BNF	100 mg/kg, ip, day3	EMND, AHH, EROD, ECOD	0.9; 41; 45; 12	39
Rainbow trout	Metyrapone	0.5 mM	AHH	0.83	18
	ANF	0.5 mM	AHH	0.19	18
	Metyrapone	0.1 mM	EMND	0.20	18
	ANF	0.1 mM	EMND	0.33	18
Rainbow trout:					
– juvenile	CdCl$_2$	0.5 mg/kg, ip, 4days	ECOD	0.90	32
– mature male	CdCl$_2$	0.5 mg/kg, ip, 4days	ECOD	0.70	32
– mature female	CdCl$_2$	0.5 mg/kg, ip, 4days	ECOD	0.70	32
Rainbow trout:					
–0.5 yr juvenile	Clophen A50	500 mg/kg, day4	PNAOD, AHH	4.3; 8.5	59
	3-MC	20 mg/kg, day4	PNAOD, AHH	13.4; 36	59
	PCN	40 mg/kg, day4	PNAOD, AHH	0.7; 1.0	59
–1.5 yr juvenile	Clophen A50	500 mg/kg, day4	PNAOD, AHH	12; 23	59
	3-MC	20 mg/kg, day4	PNAOD, AHH	22; 27	59
	PCN	40 mg/kg, day4	PNAOD, AHH	0.5; 0.5	59
–1.5-2yr mature male	Clophen A50	500 mg/kg, day4	PNAOD, AHH	5.4; 4.9	59
	3-MC	20 mg/kg, day4	PNAOD, AHH	16; 8.8	59
	PCN	40 mg/kg, day4	PNAOD, AHH	1.1; 0.4	59
–1.5-2 yr mature female	Clophen A50	500 mg/kg, day4	PNAOD, AHH	9.5; 16	59
	3-MC	20 mg/kg, day4	PNAOD, AHH	37; 18	59
	PCN	40 mg/kg, day4	PNAOD, AHH	9.5; 0.8	59

Species	Compound	Dose	Enzymes	Values	Ref.
Rainbow trout	Aroclor 5460	200 mg/kg, food, 3days	AHH, ECOD	2.0; 2.9	88
	Firemaster BP-6	200 mg/kg, food, 3days	AHH, ECOD	2.9; 1.4	88
	Halowax 2141	200 mg/kg, food, 3days	ECOD	1.4	88
	Hexachlorobutadiene	200 mg/kg, food, 3days	ECOD	0.75*	88
	Mirex	200 mg/kg, food, 3days	ECOD	0.66	88
Rainbow trout	Kepone	5 mg/kg, ip, day5	BEND, ECOD, EROD	0.80; 0.39; 0.14	82
	Mirex	25 mg/kg, ip, day5	BEND	0.72	
	Mirex	40 mg/kg, ip, day5	BEND, EROD	0.62; 1.5	
	Isosafrole	175 mg/kg, ip, day5	ECOD, EROD	5.0; 4.7	
	TCDD	1.2 ng/g, ip, day5	ECOD, EROD	16; 14	
	2,2',4,4',5,5'-hexachlorobiphenyl	150 mg/kg, ip, day5	ECOD, EROD	0.45; 0.27	
	3,3',4,4'-tetra-chlorobiphenyl	150 mg/kg, ip, day5			
Rainbow trout	BNF	70 mg/kg	ECOD, EROD	4.7; 53	26
Rainbow trout	BNF	50 mg/kg	ECOD, EROD, EMND, AHH	4.2, 93, 0.47, 14	40
Sculpin flounder	Petroleum	0.15–0.2 ppm in water	EROD, ECOD	120, 10	85
			diphenyloxazoleoxidase	3.1	
Sheepshead	Aroclor 1254	50 mg/kg, day7	AHH, ECOD, BEND	5.0; 3.8; 1.5	42
	Aroclor 1254	100 mg/kg, day7	AHH, ECOD, BEND	5.1; 5.2; 1.0	42
	Aroclor 1254	100 mg/kg, day29	AHH, ECOD, BEND	4.2; 1.7; 1.3	42
	Firemaster FF-1	15 mg/kg, day20	AHH, ECOD, BEND	6.0; 5.8; 1.0	42
Sheepshead	3-MC	20 mg/kg, day6	AHH, BEND	11; 0.9	49
	3-MC	20 mg/kg, day9	AHH, BEND	7.8; 1.0	49
	3-MC	20 mg/kg, day13	AHH, BEND	10; 0.7	49
Sheepshead	3-MC	10 mg/kg, ip, day8	AHH, ECOD	6.8; 4.1	14
	3-MC	10 mg/kg, im, day7	AHH, ECOD	5.1; 10.6	14
	3-MC	10 mg/kg, ip, day7	AHH, EROD	6.4; 6.1	14
	3-MC	5 mg/kg, im, day11	AHH, EROD	3.2; 40	14
	3-MC	20 mg/kg, day9	AHH, ECOD	3.5; 3.0	14
Sheepshead minnow	3-MC	33 mg/kg, ip, day5	EROD, ECOD, AHH	15; 9; 6	57
Southern flounder	3-MC	10 mg/kg, day8	AHH, ECOD	11.0; 24	14
Walbaum, male	Crude oil	0.3–0.8 ppm, water	AHH	5.8	89
Walbaum, female	Crude oil	0.3–0.8 ppm, water	AHH	7.5	89

nd = no detectable value

* = day n refers to the time of sampling after administration,

n day refers to the number of days a compound was administrated

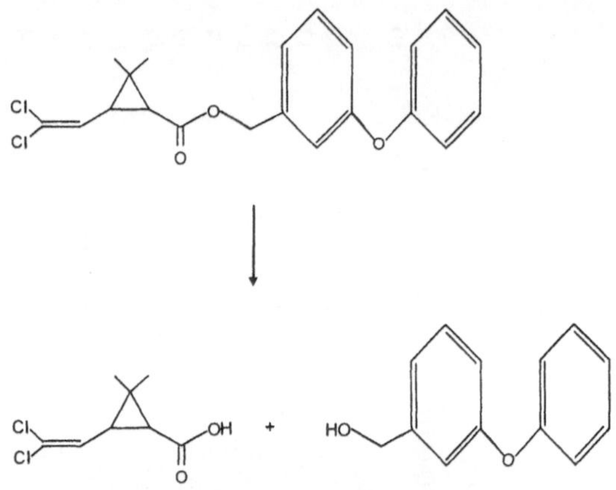

Fig. 5. Hydrolysis of an ester, cis-permethrin

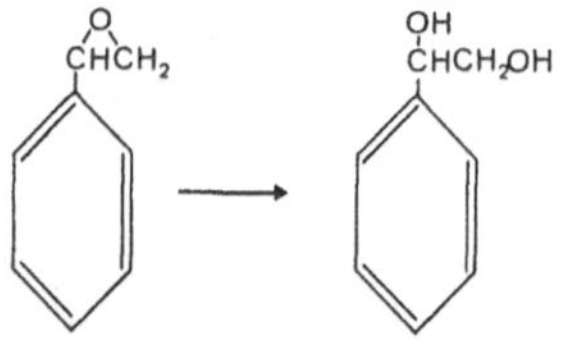

Fig. 6. Epoxidehydrase: styreneoxide to styrenedihydrodiol

arene oxides. The latter two compounds usually are the products of oxidative reactions [7].

Very little is known about esterase and amidase-activities or how to determine them in fish. Epoxidehydrase (EH) activity can be measured with e.g. styreneoxide as substrate. EH-activities range from 0.09 to 9.7 nmoles styrenedihydrodiol formed per mg protein per minute among fishes [7, 14, 15, 33, 36, 41, 49, 57, 58]. EH-activities in mammals ranged from 2.5 to 15 nmol/mg.min [36].

No induction of EH was achieved after administration of 10 or 20 mg/kg body weight of 3-methylcholanthrene or 1,2,3,4-dibenzanthracene is sheepshead, flounder, stingray and little skate [14] or after administration of 100 mg/kg body weight betanaphtoflavone or 500 mg/kg body weight Clophen A 50 in rainbow trout [60].

Reduction Reactions

Little is also known about reduction reactions in fish. Enzymes which catalyze reduction reactions comprise (a) dechlorinative, (b) ketone, (c) alkeneoxide, (d) areneoxide, (e) N-oxide, (f) nitro and (g) azoreductases (7).

Dechlorination (a) does occur in fish, but at a very low rate [7]. Ketone (b), alkeneoxide (c), areneoxide (d) and N-oxidereductases (e) have not been studied in fish [7]. Nitroreductase (f) (Fig. 7) is present both in the soluble fraction of the cell

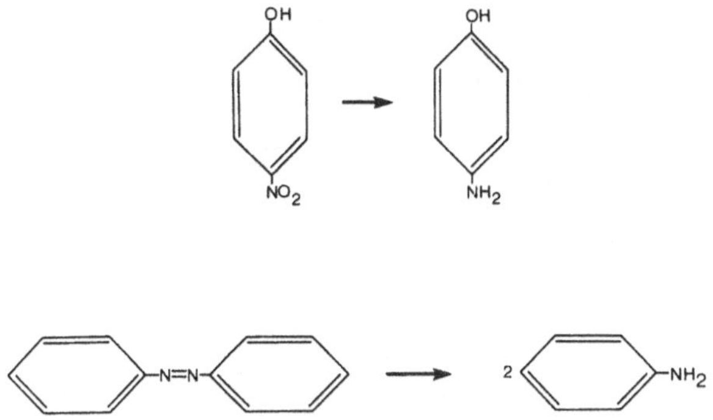

Fig. 7. Types of reductases. a) Nitroreductase; b) azoreductase

and in the microsomes and reduces a nitrocompound to the amino analogue. This enzyme is active under anaerobic conditions and has significantly reduced activity under aerobic conditions [93]. In mammals aromatic nitrocompounds can also be reduced by the bacteria which are present in the intestines [93]. It is unknown whether this reduction takes place in fish.

At least two different enzyme systems can catalyze reduction of nitrocompounds. Reduction reactions proceed faster in the presence of NADPH, rather than in the presence of NADH [7].

Azoreductase (g) (Fig. 7) cleaves azocompounds in amines, presumably in two steps. First a hydroazocompound is formed and then the N–N bond is reductively broken. The responsible enzymes are found in the cytosolic fraction of the liver cell [93]. Adamson et al. [94] measured azo and nitroreductase activities by following the formation of the cleavage product sulphanilamide from neoprontosil and the formation of p-aminobenzoic acid from p-nitrobenzoic acid, respectively.

Azoreductases were identified in the primitive fish, the elasmobranchii, while both azo and nitroreductase were found in less primitive fish, the teleosts [94]. In a comparative study it was shown that mammals have slightly higher nitroreductase activities than fish [95].

Phase II Reactions

Conjugation (phase II) reactions occur with chemicals that have functional groups such as –COOH, –OH, and –NH$_2$. These functional groups may be present in the parent compound, or may be formed during a phase I reaction [2, 7]. In the next paragraphs the steps required to form a conjugate and in some cases conjugation activities will be reviewed.

Glucuronic Acid Conjugation

Glucuronic acid conjugation (Fig. 8) first requires synthesis of uridinediphospho-glucuronic acid (UDPGA) from uridinediphospho-β-D-glucose, catalyzed by a

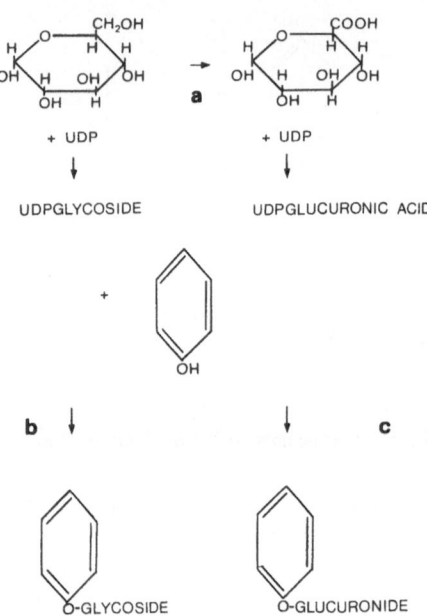

Fig. 8. Glycoside and glucuronic acid conjugation. a) UDP-glycoside dehydrogenation to UDP-glucuronic acid; b) glycoside conjugation; c) glucuronic acid conjugation

dehydrogenase in the cytosolic fraction of the cell. The transfer of the glucuronic acid from UDPGA to a substrate is catalyzed by glucuronyltransferase (UDPGA-t), present in the microsomes [2]. In addition to the glucuronic acid conjugates also glycoside conjugates can be formed (Fig. 8). The reaction between the substrate with β-D-glucose is catalyzed by uridine diphospho-β-D-glucose transferase [96].

UDPGA-t activity can be measured with several substrates of which p-nitrophenol (PNP) is most commonly used (Table 8). UDPGA-t activities among fishes range from 0.18 to 0.80 nmoles PNP conjugated per mg cytosolic protein per minute (Table 8).

UDPGA-t activities were comparable in several organs such as kidney, liver, intestine, heart and gills. In skin activity was less (Table 9). Hepatic UDPGA-t activities in some fish were higher than that in rat when PNP was used as substrate (Table 9). However, Gregus et al. found lower activities in rainbow trout than in rat when PNP and seven other substrates were used. Only when testosterone was used as substrate, activity in rainbow trout was the highest [36].

Gregus et al. also measured UDPGA-levels in livers of rat, mouse, guinea pig, rabbit, cat, dog, quail and rainbow trout. These concentrations were 408, 317, 483, 204, 344, 212, 248 and 396 nmol/g liver, respectively [36].

Bilirubine, an end product of heme catabolism, is an endogenous compound which is conjugated by UDPGA-t for 13–54% in carp, eel, aye, yellowtail, knifejaws and stingray [97].

UDPGA-t activities in carp and eel are optimal at the same pH and temperature of 8 and 37 °C, respectively. However, for stingray these optima were 8.2 and 45 °C,

Table 8. Hepatic UDPGA-transferase activities

Species	Fraction	Substrate activity	I	II	Ref.
Bream	Liver*	PNP**	2070	0.32	53
Carp	Liver	PNP	2680	0.47	53
Cod	Liver	PNP		0.55	26
Crayfish	Liver (microsomes)	PNP	0.01	0.32	27
Crayfish	Liver (cytosol)	PNP	0.01		27
Dace	Liver	PNP	2960	0.33	53
Dogfish shark	Liver	bilirubine	573		7
Eel	Liver	PNP	1320	0.18	53
Golden orfe	Liver	PNP	1900	0.25	53
Ide	Liver	PNP	3300	0.43	53
Little skate	Liver	Bilirubine	864		7
Perch	Liver	PNP	2060	0.18	53
Pike	Liver	PNP	1570	0.25	53
Rainbow trout	Liver	PNP	1720	0.23	53
Rainbow trout	Liver	TFM***	155		53
Rainbow trout	Liver	PNP		0.07–0.3	62
Rainbow trout	Liver	PNP		0.77	26
Roach	Liver	PNP	5790	0.80	53
Rudd	Liver	PNP	4840	0.72	53
Salmon	Liver	PNP	2100	0.25	53
Sea lamprey	Liver	PNP	0		53
Sea lamprey	Liver	TFM	10		7
Tench	Liver	PNP	1900	0.30	53
White bream	Liver	PNP	4320	0.55	53
Rat	Liver	PNP	46000		7

* = liver homogenate
PNP** = p-nitrophenol
TFM*** = 3-trifluoromethyl-4-nitrophenol
I = μmol/g (liver).h
II = nmol/mg.min

Table 9. UDPGA-t activities in several organs (pmol/min.mg protein)

Species	Liver	Kidney	Heart	Gills	Intestine	Ref.
Perch	93.5	—	20.8	73.9	67.6	31
Roach	108.4	158.6	36.2	235.2	75.1	31
Trout	383	60.0	89.4	70.5	245.7	31
Trout	83	—	—	—	—	37
Vendace	66.5	88.1	43.5	133.5	93.7	31
Vendace	74.3	37.7	36.0	181.3	43.0	37
Rat	58	—	—	—	—	37

— = not measured

respectively [98]. Lavaret and splake also have comparable UDPGA-t activities. UDPGA-t-activities in these two species were higher at pH 3 than at pH 6.7 [58]. These data suggest that UDPGA-transferases in different fishes are not similar.

Rainbow trout acclimated to 5 °C had lower UDPGA-t activities than trout acclimated to 15 °C, although enzyme activities were measured at the same temperature. The authors suggest that in spring when the water is cold, high levels of unconjugated steroid hormones can be maintained as a result of this low UDPGA-t activity [99]. Koivusaari et al. also observed a decreased UDPGA-t activity in rainbow trout when the environmental temperature was low, during winter [62].

UDPGA-t activities can be induced by certain compounds. For example administration of 100 mg/kg body weight Aroclor 1254 to channel catfish increased UDPGA-t-activity twofold. Lower doses of this PCB-mixture did not increase UDPGA-t-activity [24]. In winter flounder UDPGA-t-activity was increased 1.5 or 1.9-fold when these fish were pretreated with 10 mg/kg body weight 1,2,3,4-dibenzanthracene (DBA) or 100 mg/kg body weight betanaphtoflavone (BNF) [43]. Different increases of enzyme activities were seen with Clophen A50 or BNF in rainbow trout when PNP, 1-naphthol or testosterone were used as substrates for measuring UDPGA-t activities. This indicated that distinct UDPGA-t enzymes may exist within one organism [60].

Cadmium (0.5 mg/kg body weight) inhibited UDPGA-t activities in liver and kidney of rainbow trout by 20–60% [32].

Glutathione Conjugation

Glutathione, a tripeptide consisting of glutamine, cysteine and glycine, can be conjugated in the initial step of the mercapturic acid formation (7) (Fig. 9). Glutathione (GSH) reacts with electrophilic compounds and can replace hydrogen, chlorine or nitrogroups. Examples of substrates are naphthalene, 2,4-dichloronitrobenzene and pentachloronitrobenzene [93]. The mercapturic acid is formed after treatment of the glutathione conjugate with a peptidase and an acetylase [2]. Glutathione conjugation itself is catalyzed by a glutathione-transferase (GSH-t). This enzyme which is located mainly in the cytosolic fraction is also found in the microsomal fraction of the cell in liver, green glands, gills, kidney, pancreas, testes and ovaries [7, 27]. GSH-t activities can be measured with several substrates and is expressed as the rate of conjugate formed per mg of (cytosolic) protein [100]. Habig et al. [100] isolated four GSH-transferases from rat liver. Each of these enzymes was

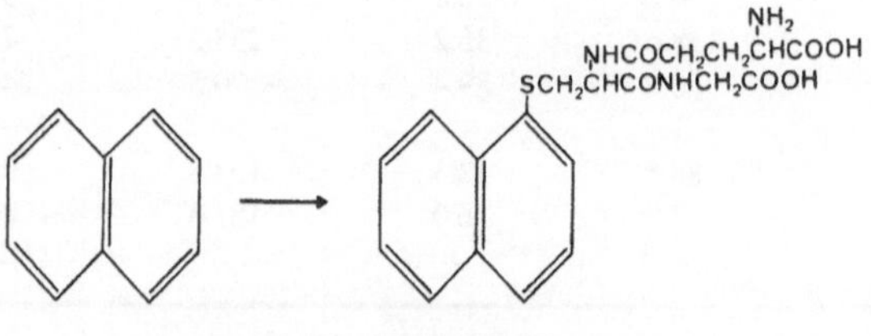

$$NH_2$$
$$NHCOCH_2CH_2CHCOOH$$
$$SCH_2CHCONHCH_2COOH$$

Fig. 9. Glutathione conjugation

identified by its reactivity with specific substrates and isolated by elution on carboxycellulose. They all had molecular weights of 45,000, but their activity of glutathione conjugation with p-nitrobenzylchloride differed. Fishes and rats have different GSH-t-activities, when different substrates are used (Table 10). Therefore they probably possess different GSH-transferases.

GSH-t activities in liver among several fishes ranged from 0.3 to 1690 nmoles substrate which is conjugated per mg cytosolic protein per minute with substrates such as 1,2-dichloro-4-nitrobenzene, styrene-7,8-oxide, octene-1,2-oxide and benzo(a)pyrene-4,5-oxide (Table 10).

In an interspecies comparison of GSH-t activities, rats and mice have the highest GSH-t activities towards five substrates, but trout has the highest GSH-t activity towards ethacrynic acid [36].

Induction of GSH-t was achieved with the administration of 500 mg/kg body weight Clophen A 50 or 100 mg/kg body weight BNF in rainbow trout [60] and with the administration of 0.75–7.5 mg/kg body weight of benzo(a)pyrene in Black Sea bass [52]. GSH-t activity which was measured with 1,2-epoxy-3-(p-nitrophenoxy)-propane as substrate was increased in channel catfish, when this fish was pretreated with 100 mg/kg body weight Aroclor 1254. When GSH-t-activity was measured with four other substrates, no increase in activity was observed [24]. This indicates that some compounds can induce only specific GSH-transferases.

No GSH-t-activity induction took place in winter flounder after pretreatment with DBA or ANF [43]. In sheepshead, southern flounder, Atlantic stingray, little skate and dogfish shark pretreated with 10 or 20 mg/kg body weight DBA or 3-MC no elevated activity was found [14.49]. Also no induction of GSH-t-activity was achieved with Aroclor 1254 (50 or 100 mg/kg body weight) or Firemaster FF1 (15 mg/kg body weight) in sheepshead [42]. In this latter study, however it was not reported which substrate was used to measure GSH-t activity. Induction by Aroclor 1254 of only one GSH-transferase of five examined was reported by Ankley et al. [24]. Hence, it is clear that the substrates which are used to express GSH-t-activity should always be reported to enable comparing activities which are induced by different compounds.

Cadmium (0.5 mg/kg body weight) inhibited GSH-t activity and UDPGA-t activity by 10–40% in rainbow trout [32]. Cyclopropenoid fatty acids in the diet reduced GSH-t activity in rainbow trout. A nonlinear dose-response relationship was observed. A dose of more than 150 mg/kg fatty acids in food decreased activity up to 50% [33]. Cyclopropenoid fatty acids induced liver cancer in rainbow trout at dietary levels as low as 45 mg/kg [101]. Whether this induced liver cancer is an effect of the decreased GSH-t activity is unknown.

Sulphate Conjugation

Hydroxy or aminocompounds can be conjugated with sulphate (Fig. 10). It requires at least three steps to form the conjugate. The first two make sulphate "active" with adenosinetriphosphate (ATP) forming 3-α-phosphoadenylsulphate (PAPS). The sulphate then can be transferred from PAPS to an acceptor molecule. This reaction

Table 10. GSH-t-activities of cytosolic liver protein

| Species | Activities (nmol/mg cytosolic protein.min) with substrates: | | | | | | | |
	A	B	C	D	E	F	G	Ref.
Atlantic stingray	5.1; 5.6	1.1; 1.7	5.0	0.5				14, 15, 49
Black drum	16	9.2	9.1	11.8				7, 49
Black Sea bass	2.1			3.3; 200				52
Bluntnose stingray	4.0		2.8					7
Channel catfish				821	26	21	0.4	24
Channel catfish				1380				25
Crayfish				394				27
Dogfish shark	15	7.8	1.1	1.5				7, 15, 49
Eel	15	2.9–3.8	15	6.7				7, 49
Fathead minnow				1592				57
Gulf killifish				1136				57
Hagfish				5.8				7
Jack crevalle	1.4			0.6				7
Japanese medaka	0.82			437				57
Killifish				45.1				7
King cobra guppy	6.3			592				57
King of Norway	2.6			2.0				7
Large skate	2.6			3.1				7
Little skate	2.4	24	0.6	0.3; 6.6				7, 14, 15, 49
Mackerel				1.9				7
Mangrove snapper	4.2	2.2						49
Nurse shark	12.1			0.2				7
Rainbow trout				87–2000	20	29	3	32, 36, 60
Red fish	6.4			10.3				7
Sailfin mollie	1.7			1119				57
Sheepshead	22.7; 26	38.5; 53	21	1.5; 6.0				7, 14, 15, 49
Sheepshead minnow	5.6			966				57
Southern flounder	1.8; 6.6			0.3; 0.8				7, 14
Spot	23.8			1.1				7
Thorny skate	7.1			2.7				7
Winter flounder	4.9	5.3	5.1	1.9; 1690				7, 15, 43, 49
Cat				2000	15	6	50	36
Dog				2000	15	29	80	36
Guinea pig				8000	145	10	43	36
Mouse				4000	140	53	62	36
Quail				2000	18	12	1	36
Rabbit				11000	20	5	10	36
Rat				2000	80	38	70	36

Substrates:
A = styrene-7,8-oxide   E = 1,2-epoxy-3-(p-nitrophenoxy)propane
B = benzo(a)pyrene-4,5-oxide   F = ethacrynic acid
C = octene-1,2-oxide   G = 1,2-dichloro-4-nitrobenzene
D = 1-chloro-2,4-dinitrobenzene
u = unknown

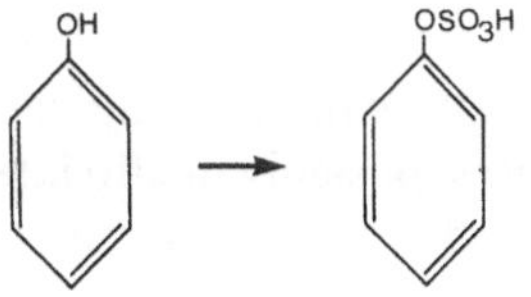

Fig. 10. Sulphate conjugation

is catalyzed by a sulphotransferase, which is located in the cytosolic fraction of the liver cell. At least two sulphotransferases have been distinguished, one catalyzing the conjugation of betahydroxysteroids and one that of phenols and estrogens [7, 93].

Rainbow trout showed less sulphotransferase activity than rat, mouse, guinea pig, rabbit, cat, dog and quail. Rabbit and guinea pig had highest activities [36].

Acetyl Conjugation

Acetylation is the conjugation of carboxylic acids with amines (Fig. 11). Prior to this the acid must be activated by forming a coenzyme A (CoA) deriyative. This activated acid can be transferred to the acceptor amine mediated by an acyltransferase [93]. Either the aminoacid or the carboxylic acid may be the xenobiotic. The aminoacid acetylation is catalyzed by a N-acetyltransferase. The carboxylic acid has to be transformed to a CoA-thioester prior to the carboxylic acid acetylation [7].

Rainbow trout acetylated β-naphthylamine and 2-aminofluorene to a high extent. In an interspecies conjugation study however, rabbits had the highest acetylation activity [36].

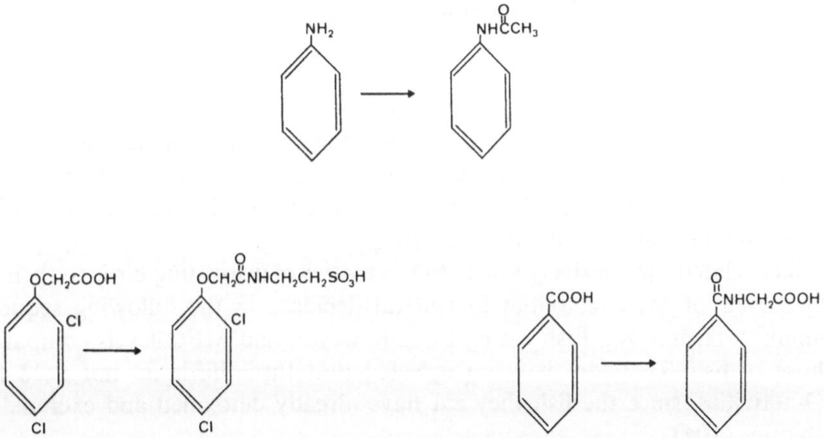

Fig. 11. Types of acetylation. a) Acetyl conjugation; b) Taurine conjugation; c) glycine conjugation

Methyl Conjugation

Methyltransferases (Fig. 12) catalyze methylconjugation in the cytosolic fraction of the cell [7]. These transferases possibly methylate nicotinamide to N-methyl-nicotinamide [93]. Initially the methyl group is transferred from S-adenosylmethionine, which is previously formed by methionine and ATP [93]. Hitherto, methyltransferase activities have not been studied in fish.

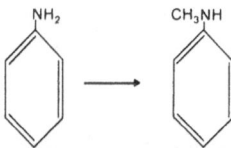

Fig. 12. Methyl conjugation

Influence on Metabolic Properties

Several factors may influence the rate of metabolism of a compound, e.g. route of administration, diet, temperature, season, hormones, age and sex of the animal, and species and strain. Some of the factors have been mentioned already. For instance the influence of route of administration was shown for sheepshead [14] in an earlier section. Other factors will be described in the following paragraphs.

Diet. The influence of diet was clearly shown by Jimenez et al. [102]. Uptake and elimination of BP was accelerated 2-fold and 10-fold, respectively, when fish were fed during the exposure to BP, compared with unfed fish. The increased elimination was partly due to an accelerated rate of metabolism of BP [102]. Feeding killifish and mullet also promoted the elimination of two biotransformation products of fenitrothion compared with not feeding these fish [103]. Takimoto et al. [103] suggested that diet stimulates bile secretion. Since the metabolites are transported from the bile to the intestine, hydrolysis of conjugates may occur in the intestinal mucus or by the intestinal microflora. At lower temperature or higher salinity of the water more conjugates were detected in the water. Under these conditions the intestinal mucus and microflora may have a reduced hydrolyzing activity, resulting in the elimination of more conjugates [103].

Walker related the relatively low MFO-activities of fish-eating birds to their diet [104]. Levels of MFO-activities in general decrease in the following sequence: mammals > birds > fish. Fish-eating birds however, had MFO-levels comparable to those of fish. Therefore it was suggested that these birds do not need high MFO-activities, since the fish they eat have already detoxified and excreted the xenobiotics [104].

Foureman et al. [43] found induction of AHH by "natural" products, such as flavonoids, safroles and indoles, which are present in the diet of e.g. herbivorous or omnivorous animals. A decreased GSH-t activity in rainbow trout was observed when cyclopropenoic acids were present in the diet at a concentration of 150 mg/kg and higher [33]. The "natural" products malvalic and sterculic acids also belong to this class of fatty acids [101].

Metabolism can thus be influenced by the food, since it contains compounds which may act as inhibitors or inducers.

Temperature and Season. In several studies no distinction could be made between temperature and season influences. DeWaide [53] found highest APND-activity in roach in the summer. This observation however, could not be due to temperature changes alone. In summer the concentration of inducers in river water is high due to a low volume of the water. The decrease of two monooxygenases, APND and AH in roach held in running tap water under laboratory conditions at a higher temperature, could be increased again by the high concentrations of inducers under environmental conditions [53].

In summer (25 °C), the extent of AHH-activity and cytochrome P-450 induction in sheepshead after 3-MC treatment was higher than in winter (11 °C) [42]. In sheepshead the course of induction also changed with season [14]. 3-MC treatment resulted in a maximum induction of AHH-activity after 3 days in summer (26 °C), whereas AHH-activity returned to control values after 14 days. In winter (14 °C) however, maximum induction of AHH-activity appeared after 8 days, whereas control values were not reached even after 28 days [14].

Induced AHH-activity was found in channel catfish after BP treatment. Induction was found to be highest in channel catfish held at 26 °C, lowest in fish at 15 °C and intermediate in fish at 19 ° after 2 days. After 4 days however, there was the same level of induction at the three temperatures [84].

The influence of temperature on metabolism was also clearly shown by Jimenez et al. [102]. Uptake rates of BP in bluegill sunfish which were 5.8 times higher when the fish were acclimated at 23 °C than when they were acclimated at 13 °C. Elimination rates and the amounts of BP which were metabolized were higher at the higher temperature [102].

ECOD and AHH-activities in cold-acclimated (5 °C) rainbow trout were 1.5 and 3-fold higher, respectively than those activities in warm-acclimated trout (15°C). These activities were measured at assay temperatures of 5, 10 and 15 °C [96]. It was concluded that metabolism thus may proceed at similar rates at different body temperatures [60].

EROD-activities of cold-acclimated (5 °C) rainbow trout were also higher when measured at 17 °C than those from warm-acclimated (15 °C) trout [87]. EROD-activities of cold-acclimated fish measured at 5 °C, however were only slightly lower than those of warm-acclimated fish measured at 17 °C.

Förlin et al. [87] suggest that the phospholipid content of the microsomes has a large influence on the microsomal enzyme activities. Livers from cold-acclimated fish contain a higher concentration of unsaturated fatty acids than livers from warm-acclimated fish. When MFO-activities are measured at the same assay temperature, unsaturated fatty acids will facilitate the mobility of the xenobiotics in the membranes more than saturated fatty acids do. The mobility of MFOs, which are located in the endoplasmic reticulum, are also facilitated. At the same assay temperatures MFO-activities will thus be higher in liver from cold-acclimated fish than from warm-acclimated fish.

Apart from influence on enzymatic activities, temperature has an influence on other properties as well. In fish and in mammals the ratio liver weight to body weight decreased with increasing temperature [53, 62]. Koivusaari et al. [62] observed decreased MFO-activities per mg liver protein in rainbow trout in summer, but oxidation activity per liver remained unchanged due to liver growth.

As mentioned earlier UDPGA-t-activity appeared to be lower in cold-acclimated rainbow trout than in warm-acclimated trout [62, 99], but a reliable explanation for this is still lacking.

Another aspect of the influence of temperature on metabolism is the optimum temperature for measuring enzyme activities. Optimum temperatures for AHH and ECOD-activities were 20 and 30 °C, respectively in rainbow trout and vendace [37]. Optimum temperatures for AH and APND-activities in bream, white bream, orfe, roach, eel and perch were 35–40 °C, in tench and pike 30–40 °C, in salmon and sea lamprey 30–35 °C, and in rainbow trout 25–30 °C [53]. In mammals optimum temperatures were about 10°C higher [53]. Optimum temperature for AHH-activity was 40–45 °C in channel catfish [84] and for EMND-activity in guppy was 35 °C [28].

In summary, temperature and/or season has a great influence on metabolism. At lower temperatures metabolism usually proceeds at lower rates than at higher temperatures. At lower temperatures fish have higher MFO-capacities to oxidize compounds, but lower capacities to conjugate compounds. It is not yet clear whether these different metabolizing capacities are determined by hormonal regulation or by physico-chemical properties such as phospholipid composition of membranes.

Hormones. The influence of hormones in fish metabolism was demonstrated in a few studies. Testosterone (3 mg/kg) and estradiol (3 mg/kg) increased (40%) and decreased (20%) cytochrome P-450 concentrations, respectively in male and female juvenile brook trout [20]. A protein with a molecular weight of 56,000 decreased concomitantly with cytochrome P-450 by administration of estradiol-17β [20]. This compound is the principle female sex steroid hormone [87] and is produced by the hypophyse [38]. In juvenile rainbow trout estradiol-17β decreased MFO-activities in a direct way [38]. After administration of estradiol-17β to hypophysectomized and sham-operated rainbow trout the same suppressive effects on MFO-activities were observed [38]. Cytochrome P-450 levels, however were also decreased, which resulted in undisturbed levels of MFO-activities per nmoles of cytochrome P-450 [38]. The low cytochrome P-450 concentration and MFO-activities in maturing female fish served to maintain high levels of estradiol-17β in the liver during the vitellogenin synthesis, which is associated with reproduction [38, 105]. Concentrations of steroid hormones in blood increase as much as 10-fold at spawning times [20] and thus may influence MFO-activities significantly.

AHH-activity in walbaum decreased significantly in male and female fish from maximal values before gonad maturation to minimum values two weeks following spawning [89].

Sivarajah et al. [106] found a nonlinear correlation between decreased plasma hormone levels of androgens, estrogens and corticoids and increased enzyme activities (APND and UDPGA-t).

In conclusion, hormones such as testosterone and estradiol-17β seem to play a significant role in the regulation of the sex differences in the cytochrome P-450 mediated metabolism [105] and conjugation [106]. Much work, however has to be carried out to elucidate influences of several hormones on biotransformation.

Age and Sex. The influence of age and sex was demonstrated in only a few studies. Inhibition of ECOD-activity by $CdCl_2$ in rainbow trout was higher in mature male and female fish than in juvenile fish [32].

Induction of PNAOD-activity by PCN was higher in 1.5 to 2 year old mature male and female rainbow trout than in 0.5 to 1.5 year old juvenile fish [59].

Androstenedione-6β-hydroxylase-activity was lower in yearling and 1 year old juvenile rainbow trout than in adult trout. Mature female fish however had the lowest activity [83]. Higher induction of androstenedione-6β-hydroxylase by Clophen A50 (500 mg/kg) or 3-MC (20 mg/kg) was observed in maturing female rainbow trout than in male trout [83].

In general MFO-activities and cytochrome P-450 concentration are lower in maturing female than in male fish [83]. Induced androstenedione-6β-hydroxylase activities in female rainbow trout were found after 3-MC treatment, however this activity was equal to the androstenedione-6β-hydroxylase-activity in male fish which were not pretreated with 3-MC [83]. After spawning differences in MFO-activities and cytochrome P-450 concentration between male and female fish disappeared [62].

In summary from the few data available, adult fish seem to have higher metabolic activities than juvenile fish, whereas maturing female fish have the lowest levels.

Species and Strain. The influence of species and strain can be clarified with the aid of the Tables 3 to 5, 9 and 10, showing different enzyme activities among fishes. Enzyme activities can vary more than 100-fold.

Even within one species enzyme activities can vary significantly for several strains [107]. In six distinct rainbow trout strains for instance AHH-activity varied maximally 57-fold, AH-activity 2-fold and nitroreductase-activity 2.5-fold.

Walker et al. [104, 108] observed a decline of enzyme activities with increasing body size in mammals. A good explanation for this, however was lacking. Whether or not this phenomenon holds true for fish is unknown.

Biotransformation of Xenobiotics in Fish

In the previous sections it was shown that fish posses the capabilities to bio-transform organic compounds. Only selected compounds were used in *in vitro* tests to measure the activities of several enzymes in fish. In the next sections the biotransformation of other organic compounds in fish will be reviewed.

As has been mentioned before, the route of administration can be important in the quantitative and qualitative metabolism of xenobiotics. Therefore it is difficult to compare results among studies which applied different exposure regimes.

Especially experiments done with model ecosystems (ME) are difficult to compare with all other *in vivo* and *in vitro* experiments. These model ecosystems are always constructed with a variety of biotic species. The advantage of such systems is that more information can be obtained about the interrelationship between excretion of metabolites from one organism and further biotransformation of these compounds by other organisms. Consequently the major problem is to identify from which organism the metabolites originate which are found in the water or the organisms. Hence, experiments with single organisms are usually required to enable interpretation of the results.

Oxidations

Many organic compounds can be oxidized by fish (Table 11). From some interspecies studies it is clear that oxidation takes place at different rates in different species. For example in the study of Chin et al. [123] carbaryl was oxidized to naphthalenediol for 28.8, 11.5, 14.1, 67.1 and 24.2%, respectively in bluegill sunfish, channel catfish, goldfish, kissing gourami and yellow perch. These species differences have also been found by Thornton et al. [118], Lee et al. [121]. Varanasi et al. [9] and Ludke et al. [110].

In some in vitro tests it was demonstrated that aniline could be oxidized to aminophenol [10], and benzo(a)pyrene predominantly to hydroxylated compounds and quinones [21, 55, 107, 119, 184–192].

Further, dodecylcyclohexane was oxidized by rainbow trout to phenylacetic acid, probably via the oxidation of dodecylcyclohexaneacetic acid, which is a oxidation product of the parent compound [161]. This means that a saturated hydrocarbon can be transformed to an unsaturated compound in fish.

Reductions

In fish it is demonstrated that only a small number of compounds underwent reduction reactions (Table 12). For instance a DDT biotransformation product, DBP, underwent ketoreduction and some nitrocompounds underwent nitroreduction.

Glucuronide Conjugates

A large number of compounds can be glucuronidated by fish (Table 13). Among these compounds there are probably many which have been oxidized first and have subsequently been glucuronidated, e.g. naphthalene and benzo(a)pyrene.

In rainbow trout it was shown that BNF-pretreatment increased the amounts of 2-methylnaphthalene which were glucuronidated [168]. Whether this finding was the result of an increased glucuronic acid transferase-activity or an increased MFO-activity, is unknown.

Table 11. Oxidation reactions

Fish	Compound	Exposure	Metabolites	%	Ref.
Atlantic salmon	Aldrin	im, 5 mg/kg	Dieldrin	u	109
Bluegill sunfish	Aldrin	in vitro	Dieldrin	u	110
Brook trout	Aldrin	u	Dieldrin	u	111
Golden shiner	Aldrin	in vitro	Dieldrin	u	110
Goldfish	Aldrin	u	Dieldrin	u	112
Goldfish	Aldrin	u	Dieldrin, 9-keto-dieldrin, 9-hydroxy-dieldrin	u	113
Green sunfish	Aldrin	in vitro	Dieldrin	u	110
Green sunfish	Aldrin	u	Dieldrin, 9-keto-dieldrin, 9-hydroxy-dieldrin, dihydroxy-trans-aldrin	u	114
Mosquitofish	Aldrin	in vitro	Dieldrin	u	110
Mosquitofish	Aldrin	u	Dieldrin	u	112
Mosquitofish	Aldrin	u	Dieldrin	u	115
Mosquitofish	Aldrin	u	Dieldrin, hydroxy-aldrin	u	116
Rainbow trout	Aldrin	u	Dieldrin	u	117
Yellow bullhead catfish	Aldrin	in vitro	Dieldrin	u	110
Mosquitofish	Anisole	u	o- and p-hydroxyanisole	u	113
Mosquitofish	Anisole	u	Phenol	u	116
Bluegill sunfish	[³H]-BP	in vitro	9,10-dihydrodiol-BP	15.8	118
			7,8-dihydrodiol-BP	15.0	
			9-hydroxy-BP	26.5	
			BP-quinone	5.6	
California killifish	BP	in vitro	9,10-dihydrodiol-BP	19	23
			7,8-dihydrodiol-BP	20	
			1,6- and 3,6-BP-quinones	8	
			1-, 3- and 7- hydroxy-BP	52	
Carp	BP	u	Hydroxy-BP	u	119
Coho salmon	BP	in vitro	9,10-dihydrodiol-BP	u	120
			7,8-dihydrodiol-BP	u	
			3-hydroxy-BP	u	

Table 11. (continued)

Fish	Compound	Exposure	Metabolites	%	Ref.
English sole	BP	in vitro	9,10-dihydrodiol-BP	u	120
			7,8-dihydrodiol-BP	u	
			3-hydroxy-BP	u	
Fathead minnow	[³H]-BP	in vitro	9,10-dihydrodiol-BP	12.5	118
			7,8-dihydrodiol-BP	9.7	
			3-hydroxy-BP	9.8	
			BP-quinone	6.4	
Mudsucker	[³H]-BP	Water, 1 µg/l (6 µg/l)	7,8-dihydrodiol-BP	>50	121
			6-hydroxy-BP	10	
Rainbow trout	[³H]-BP	in vitro	9,10-dihydrodiol-BP	3.9	118
			9-hydroxy-BP	8.8	
			BP-quinone	4.6	
			7,8-dihydrodiol-BP	2.9	
Sand dab	[³H]-BP	Water, 1 µg/l (6 µg/l)	7,8-dihydrodiol-BP	>50	121
			6-hydroxy-BP	10	
Sand goby	BP	Water, 1 µg/l (6 µg/l)	7,8-dihydrodiol-BP	u	121
Sculpin	[³H]-BP	Water, 1 µg/l (6 µg/l)	7,8-dihydrodiol-BP	>50	121
			6-hydroxy-BP	10	
Speckled sanddab	BP	in vitro	9,10-dihydrodiol-BP	21	23
			7,8-dihydrodiol-BP	26	
			1-, 3- and 7-hydroxy-BP	53	
Starry flounder	BP	in vitro	9,10-dihydrodiol-BP	u	120
			7,8-dihydrodiol-BP	u	
			3-hydroxy-BP		
Rainbow trout	n-butylbenzene	Food, 25 µl	1-phenyl-1-butanol	u	122
Bluegill sunfish	Carbaryl	in vitro	1-naphthol	u	123
			dihydrodiolcarbaryl	28.8	
Channel catfish	Carbaryl	in vitro	1-naphthol	2.8	123
			dihydrodiolcarbaryl	11.5	

Species	Compound	Condition	Metabolite	Value	Ref.
Goldfish	Carbaryl	in vitro	1-naphthol	u	123
			dihydrodiolcarbaryl	14.1	
Kissing gourami	Carbaryl	in vitro	1-naphthol	2.2	123
			dihydrodiolcarbaryl	67.1	
Rainbow trout	Carbaryl	Water, 0.25 mg/l	5,6-dihydrodiol-carbaryl	u	124
Yellow perch	Carbaryl	in vitro	1-naphthol	2.2	123
			dihydrodiolcarbaryl	24.2	
Sepat siam	[^{14}C]-carbofuran	in vitro	Carbofuranphenol	1.2	125
			3-ketophenol	0.2	
			3-ketocarbofuran	0.2	
			3-hydroxycarbofuran	4.9	
			N-hydroxymethylcarbofuran	12.9	
Bluegill sunfish	cis-[^{14}C]-chlordane	Water, 5 µg/l	Chlordane-chlorohydrin	u	126
			heptachlordiol		
Goldfish	cis-[^{14}C]-chlordane	Water, 26 µg/l	Chlordane-chlorohydrin	0.06	127
			monohydroxychlordane	0.06	
Channel catfish	[^{14}C]-chlordecone	Food, 1.36 mg/kg	Chlordeconealcohol	u	128
Goldfish	Chloroalkylene-9	Water, 1 mg/l	Hydroxylated compounds	u	129
			sidechain- and aromatic oxidation products	u	129
Guppy	Chloroanisoles	Water, 350-2595 µg	Chlorophenols	u	130
Winter skate	4-chlorodiphenylether	iv, 68 nmol/kg	4'-hydroxy-4-chlorodiphenylether	u	131
Goldfish	4-chloro-4'-isopropyl-biphenyl	u	2-(4'-chlorobiphenyl)-propanoic acid	u	132
Mosquitofish	Chlorpyrifos	u	3,5,6-trichloropyridinol	u	113
Bitterling	[^{14}C]-m-cresol	Water, 4 mg/l	m-hydroxybenzoic acid	u	133
Bream	[^{14}C]-m-cresol	Water, 5 mg/l	m-hydroxybenzoic acid	u	133
Crucian carp	[^{14}C]-m-cresol	Water, 3 mg/l	m-hydroxybenzoic acid	u	133
Goldfish	[^{14}C]-m-cresol	Water, 15 mg/l	m-hydroxybenzoic acid	u	133
Gudgeon	[^{14}C]-m-cresol	Water, 8 mg/l	m-hydroxybenzoic acid	u	133
Minnow	[^{14}C]-m-cresol	Water, 4 mg/l	m-hydroxybenzoic acid	u	133
Perch	[^{14}C]-m-cresol	Water, 5 mg/l	m-hydroxybenzoic acid	u	133
Roach	[^{14}C]-m-cresol	Water, 8 mg/l	m-hydroxybenzoic acid	u	133
Rudd	[^{14}C]-m-cresol	Water, 8 mg/l	m-hydroxybenzoic acid	u	133
Three-spined stickleback	[^{14}C]-m-cresol	Water, 4 mg/l	m-hydroxybenzoic acid	u	133

Table 11. (continued)

Fish	Compound	Exposure	Metabolites	%	Ref.
Tench	[14C]-m-cresol	Water, 15 mg/l	m-hydroxybenzoic acid	u	133
Rainbow trout	DBH	ip, 5 mg/kg	DBP	6.5	134
Bluegill sunfish	2,4-D-butoxyethylester	u	2,4-D	u	135
Channel catfish	2,4-D-butoxyethylester	u	2,4-D	u	135
Rainbow trout	2,4,-D-n-butylester	u	2,4-D	100	136
Rainbow trout	DDA	ip, 5 mg/kg	DBP	70	134
Mosquitofish	DDD	u	DDMU	u	137
Rainbow trout	DDD	ip, 5 mg/kg	DDMU	u	134
Rainbow trout	DDD	ip, 5 mg/kg	DDMU	0.6	134
Rainbow trout	DDMS	ip, 5 mg/kg	DDNU	0.5	134
Rainbow trout	DDOH	ip, 5 mg/kg	DBP	70	134
Atlantic salmon	DDT	Water, 2.0 ppm	DDD, DDE	u	138
Atlantic salmon	DDT	im, 5 mg/kg	DDE	u	109
Brook trout	DDT	u	DDD, DDE	u	139
Brook trout	DDT	im, 150 µg	DDE	u	140
Dogfish shark	DDT	iv, 0.5 mg/kg	DDA	u	141
Mosquitofish	DDT-analogs	ME	Sulfoxidation-, O-dealkylation-, sidechain oxidation- and dehydro-dechlorination compounds	u	142
Mosquitofish	DDT	Water, 0.22 µg/l	DDD	9.8*	143
			DDE	53.9*	143
Rainbow trout	DDT	ip, 5 mg/kg	DDD	u	134
			DDE	94	134
Sole	DDT	Oral 1.7, 17 or 35 µg	DDD, DDE	u	144
Winter flounder	DDT	u	DBP, DDA, DDD, DDE	u	145
Channel catfish	DEHP	Water, 1 µg/l	MEHP	66	146
			phthalic acid	4	
Fathead minnow	DEHP	Water, 1.9-62 µg/l	MEHP	37.1	146
			phthalic acid	5.2	
Guppy	DEHP	u	Phthalic anhydride, phthalic acid	u	147

Fish	Chemical	Exposure	Product	Value	Ref.
Mosquitofish	DEHP	u	MEHP, phthalic anhydride, phthalic acid	u	113, 147
Rainbow trout	DEHP	Water, 0.5 µg/l	MEHP	0.5	148, 149
Mosquitofish	2,4-D-ethylthioester	in vitro	2,4-D, butanethiol	u	150
Channel catfish	di-n-butylphthalate	in vitro	Mono-n-butylphthalate	u	146
Guppy	2,5-dichlorobiphenyl	Water, 500 µg/l	2,5-dichloro-4-hydroxybiphenyl	u	151
			2,5-dichloro-3'-hydroxybiphenyl	u	
			2,5-dichloro-2'-hydroxybiphenyl	u	
			2,5-dichloro-3',4'-dihydroxybiphenyl	u	
			2,5-dichloro-dihydroxybiphenyl	u	
			2,5-dichlorobenzoic acid	u	
Guppy	4,4'-dichlorobiphenyl	Water, 31 µg/l	4,4'-dichloro-3-hydroxybiphenyl	15	151
			4,3'-dichloro-4-hydroxybiphenyl	1.5	
Goldfish	2,2'-dichlorobiphenyl	u	2-hydroxydichlorobiphenyl	u	152
Goldfish	2,8-dichlorodibenzo-p-dioxin	Water, 5 µg/l	3-hydroxy-2,8-dichlorodibenzo-p-dioxin	u	153
Goldfish	2,5-dichloro-4'-isopropyl-biphenyl	u	2-(2',5'-dichloro-4-biphenyl)-propanoic acid	u	132
Bluegill sunfish	[^{14}C]-dieldrin	Water, 50 µg/l	Pentachloroketone	8.17	154
			aldrin-trans-diol	8.04	
			aldrin-carboxylic acid	0.34	
Bluegill sunfish	2,4-D-dimethylamine salt	u	2,4-D	u	155
Channel catfish	2,4-D-dimethylamine salt	u	2,4-D	u	155, 156
Carp	Dinitramine	Water, 1 mg/l	N^3-ethyl-2,4-dinitro-6-trifluoromethyl-m-phenylenediamine	u	157
			2,4-dinitro-6-trifluoromethyl-m-phenylenediamine		
Mosquitofish	Di-n-octylphthalate	in vitro	Mono-n-octylphthalate, phthalic acid	12	158
Cunner	2,5-diphenyloxazole	u	2-(4-hydroxyphenyl)-5-phenyloxazole	u	159
			5-(4-hydroxyphenyl)-2-phenyloxazole	u	
Rainbow trout	[^{3}H]-dodecylcyclohexane	Food, 5 mg	3-dodecylcyclohexane	u	160
			4-dodecylcyclohexane	u	
			cyclohexyldodecane-2-ol	u	
			cyclohexyldodecanoic acid	u	

Table 11. (continued)

Fish	Compound	Exposure	Metabolites	%	Ref.
Rainbow trout	[³H]-dodecylcyclohexane	Food, 5 mg	1-hydroxycyclohexylacetic acid	2.1	161
			3-hydroxycyclohexylacetic acid	u	
			4-hydroxycyclohexylacetic acid	u	
			cyclohexylacetic acid	0.7	
			phenylacetic acid	u	
Catfish	Endosulfan	Water, 0.5 µg/l	Endosulfanalcohol	u	162
			endosulfanether	u	
			endosulfan-α-hydroxyether	u	
			endosulfanlactone		
Macrognathus aculeatum					
Bluegill sunfish	Endosulfan	Water, 0.5 µg/l	Endosulfanether	u	163
	[¹⁴C]-endrin	Water, 1 µg/l	12-anti-hydroxyendrin	3	164
			12-syn-hydroxyendrin	1	
Mosquitofish	Ethoxychlor	u	hydroxy-ethoxychlor, hydroxychlor, ethoxychlorethylene	u	165
Killifish	Fenitrothion	Water, 0.1 µg/l	3-methyl-4-nitrophenol	0.3–3.1	103
			fenitrooxon	0.5–0.7	
			desmethylfenitrothion	0.1–0.8	
			desmethylfenitrooxon	0.1–0.2	
Mullet	Fenitrothion	Water, 0.1 µg/l	3-methyl-4-nitrophenol	0.3–0.9	103
			fenitrooxon	0.1–0.2	
			desmethylfenitrothion	48.7–55.2	
			desmethylfenitrooxon	0.1–0.4	
Tilapia mossambica	Fenitrothion	200 mg/kg	Fenitrooxon, 3-methyl-4-nitrophenol, desmethyl-N-acetylaminofenitrothion, desmethyl-N-acetyl-aminofenitrooxon	u	4
Goldfish	[¹⁴C]-heptachlor	ip, 0.87 µg/fish	1-hydroxy-2,3-epoxychlordene	4.6	166
			1-hydroxychlordene	1.0	
			heptachlor-2,3-exo-epoxide	4.3	
Green sunfish	Hexachlorobenzene	u	Pentachlorophenol	u	167

Species	Compound	Exposure	Metabolites		Ref.
Rainbow trout	n-hexylbenzene	Food, 25 μl	4-phenyl-1-butanol	u	122
			6-phenyl-1-hexanol	u	
			1-phenyl-1-hexanol	u	
			6-phenylhexanoic acid	u	
Rainbow trout	Lauric acid	in vitro	Hydroxylauric acid	u	79
Mosquitofish	Methiochlor	Water, 73 μg/l	Sulfoxidation products	12.3	143
Green sunfish	Methoxychlor	u	Hydroxymethoxychlor, hydroxychlor, hydroxychlorethylene, bis-(p-hydroxyphenyl) acetic acid	u	114
Mosquitofish	Methylchlorpyrifos	u	3,5,6-trichloropyridinol	u	113
Mosquitofish	Methyl-ethoxychlor	ME	Methyl-hydroxychlor, ethoxychloracetic acid	u	142
Rainbow trout	2-methyl-[8-^{14}C]-naphthalene	Water, 0.5 mg/l	1,2-dihydrodiol-3-methylnaphthalene	12	168
			1,2-dihydrodiol-6-methylnaphthalene	40	
			1,2-dihydrodiol-7-methylnaphthalene	41	
			2-hydroxy-methylnaphthalene	6	
Crayfish	[^{14}C]-methylparathion	Water, 0.01–0.1 mg/l	Methylparaoxon	0.2	169
			p-nitrophenol	43.6	
			desmethylparathion	4.8	
Coho salmon	[^{14}C]-naphthalene	Food, 0.13 mg	1-naphthol	7–61	170
			naphthalenediol	5–26	
English sole	1-[^{3}H]-naphthalene	Food, 0.29 μmol	1,2-dihydrodiolnaphthalene	35–85	171
			1- and 2-naphthol	5	
Mudsucker	[^{14}C]-naphthalene	Water, 32 μg/l (and 29 mg/l)	1,2-dihydrodiolnaphthalene	>50	121
			hydroxynaphthalene	7	
Rainbow trout	Naphthalene	Food, 73 mg/kg	1-naphthol	21–33	172
Rock sole	1-[^{3}H]-naphthalene	Food, 0.28 μmol	1- and 2-naphthol	1–16	9
Sand dab	[^{14}C]-naphthalene	Water, 32 μg/l (and 29 mg/l)	1,2-dihydrodiolnaphthalene	35–39	121
			hydroxynaphthalene	>50	
Sculpin	[^{14}C]-naphthalene	Water, 32 μg/l (and 29 mg/l)	1,2-dihydrodiolnaphthalene	7	121
			hydroxynaphthalene	>50	
Starry flounder	1-[^{3}H]-naphthalene	Food, 0.28 μmol	1,2-dihydrodiolnaphthalene	u	173
			1- and 2-naphthol	u	
Starry flounder	1-[^{3}H]-naphthalene	Food, 0.28 μmol	1,2-dihydrodiolnaphthalene	4–40	9
			1- and 2-naphthol	1.5–11	

Table 11. (continued)

Fish	Compound	Exposure	Metabolites	%	Ref.
Crayfish	[14C]-p-nitroanisole	Water, 0.01–0.1 mg/l	p-nitrophenol	31.7	169
Mosquitofish	Nitrobenzene	u	Nitrophenols, aminophenols	u	116
Crayfish	[14C]-p-nitro-m-cresole	Water, 0.01–0.1 mg/l	5-hydroxy-2-nitrobenzaldehyde	20.3	169
Rainbow trout	n-octylbenzene	Food, 25 µl	6-phenyl-1-hexanol	u	122
			8-phenyl-1-octanol	u	
			1-phenyl-1-octanol	u	
			1-phenyloctanoic acid	u	
Bluegill sunfish	Parathion	in vitro	Paraoxon	u	110
Brook trout	Parathion	u	Paraoxon	u	174
Brown trout	Parathion	u	Paraoxon	u	174
Golden shiner	Parathion	in vitro·	Paraoxon	u	110
Green sunfish	Parathion	in vitro	Paraoxon	u	110
Mosquitofish	Parathion	in vitro	Paraoxon	u	110
Yellow bullhead catfish	Parathion	in vitro	Paraoxon	u	110
Rainbow trout	[14C]-pentachloroanisole	Water, 24 µg/l	Pentachlorophenol	89	3
Carp	[14C]-cis-permethrin	in vitro	Hydroxy-cis-permethrin	29.3	175
Rainbow trout	[14C]-cis-permethrin	in vitro	Hydroxy-cis-permethrin	29.3	175
Carp	[14C]-trans-permethrin	in vitro	Hydroxy-trans-permethrin	13.3	175
Rainbow trout	[14C]-trans-permethrin	in vitro	Hydroxy-trans-permethrin	13.3	175
Mosquitofish	Phthalic anhydride	u	Phthalic acid	u	113, 116
Mosquitofish	Prolan	ME	1,1-bis-(p-chlorobiphenyl)-2-propane	3.2	176
			1,1-bis-(p-chlorobiphenyl)-acetic acid	5.6	
Mosquitofish	Propoxur	u	Isopropoxyphenol, 2-isopropoxyphenyl-N-hydroxymethylcarbamate, 2-isopropoxy-phenylcarbamate	u	113
Rainbow trout	1,3,6,8-TCDD	Water, 130–290 ng/l	Hydroxy-1,3,6,8-TCDD	70	177
Rainbow trout	[14C]-2,2',5,5'-tetrachloro-biphenyl	Water, 0.5 µg/l	Hydroxy-2,2',5,5'-tetrachlorobiphenyl	u	178

Species	Parent compound	Conditions	Metabolites	*	Ref
Mosquitofish	Tetrachloro-DDT	ME	Dicofol	u	179
Bluegill sunfish	[14C]-thidiazuron	Water, 0.1 µg/l	2-hydroxythidiazuron	u	180
Guppy	2,4,5-trichlorobiphenyl	Water, 8.5–15.8 µg/l	2,4,5-trichloro-4'-hydroxybiphenyl	u	181
			trichlorohydroxybiphenyl	u	
			trichlorodihydroxybiphenyl	u	
			trichlorohydroxymethoxybiphenyl	u	
			trichloromethoxybiphenyl	u	
Guppy	2,4,5-trichlorobiphenyl + 2,2',5,5'-tetrachlorobiphenyl	Water, 8.5–15.8 µg/l; 3.8–44 µg/l	Two trichlorohydroxybiphenyls	u	181
			Two trichlorohydroxymethoxybiphenyls	u	
			Tetrachlorohydroxybiphenyl	u	
Guppy	2,4,5-trichlorobiphenyl + 3,3',5,5'-tetrachlorobiphenyl	Water, 14–36 µg/l; 0.7–0.74 µg/l	Two trichlorohydroxybiphenyls	u	181
			Trichlorohydroxybiphenyl	u	
			Trichloromethoxybiphenyl	u	
			trichlorohydroxymethoxybiphenyl	u	
Goldfish	2,2',5-trichlorobiphenyl	in vitro	Hydroxytrichlorobiphenyl	u	182
Goldfish	2,4',5-trichlorobiphenyl	Water, 1 mg/l	Hydroxytrichlorobiphenyl	0.9	129
Topmouth gudgeon	2,4,6-trichlorophenyl-4'-nitrophenylether	u	Nitrophenol	u	183
Green sunfish	Trifluralin	u	$C_6H_2(NO_2)_2NH(C_3H_7)$	u	114

u = unknown

* = percentage of metabolites in organ of the total amount of the parent and its metabolites

Table 12. Reduction reactions in fish

Fish	Compound	Exposure	Metabolites	%	Ref.
Rainbow trout	DBP	ip, 5mg/kg	DBH	8.5	134
Bullhead	EPN	u	Amino-EPN	u	193
Sunfish	EPN	u	Amino-EPN	u	193
Mosquitofish	Nitrobenzene	u	Aniline	u	116
Fathead minnow	[^{14}C]-p-nitrophenol	Water, 4.1mg/l (and 44.1mg/l)	Aminophenol	0.6	194
Bullhead	Parathion	u	Aminoparathion, aminoparaoxon	u	193
Sunfish	Parathion	u	Aminoparathion, aminoparaoxon	u	193
Golden orfe	[^{14}C]-pentachloronitrobenzene	Water, 5μg/l	Pentachloroaminobenzene	0.8	195
Mosquitofish	Prolan	ME	1,1-bis-(p-chlorophenyl)-2-aminopropane	4.9	176

u = unknown

Table 13. Glucuronide conjugates formed in fish

Species	Compound	Exposure	Organ	β-glucuronidase treatment	%*	Ref.
Rainbow trout	Abietic acid	Water	Bile, blood	+	84	196
Bluegill sunfish	BP	in vitro	Liver	+	60	118
Fathead minnow	BP	in vitro	Liver	+	20	118
Rainbow trout	BP	in vitro	Liver	+	15	118
Bluegill sunfish	Carbaryl	u	Liver	u	19.2	123
Channel catfish	Carbaryl	u	Liver	u	13.8	123
Goldfish	Carbaryl	u	Liver	u	23.3	123
Kissing gourami	Carbaryl	u	Liver	u	31.6	123
Rainbow trout	[1-14C]-carbaryl	0.25 mg/l	u	+	u	124
Yellow perch	Carbaryl	u	Liver	u	28.0	123
Channel catfish	[14C]-chlordecone	1.36 mg/kg Food / 0.20 mg/kg im	Bile, gall bladder	+	19	128
Bitterling	[14C]-o-chlorophenol	3 mg/l	Water	Bile, water	+	u	133
Bream	[14C]-o-chlorophenol	3 mg/l	Water	Bile, water	+	u	133
Crucian carp	[14C]-o-chlorophenol	10 mg/l	Water	Bile, water	+	u	133
Goldfish	[14C]-o-chlorophenol	10 mg/l	Water	Bile, water	+	u	133
Gudgeon	[14C]-o-chlorophenol	2 mg/l	Water	Bile, water	+	u	133
Guppy	[14C]-o-chlorophenol	3 mg/l	Water	Bile, water	+	u	133
Minnow	[14C]-o-chlorophenol	4 mg/l	Water	Bile, water	+	u	133
Perch	[14C]-o-chlorophenol	4 mg/l	Water	Bile, water	+	u	133
Roach	[14C]-o-chlorophenol	5 mg/l	Water	Bile, water	+	u	133
Rudd	[14C]-o-chlorophenol	7 mg/l	Water	Bile, water	+	u	133
Three-spined stickleback	[14C]-o-chlorophenol	5 mg/l	Water	Bile, water	+	u	133
Tench	[14C]-o-chlorophenol	3 mg/l	Water	Bile, water	+	u	133
Bitterling	[14C]-m-cresol	4 mg/l	Water	Bile	+	u	133
Bream	[14C]-m-cresol	3 mg/l	Water	Bile	+	u	133
Crucian carp	[14C]-m-cresol	15 mg/l	Water	Bile	+	u	133
Goldfish	[14C]-m-cresol	15 mg/l	Water	Bile	+	u	133
Gudgeon	[14C]-m-cresol	8 mg/l	Water	Bile	+	u	133

Table 13. (continued)

Species	Compound	Exposure	Organ.	β-glucuronidase treatment	%	Ref.
Minnow	[14C]-m-cresol	4 mg/l	Water	Bile	+	u	133
Perch	[14C]-m-cresol	5 mg/l	Water	Bile	+	u	133
Roach	[14C]-m-cresol	8 mg/l	Water	Bile	+	u	133
Rudd	[14C]-m-cresol	8 mg/l	Water	Bile	+	u	133
Three-spined stickleback	[14C]-m-cresol	4 mg/l	Water	Bile	+	u	133
Tench	[14C]-m-cresol	5 mg/l	Water	Bile	+	u	133
Rainbow trout	DEHP	0.5 mg/l	Water	Bile	+	74	149
Rainbow trout	Dehydroabietic acid	u	Water	Bile, blood	+	91	196
Rainbow trout	2',5-dichloro-4'-nitro-salicylanilide (Bayer 73)	0.05 mg/l	Water	u	+	u	197
Rainbow trout	[3H]-dodecylcyclohexane	5 mg	Food	Urine	+	0.2	161
Killifish	Fenitrothion	0.1 mg/l	Water	Water	+	16.6–26.7	103
Mullet	Fenitrothion	0.1 mg/l	Water	Water	+	15.0–21.7	103
Cunner	No.2 fuel oil	1–40 mg/l	Water	Gall bladder/bile	–	u	198
Rainbow trout	Isopimaric acid	u	Water	Bile, blood	+	100	196
Rainbow trout	2-methyl-[8-14C]naphthalene	3.52 nmol/kg	ip	Bile	+	21.8	168
Rainbow trout	BNF + 2-methyl-[8-14C]-naphthalene	3.52 nmol/kg	ip	Bile	+	34.8	168
Crayfish	[14C]-methylparathion	0.1–0.01 mg/l	Water	Bile	u	2.3	169
Coho salmon	[14C]-naphthalene	u	Food, ip	Brain	–	35**	170
Coho salmon	[14C]-naphthalene	u	Food, ip	Liver		25**	170
Coho salmon	[14C]-naphthalene	u	Food, ip	Gall bladder		75**	170
Coho salmon	[14C]-naphthalene	u	Food, ip	Heart		13**	170
Coho salmon	[14C]-naphthalene	u	Food ip	Flesh		14**	170
English sole	[3H]-naphthalene	u	Food	Testes	u	37	171
English sole	[3H]-naphthalene	u	Food	Liver	u	50	171
English sole	[3H]-naphthalene	u	Food	Blood	u	35	171
Rainbow trout	Naphthalene	73 mg/kg	u	u	u	7	172
Rainbow trout	Naphthalene	7.3 mg/kg	u	u	u	66	172

Species	Chemical	Concentration	Administration	Tissue		Value	Ref.
Rock sole	1-[^{3}H]-naphthalene		Food, ip	Bile	—	80**	9
Rock sole	1-[^{3}H]-naphthalene		Food, ip	Liver	—	13**	9
Starry flounder	1-[^{3}H]-naphthalene		Food, ip	Bile	—	82***	9
Starry flounder	1-[^{3}H]-naphthalene		Food, ip	Liver	—	32**	9
Starry flounder	1-[^{3}H]-naphthalene	0.28 µmol	Food	u	u	u	173
Bitterling	[^{14}C]-1-naphthol	3 mg/l	Water	Bile, water	+	u	133
Bream	[^{14}C]-1-naphthol	3 mg/l	Water	Bile, water	+	u	133
Crucian carp	[^{14}C]-1-naphthol	5 mg/l	Water	Bile, water	+	u	133
Goldfish	[^{14}C]-1-naphthol	5 mg/l	Water	Bile, water	+	u	133
Gudgeon	[^{14}C]-1-naphthol	4 mg/l	Water	Bile, water	+	u	133
Guppy	[^{14}C]-1-naphthol	2 mg/l	Water	Bile, water	+	u	133
Minnow	[^{14}C]-1-naphthol	2 mg/l	Water	Bile, water	+	u	133
Perch	[^{14}C]-1-naphthol	4 mg/l	Water	Bile, water	+	u	133
Roach	[^{14}C]-1-naphthol	3 mg/l	Water	Bile, water	+	u	133
Rudd	[^{14}C]-1-naphthol	3 mg/l	Water	Bile, water	+	u	133
Three-spined stickleback	[^{14}C]-1-naphthol	2 mg/l	Water	Bile, water	+	u	133
Tench	[^{14}C]-1-naphththol	3 mg/l	Water	Bile, water	+	u	133
Crayfish	[^{14}C]-nitroanisole	0.1–0.01 mg/l	Water	Bile	u	0.2	169
Rainbow trout	[^{14}C]-pentachloroanisole	24 µg/l	Water	Fat, muscle	+	83	3
Rainbow trout	Pentachlorophenol	29 µg/l	Water	Bile	+	72	199
Goldfish	[^{14}C]-cis-permethrin and [^{14}C]-trans-permethrin	u	u	u	u	u	175
Rainbow trout	[^{14}C]-cis-permethrin and [^{14}C]-trans-permethrin	u	u	Bile	u	u	175
Bream	[^{14}C]-phenol	6 mg/l	Water	Water	+	19.7	200
Golden orfe ide	[U-^{14}C]-phenol	20 mg/l	Water	Water	—	4	201
Goldfish	[U-^{14}C]-phenol	20 mg/l	Water	Water	—	2.1	201
Goldfish	[U-^{14}C]-phenol	0.2 mg/l	Water	Water	—	2.1	201
Goldfish	[U-^{14}C]-phenol	2 mg/l	Water	Water	—	4.3	201
Goldfish	[U-^{14}C]-phenol	20 mg/l	Water	Water	—	5.4	201
Goldfish	[^{14}C]-phenol	10 µg/kg	ip	u	u	u	202
Phenol	Phenol	u	u	Bile	u	>10	203
Perch	[^{14}C]-phenol	4 mg/l	Water	Water	+	25.5	200
Rainbow trout	[U-^{14}C]-phenol	20 mg/l	Water	Water	—	30.19	201

Table 13. (continued)

Species	Compound	Exposure		Organ.	β-glucuronidase treatment	%	Ref.
Roach	[14C]-phenol	6 mg/1	Water	Water	+	9.6	200
Rudd	[14C]-phenol	6mg/1	Water	Water	+	27	200
Dogfish shark	Phenol red	u	u	Bile	u	24	204
Little skate	Phenol red	u	u	Bile	u	65	204
Rainbow trout	Pimaric acid	u	Water	Bile, blood	+	100	196
Rainbow trout	Sandaracopimaric acid	u	Water	Bile, blood	+	100	196
Rainbow trout	[14C]-2,2',5,5'-TCB	0.5 mg/1	Water	Bile	+	u	178
Rainbow trout	1,3,6,8-TCDD	130–230 mg/1	Water	Bile	+	70	177
Yellow perch	[1,6-3H]-2,3,7,8-TCDD	60 µg/kg	ip	Bile	+	u	205
Yellow perch	[1,6-3H]-2,3,7,8-TCDD	60 µg/kg	ip	Bile	+	u	206
Rainbow trout	3,4,5,6-tetrachloro guaiacol	u	Water	Bile, blood	+	68	199
Rainbow trout	2,3,4,6-tetrachlorophenol	u	Water	Bile, blood	+	69	199
Rainbow trout	2,3,4,6-tetrachlorophenol	71 µg/1	Water	Bile	+	81	199
Rainbow trout	3,4,5-trichloro guaiacol	u	Water	Bile, blood	+	75	199
Rainbow trout	2,4,6-trichlorophenol	u	Water	Bile, blood	+	90	199
Rainbow trout	2,4,6-trichlorophenol	50 µg/1	Water	Bile	+	84	199
Rainbow trout	4,5,6-trichlorophenol	u	Water	Bile, blood	+	82	199
Rainbow trout	3,4,5-trichlorosyringol	u	Water	Bile, blood	+	68	199
Largemouth bass	TFM	1.0 g/1	Water	Muscle, head	+	u	205
Rainbow trout	TFM	0.5 mg/1	Water	Bile	+	3	207
Rainbow trout	TFM	1.0 mg/1	Water	Bile	+	u	208
Rainbow trout	TFM	u	in vitro	Liver	+	u	208
Rainbow trout	TFM	1 mg/1	Water	Blood, bile	+	u	5
Sea lamprey	TFM	u	in vitro	Liver	+	u	208

* = percentage of total dose (u = unknown percentage)

** = percentage of glucuronides in organ of total amount of parent and metabolised compound

u = unknown

It is interesting that bluegill sunfish is able to glucuronidate benzo(a)pyrene *in vitro* to a 3 and 4 times greater extent than fathead minnow and rainbow trout [118]. This shows clearly that species differences exist in xenobiotic metabolism. In another study in an *in vitro* test with 1-naphthyl-N-methylcarbamate (carbaryl) [123] and in *in vivo* experiments with phenol [201] and phenol red [204] these species differences were also observed.

That glucuronidation does not occur to the same extent in brain, liver, gall bladder, heart and flesh of Coho salmon is shown by Roubal et al. [170]. The highest percentage of naphthalene glucuronides were found in the gall bladder.

The influence of dose was demonstrated by Krahn et al. [172]. A higher percentage of the administrated dose of naphthalene was conjugated at a lower dose (7.3 mg/kg) than at a higher dose (73 mg/kg) in rainbow trout. A slightly increased percentage of phenol was glucuronidated in goldfish with increasing phenol concentration in the water [201]. Consequently there is no simple relationship between the amount of a compound which is glucuronidated and the dose at which it is administered.

Finally, among the compounds which were glucuronidated, there are many phenolic agents, which is in good agreement with Table 2.

Glutathione Conjugates

Hitherto, glutathione conjugation (Table 14) has been examined in fish studies only with naphthalene. The scarce information reveals however, that relatively high percentages of glutathione conjugates were present in brain, flesh, gall bladder and liver [170]. This interesting observation indicates that in these extrahepatic organs glutathione conjugation may take place, as well as in the liver. The glutathione conjugation of naphthalene is in agreement with Table 2.

Table 14. Glutathione conjugates formed in fish

Fish	Compound	Exposure	Organ	%	Ref.
Coho salmon	[^{14}C]-naphthalene	Food, ip	Brain	11*	170
Coho salmon	[^{14}C]-naphthalene	Food, ip	Liver	22*	170
Coho salmon	[^{14}C]-naphthalene	Food, ip	Gall bladder	9*	170
Coho salmon	[^{14}C]-naphthalene	Food, ip	Flesh	6*	170
Coho salmon	[^{14}C]-naphthalene	Food, ip	Liver	14*	170
Rock sole	1-[^{3}H]-naphthalene	Food, ip	Bile	11*	9
Rock sole	1-[^{3}H]-naphthalene	Food, ip	Liver	14*	9
Starry flounder	1-[^{3}H]-naphthalene	Food, ip	Bile	9*	9
Starry flounder	1-[^{3}H]-naphthalene	Food, ip	Liver	10*	9
Starry flounder	1-[^{3}H]-naphthalene	Food; 0.28 μmol	u	u	173

* = percentage of total dose
u = unknown

Sulphate Conjugates

Sulphate conjugation has been found for carbaryl, endosulfan, methylparathion, naphthalene and (chloro)phenols (Table 15). This is in good agreement with Table 2 for the phenolic compounds. Naphthalene is probably oxidized before it is conjugated with sulphate.

From the dose dependent percentage of phenol conjugation in goldfish it was concluded by Nagel et al. [201] that sulphate conjugation may be limited. This may either be due to a limited pool of PAPS from which the sulphate is transferred, or to

Table 15. Sulphate conjugates formed in fish

Species	Compound	Exposure	Organ	%	Ref.
Bluegill sunfish	Carbaryl	in vitro	u	5.1	123
Catfish	Carbaryl	in vitro	u	3.1	123
Goldfish	Carbaryl	in vitro	u	1.2	123
Kissing gourami	Carbaryl	in vitro	u	1.7	123
Perch	Carbaryl	in vitro	u	4.9	123
Bitterling	[^{14}C]-o-chlorophenol	3 mg/1, water	Bile, water	u	133
Bream	[^{14}C]-o-chlorophenol	3 mg/1, water	Bile, water	u	133
Crucian carp	[^{14}C]-o-chlorophenol	10 mg/1, water	Bile, water	u	133
Goldfish	[^{14}C]-o-chlorophenol	10 mg/1, water	Bile, water	u	133
Gudgeon	[^{14}C]-o-chlorophenol	2 mg/1, water	Bile, water	u	133
Guppy	[^{14}C]-o-chlorophenol	3 mg/1, water	Bile, water	u	133
Minnow	[^{14}C]-o-chlorophenol	4 mg/1, water	Bile, water	u	133
Perch	[^{14}C]-o-chlorophenol	4 mg/1, water	Bile, water	u	133
Roach	[^{14}C]-o-chlorophenol	5 mg/1, water	Bile, water	u	133
Rudd	[^{14}C]-o-chlorophenol	7 mg/1, water	Bile, water	u	133
Three-spined stickleback	[^{14}C]-o-chlorophenol	5 mg/1, water	Bile, water	u	133
Tench	[^{14}C]-o-chlorophenol	3 mg/1, water	Bile, water	u	133
Bitterling	[^{14}C]-m-cresol	4 mg/1, water	Bile, water	u	133
Bream	[^{14}C]-m-cresol	3 mg/1, water	Bile, water	u	133
Crucian carp	[^{14}C]-m-cresol	15 mg/1, water	Bile, water	u	133
Goldfish	[^{14}C]-m-cresol	15 mg/1, water	Bile, water	u	133
Gudgeon	[^{14}C]-m-cresol	8 mg/1, water	Bile, water	u	133
Guppy	[^{14}C]-m-cresol	3 mg/1, water	Bile, water	u	133
Minnow	[^{14}C]-m-cresol	4 mg/1, water	Bile, water	u	133
Perch	[^{14}C]-m-cresol	5 mg/1, water	Bile, water	u	133
Roach	[^{14}C]-m-cresol	8 mg/1, water	Bile, water	u	133
Rudd	[^{14}C]-m-cresol	8 mg/1, water	Bile, water	u	133
Three-spined stickleback	[^{14}C]-m-cresol	4 mg/1, water	Bile, water	u	133
Tench	[^{14}C]-m-cresol	15 mg/1, water	Bile, water	u	133
Catfish, *H. fossilis*	Endosulfan	0.5 µg/1, water	Several tissues	u	162
Catfish, *M. carasius*	Endosulfan	0.5 µg/1, water	Several tissues	u	162
Catfish, *M. vittatus*	Endosulfan	0.5 µg/1, water	Several tissues	u	162
Macrognathus aculeatum	Endosulfan	0.5 µg/1, water	Liver, kidney	u	163
Crayfish	[^{14}C]-methylparathion	0.1–0.01 mg/1, water	Soft tissues and water	5.8	169

Table 15. (continued)

Species	Compound	Exposure	Organ	%	Ref.
Rainbow trout	Naphthalene	73 mg/kg u	u	19	172
Rainbow trout	Naphthalene	7.3 mg/kg u	u	13	172
Rock sole	1-[^{3}H]-naphthalene	Food, i.p.	Bile	1	9
Rock sole	1-[^{3}H]-naphthalene	Food, i.p.	Liver	14	9
Starry flounder	1-[^{3}H]-naphthalene	Food, i.p.	Bile	1.4	9
Starry flounder	1-[^{3}H]-naphthalene	Food, i.p.	Liver	2	9
Starry flounder	1-[^{3}H]-naphthalene	0.28 μmol, food	u	u	173
Bitterling	[^{14}C]-1-naphthol	3 mg/1, water	Bile, water	u	133
Bream	[^{14}C]-1-naphthol	3 mg/1, water	Bile, water	u	133
Crucian carp	[^{14}C]-1-naphthol	5 mg/1, water	Bile, water	u	133
Goldfish	[^{14}C]-1-naphthol	5 mg/1, water	Bile, water	u	133
Gudgeon	[^{14}C]-1-naphthol	4 mg/1, water	Bile, water	u	133
Guppy	[^{14}C]-1-naphthol	2 mg/1, water	Bile, water	u	133
Minnow	[^{14}C]-1-naphthol	2 mg/1, water	Bile, water	u	133
Perch	[^{14}C]-1-naphthol	4 mg/1, water	Bile, water	u	133
Roach	[^{14}C]-1-naphthol	3 mg/1, water	Bile, water	u	133
Rudd	[^{14}C]-1-naphthol	3 mg/1, water	Bile, water	u	133
Three-spined stickleback	[^{14}C]-1-naphthol	2 mg/1, water	Bile, water	u	133
Tench	[^{14}C]-1-naphthol	3 mg/1,water	Bile, water	u	133
Crayfish	[^{14}C]-nitroanisole	0.1–0.01 mg/1, water	Soft tissues and water	5.8	169
Carp	[^{14}C]-penta-chlorophenol	in vitro	Liver	u	209
Goldfish	Pentachlorophenol	Water	u	u	210
Goldfish	Pentachlorophenol	Water	Water	u	211–213
Goldfish	[^{14}C]-penta-chlorophenol	in vitro	Liver	u	209
Rainbow trout	[^{14}C]-penta-chlorophenol	in vitro	Liver	u	209
Rainbow trout	Pentachlorophenol	29 μg/1, water	Bile	25	199
Bream	[^{14}C]-phenol	6 mg/1, water	Medium	34	200
Fathead minnow	[^{14}C]-phenol	4 mg/1, water	Medium	19	200
Golden orfe ide	[^{14}C]-phenol	20 mg/1, water	Medium	22	201
Goldfish	[^{14}C]-phenol	0.2 mg/1, water	Medium	32	201
Goldfish	[^{14}C]-phenol	2 mg/1, water	Medium	18	201
Goldfish	[^{14}C]-phenol	20 mg/1, water	Medium	14	201
Goldfish	[^{14}C]-phenol	20 mg/1, water	Medium	33	201
Goldfish	[^{14}C]-phenol	10 mg/1kg, i.p.	u	u	202
Goldfish	[^{14}C]-phenol	20 mg/1, water	Medium	45	200
Guppy	[^{14}C]-phenol	3 mg/1, water	Medium	42	200
Perch	[^{14}C]-phenol	4 mg/1, water	Medium	5	200
Rainbow trout	[^{14}C]-phenol	20 mg/1, water	Medium	19	201
Roach	[^{14}C]-phenol	6 mg/1, water	Medium	35	200
Rudd	[^{14}C]-phenol	6 mg/1, water	Medium	23	200
Tench	[^{14}C]-phenol	4 mg/1, water	Medium	38	200
Rainbow trout	2,3,4,6-tetrachlorophenol	71 μg/1, water	Bile	25	199
Rainbow trout	2,4,6-trichlorophenol	50 μg/1, water	Bile	9	199

u = unknown

Table 16. Acetylation conjugates formed in fish

Fish	Compound	Exposure	Organ	Conjugate	%	Ref.
Dogfish shark	o-, m-, p-aminobenzoic acid	i.m.	Urine	Glycine	u	214
Dogfish shark	p- and m-aminobenzoic acid	in vitro	u	Acetyl/glycine	u	214
Dogfish shark	p- and m-aminobenzoic acid	in vitro	Kidney	Acetyl/glycine	u	214
Lophius americanus	p- and m-aminobenzoic acid	in vitro	u	Acetyl/glycine	u	214
Winter flounder	p- and m-aminobenzoic acid	in vitro	Kidney	Acetyl/glycine	u	214
Winter flounder	p- and m-aminobenzoic acid	in vitro	u	Acetyl/glycine	u	214
Mosquitofish	Aniline	Water, ME	Whole	Acetyl	u	113
Mosquitofish	Benzoic acid	Water, ME	Whole	Glycine	u	113, 116
Bluegill sunfish	2,4-D	Water, 2 mg/l	Fillet, head, viscera	u	u	215
Dogfish shark	2,4-D	in vitro	Urine	Taurine	>90	216
Winter flounder	2,4-D	in vitro	Urine	Taurine	>90	216
Rainbow trout	[³H]-dodecylcyclohexane	Food, 5 mg	Urine	Taurine	3.6	161
Tilapia mossambica	Fenitrothion	200 mg/kg	Liver, kidney, brain	Acetyl	u	4
Channel catfish	p-nitrobenzoic acid.	in vitro	Liver, kidney, brain	Acetyl	u	4
Flathead catfish	p-nitrobenzoic acid	in vitro	Liver, kidney, brain	Acetyl	u	4
Mosquitofish	p-nitrobenzoic acid	Water, ME	Whole	Acetyl	u	116
Rainbow trout	p-nitrobenzoic acid	in vitro	u	Acetyl	u	116
Dogfish shark	Phenyl acetic acid	in vitro	Urine	Taurine	>90	216
Winter flounder	Phenyl acetic acid	in vitro	Urine	Taurine	>90	216
Rainbow trout	Tricaine (MS 222)	Water, 100 mg/l	Urine	Acetyl	11–20	217

u = unknown

a higher activity of glucuronyltransferase than sulphotransferase at higher phenol concentrations [201].

All fishes in Table 15 conjugate phenol with sulphate at percentages between 5 and 45%, carbaryl (in vitro) between 1.2 and 5.1% [120] and naphthalene between 1 and 19% of the dose [9, 172]. This indicates that there exists a species difference in sulphate conjugation as has also been observed for other conjugations. This means that different compounds are not conjugated to similar extents.

Acetyl, Methyl and Glycoside Conjugates

Only three conjugates have been found which can be acetylated with xenobiotics: acetyl, taurine and glycine (Table 16). However, no quantitative data are available, except of the taurine conjugation of two biotransformation products of dodecylcyclohexane in rainbow trout [161].

Compounds which were found to be acetylated were benzoic acids, phenols and aromatic amines, which is in agreement with Table 2.

Very little information is also available about methyl and glycoside conjugation (Tables 17 and 18), although these conjugations may be very important to the biotransformation of some compounds. For instance quintozene (pentachloronitro-benzene) was methylated for 67% in golden orfe [195]. In addition, p-nitroanisole, p-nitrocresol and methylparathion were conjugated with glycosides in crayfish for 57.9, 15.0 and 30.7%, respectively [169].

Table 17. Methyl conjugates formed in crayfish

Species	Compound	Exposure	Organ	%	Ref.
Mosquitofish	Aniline	Water, ME	Whole	u	113, 116
Golden orfe	[^{14}C]-pentachloronitrobenzene	5 μg/l, Water	u	67	195
Goldfish	2, 4′, 5-trichlorobiphenyl	1 mg/l, water	u	u	129

u = unknown

Table 18. Glycoside conjugates formed in fish

Fish	Compound	Exposure	Organ	%	Ref.
Crayfish	[^{14}C]-p-nitroanisole	0.1–0.01 mg/l, water	Water	57.9	169
Crayfish	[^{14}C]-p-nitro-cresol	0.1–0.01 mg/l, water	Water	15.0	169
Crayfish	[^{14}C]-methylparathion	0.1–0.01 mg/l, water	Water	30.7	169

Other Reactions

Compounds which were biotransformed in fish, but of which no metabolites were identified, are summarized in Table 19. Many of the biotransformation products could be characterized as polar compounds from thin layer chromatography.

Table 19. Other reactions

Fish	Compound	Exposure	Metabolites	%	Ref.
Green sunfish	Aldrin	u	u	u	114
Lake trout	Aldrin	u	u	u	190
Mosquitofish	Aniline	u	Polar	u	113, 116
Mosquitofish	Anisole	u	Polar	u	113, 116
Coho salmon	[14C]-anthracene	ip, 28 µg	u	290	170
Fathead minnow	[14C]-benz(a)acridine	Water, 78 µg/l	u	90	218
Coho salmon	[14C]-benzene	ip, 25 µg	u	0–71	170
Mosquitofish	Chlorpyrifos	u	Polar	u	113
Rainbow trout	DBP	ip, 5 mg/kg	u	8.8	134
Rainbow trout	DDA	ip, 5 mg/kg	u	2.4	134
Mosquitofish	DDD	u	Polar	u	137
Rainbow trout	DDD	ip, 5 mg/kg	u	0.2	134
Green sunfish	DDE	u	Polar	u	219
Mosquitofish	DDE	u	Polar	u	137, 220
Green sunfish	DDT	u	Polar	u	219
Mosquitofish	DDT	ME, 4 µg/l	Polar	1.5	143
Mosquitofish	DDT	ME	Polar	1.5	137
Sole	DDT	Food, 1.7, 17 or 35 µg	Polar	<20	144
Channel catfish	DEHP	Water, 1 µg/l	u	2	144
Guppy	DEHP	u	u	u	147
Rainbow trout	DEHP	Water, 0.5 mg/l	2 polar compounds	21.5	148
Rainbow trout	DEHP	u	2 polar compounds	u	221
Carp	Diazinon	in vitro	u	1–30	222
Winter skate	Diazinon	in vitro	u	0–61	222
Channel catfish	Di-n-butylphthalate	in vitro	3 unidentified	u	146
Mosquitofish	2,6-diethylaniline	u	Polar	u	113, 116
Bluegill sunfish	2,4-D dimethylamine salt	Water, 2 mg/l	Incorporation in fish, CO_2	u	215
Mosquitofish	Di-n-octylphthalate	u	u	u	113
Mosquitofish	Di-n-octylphthalate	ME	Polar	21.5	158
Lake trout	2,5-diphenyloxazole	u	u	u	21, 191

Species	Compound	Dose	Metabolite	Value	Ref.
Carp	Fenitrothion	in vitro	u	0–15	222
Tilapia mossambica	Fenitrothion	200 mg/kg	3-methyl-4-nitrophenol	u	4
Winter skate	Fenitrothion	in vitro	u	0–20	222
Mosquitofish	Hexachlorobenzene	u	Polar	u	113, 116
Carp	Malathion	in vitro	u	61	222
Winter skate	Malathion	in vitro	u	30–77	222
Mosquitofish	Methiochlor	ME, 73 µg/l	Polar	12.1	143
Mosquitofish	Methoxychlor	ME, 1.6 µg/l	u	48.5	143
Mosquitofish	Methoxychlor	u	Polar	u	137, 223
Mosquitofish	Methylchlorpyrifos	u	Polar	u	113
Rainbow trout	[14C]-methylnaphthalene	Water, 0.5 mg/l	u	<1–70	224
Coho salmon	[14C]-naphthalene	ip, 0.13 µg	u	1–72	170
Rainbow trout	[14C]-naphthalene	Water, 0.5 mg/l	u	<1–70	224
Rainbow trout	Naphthalene	Food, 0.9 µmol	Polar	u	225
		ip, 9.2 mg/kg	Polar	u	
		Water, 72 mg/l	Polar	u	
			Polar	u	
Mosquitofish	Nitrobenzene	u	Polar	u	116
Fathead minnow	[14C]-p-nitrophenol	Water, 4.1 or 44.1 mg/l	u	95.9	194
Brook trout	n-octano-[1-14C]-hydroxyamic acid	Water, 7 mg/l	u	25–50	226
Green sunfish	2,2':4,5,5'-pentachlorobiphenyl	u	Polar	u	219
Mosquitofish	2,2':4,5,5'-pentachlorobiphenyl	u	Polar	u	113
Golden orfe	[14C]-pentachloronitrobenzene	Water, 5 µg/l	Substitution nitro group by -SH or -OH, then conjugation	79.6	195
Mosquitofish	Pentachlorophenol	u	Polar	u	116
Carp	[14C]-cis-permethrin	in vitro	Hydrolysis product	3.5	175
Carp	[14C]-trans-permethrin	in vitro	Hydrolysis product	62	175
Rainbow trout	[14C]-cis-permethrin	in vitro	Hydrolysis product	3.5	175
Rainbow trout	[14C]-trans-permethrin	in vitro	Hydrolysis product	62	175
Fathead minnow	[14C]-phenol	Water, 2.5 or 32.7 mg/l	u	91.3	194
English sole	Poly-aromatic hydrocarbons	u	Hydroxy- and quinone-derivatives	u	227
Rainbow trout	Pristane	Food, 0.1 mg	u	2.6	228

Table 19. (continued)

Fish	Compound	Exposure	Metabolites	%	Ref.
Mosquitofish	Prolan	ME	u	6.5	176
Mosquitofish	Prolan	ME	Polar	76.6	176
Mosquitofish	Propoxur	u	Polar	u	113
Rainbow trout	[1,6-^{3}H]-2,3,7,8-TCDD	ip, 60 µg/kg	3 polar compounds	u	11
Yellow perch	[1,6-[^{3}H]-2,3,7,8-TCDD	ip, 60 µg/kg	4 polar compounds	u	206
Green sunfish	2,2',5,5'-tetrachlorobiphenyl	u	Polar	u	219
Mosquitofish	2,2',5,5'-tetrachlorobiphenyl	u	Polar	u	113, 220
Rainbow trout	2,2',5,5'-tetrachlorobiphenyl	u	u	u	229
Bluegill sunfish	Thidiazuron	Water, 0.1 mg/l	Phenylurea	u	180
Channel catfish	Thidiazuron	Water, 0.1 mg/l	Phenylurea	u	180
Carp	[U-^{14}C]-1,2,4-trichlorobenzene	Water, 18 µg/l	Polar	60	230
Rainbow trout	[U-^{14}C]-1,2,4-trichlorobenzene	Water, 18 µg/l	Polar	60	230
Bullhead	[^{14}C]-2,2',5-trichlorobiphenyl	in vitro	2 polar compounds	0.25	182
Channel catfish	2,2',5-trichlorobiphenyl	in vitro	Polar	u	182
Goldfish	[^{14}C]-2,2',5-trichlorobiphenyl	in vitro	2 polar compounds	7.7	182
Green sunfish	2,2',5-trichlorobiphenyl	u	Polar	u	219
Mosquitofish	2,2',5-trichlorobiphenyl	u	Polar	u	113, 220
Rainbow trout	[^{14}C]-2,2',5-trichlorobiphenyl	in vitro	Polar	u	182
Rainbow trout	2,2',5-trichlorobiphenyl	in vitro	Polar compounds	0.25	182
Fathead minnow	[^{14}C]-2,4,5-trichlorophenol	Water, 4.8 or 49.3 mg/l	u	21.4	194

u = unknown

Table 20. Unknown conjugates formed in fish

Fish	Compound	Exposure	Organ	%	Ref.
Mosquitofish	Aldrin	Water, ME	Whole	u	116
Bluegill sunfish	cis-[14C]-chlordane	Water, 5 μg/l	u	u	126
Mosquitofish	Chloro-methylchlor	Water, ME	Whole	u	142
Perch	Chlorophenols	u	Bile	99	231
Roach	Chlorophenols	u	Bile	99	231
Rainbow trout	Chlorophenols	Water	Bile, blood	>96	198
Rainbow trout	Chlorophenols	Water, 0.5–5 μg/l	Blood	97	232
Mosquitofish	DDT-analogs	Water, ME	u	u	142
Channel catfish	DEHP	Water, 1 mg/l	Whole	14	146
Fathead minnow	DEHP	Water	Whole	u	146
Bluegill sunfish	[14C]-endrin	Water, 1 μg/l	u	u	20
Mosquitofish	Ethoxychlor	Water, ME	Whole	u	165
Cunner	No. 2 fuel oil	Water, 1–40 mg/l	Bile, gall bladder	u	231
Goldfish	[14C]-heptachlor	0.868 μg/g i.m.	Water, fish, faeces	10.2	166
Mosquitofish	Methoxy-methiochlor	Water, ME	Whole	u	142
Mosquitofish	Methylchlor	Water, ME	Whole	u	165
Mosquitofish	Methyl-ethoxychlor	Water, ME	Whole	u	142
Coho salmon	[14C]-naphthalene	Food, i.p.	Brain	17*	170
Coho salmon	[14C]-naphthalene	Food, i.p.	Liver	2*	170
Coho salmon	[14C]-naphthalene	Food, i.p.	Gall bladder	2*	170
Coho salmon	[14C]-naphthalene	Food, i.p.	Flesh	26*	170
Mudsucker	[14C]-naphthalene	u	Gall bladder	u	121
Sand dab	[14C]-naphthalene	u	Gall bladder	u	121
Sculpin	[14C]-naphthalene	u	Gall bladder	u	121
Golden orfe	[14C]-pentachloronitrobenzene	Water, 5 μg/l	u	70	195
Bluegill sunfish	cis-[14C]-photochlordane	Water, 5 μg/l	u	u	126
Mosquitofish	Phthalic anhydride	Water, ME	Whole	99	113, 116
Perch	Resin acids	u	Bile	99	231
Rainbow trout	Resin acids	Water	Bile, blood	>99	196
Rainbow trout	Resin acids	Water, 3.5–13 μg/l	Blood	95	232
Roach	Resin acids	u	Bile	99	231
Largemouth bass	TFM	Water, 1.0 g/l	u	u	205

u = unknown

* = percentage of metabolites in organ of the total amount of parent compound and metabolites

Table 20 summarizes compounds which were believed to be conjugated, since hydroxylated compounds were found after e.g. acid hydrolysis. The nature of the conjugate however has still to be elucidated. These conjugates may be glucuronides, glutathiones or other conjugates which have been mentioned already.

General

From the biotransformation reactions presented in Tables 11 to 20 some general remarks can be made and some degradation pathways of some well investigated compounds can be deduced.

An illustrative example of several degradable groups in one compound is fenitrothion. In Fig. 13 the biotransformation products of fenitrothion in *Tilapia mossambica* (4) are shown. The following reactions take place: oxidative desulphuration with fenitrooxon as the product, O-demethylation with desmethyl-fenitrothion as the product, phosphor oxygen deesterification, which resulted in the hydrolysis to 3-methyl-4-nitrophenol and nitroreduction followed by acetylation to N-acetylamino-fenitrothion.

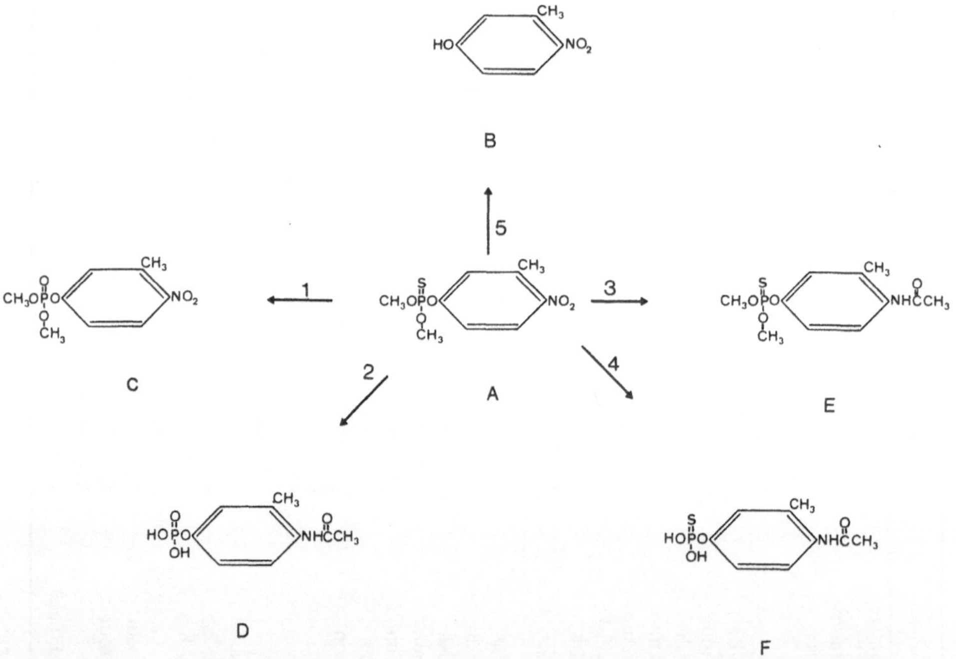

Fig. 13. Biotransformation of fenitrothion in *Tilapia mossambica* (4). A: fenitrothion; B: 3-methyl-4-nitrophenol; C: fenitrooxon; D: desmethyl-N-acetylaminofenitrooxon; E: N-acetylaminofenitrothion; F: desmethyl-N-acetylaminofenitrothion; 1: oxidative desulphuration; 2: oxidative desulphuration, followed by oxidative demethylation; 3: nitroreduction, followed by acetylation; 4: nitroreduction, followed by acetylation, followed by oxidative demethylation; 5: phosphoroxygen deesterification (hydrolysis)

Since the use of the insecticide DDT led to accumulation in several species, e.g. fish, other compounds have been synthesized which should have the same acetylcholinesterase inhibition activity as DDT, but should be more degradable. DDT-isosteric compounds such as prolan, methiochlor, methoxychlor and methyl-ethoxychlor were therefore tested in biotransformation experiments. From these experiments it appeared that these DDT-isosteric compounds were indeed better degradable than DDT [142]. The higher degradability was due to an easier oxidation or reduction of e.g. the aliphatic or nitrogroups of these compounds. The resulting hydroxy or aminoderivatives are good substrates for conjugation and subsequently could be readily excreted, which prohibits accumulation.

To reduce the persistence of compounds with comparable dielectric and physical properties as PCBs (polychlorinated biphenyls), isopropyl-polychlorinated biphenyls such as Chloroalkylene-9 were synthesized. Although the isopropyl group is expected to be readily degradable, hitherto no evidence was found whether or not fish do degrade these compounds better than PCBs [129, 132]. The hydroxylated metabolites of these isopropyl-polychlorinated biphenyls which are found in the water could have been formed by microorganisms in the water.

Finally, an important number of biotransformation studies dealt with the insecticides aldrin, dieldrin, endrin, chlordane and heptachlor.

Concluding Remarks

Information on the capacities of the fish to metabolize xenobiotics and the information derived from biotransformation studies are difficult to compare. For many compounds no comparisons can be made, since very little information is available about the extents of metabolism on the one hand and MFO-activities, cytochrome P-450 concentrations and conjugating activities on the other hand.

For only a few compounds comparisons can be made for a selected number of fish between metabolizing capacities and "proven" metabolism of xenobiotics. For instance bluegill sunfish, carp and rainbow trout have comparable cytochrome P-450 levels (Table 5) and have AHH-activities in the order: bluegill sunfish > carp > rainbow trout. The capacities to oxidize benzo(a)pyrene (AHH-activity), are however not in the same order as the extent of metabolism of benzo(a)pyrene in these fish (Table 11).

When phenolglucuronidation (Table 13) and the ability to glucuronidate p-nitrophenol (Table 8) is compared, also no conclusions can be drawn based on the limited number of data which are available. For instance roach has the highest activity to conjugate p-nitrophenol, but glucuronidates phenol to almost the lowest extent among the fish from which both the UDPGA-t-activities and phenol-glucuronidation data were available.

From the three fish whose cytochrome P-450 concentrations and the extent of carbaryl-oxidation are known, kissing gourami oxidized carbaryl in the highest extent and had the highest cytochrome P-450 concentration. In this case a high amount of carbaryl which was oxidized seems to parallel a high cytochrome P-450 level.

References

1. Brodie B B, Maickel R G (1962) Proc. First Int. Pharmacol. Meeting 6: 299
2. Lech J J, Vodicnik M J (1985) In: Fundamentals of Aquatic Toxicology (Rand G M, Petrocelli S R (eds). Hemisphere, Washington, p 526
3. Glickman A H, Statham C N, Wu A, Lech J J (1977) Toxicol. Appl. Pharmacol. 41: 649
4. Anjum F, Qadri S S H (1986) Bull. Environ. Contam. Toxicol. 36: 140
5. Lech J J (1974) Biochem. Pharmacol. 23: 2403
6. Lech J J, Vodicnik M J, Elcombe C R (1982) In: Weber L J (ed) Aquatic Toxicology. Raven, New York, p 107
7. Bend J R, James M O (1978) Biochem. Biophys. Persp., Mar. Biol. 4: 125
8. Williams R T (1947) In: Detoxication mechanisms. Wiley, New York
9. Varanasi U, Gmur D J, Treseler P A (1979) Arch. Environ. Contam. Toxicol. 8: 673
10. Buhler D R, Rasmusson M E (1968) Comp. Biochem. Physiol, 25: 223
11. Kleeman J M, Olson J R, Chen S M, Peterson R E (1986) Toxicol. Appl. Pharmacol. 83: 391
12. Omura T, Sato R (1964) J. Biol. Chem. 239: 2370
13. Johannesen K A M, DePierre J W (1978) Anal. Biochem. 86: 725
14. James M O, Bend J R (1980) Toxicol. Appl. Pharmacol. 54: 117
15. Stegeman J J (1981) In Gelboin H V, Ts'O P O P (eds) Polycyclic hydrocarbons and cancer. Vol. 3, Academic, New York, p 1
16. Ahokas J T (1979) ACS Symp. Series 99 (Pesticides and Xenobiotic Metabolism in Aquatic Organisms), 279
17. Stegeman J J, Kloepper-Sams P J (1987) Environ. Health Persp. 71: 87
18. Funari F, Zoppini A, Verdino A, De Angelis G, Vittozzi L (1987) Ecotoxicol. Environ. Safety 13: 24
19. Fabacher D L (1982) Comp. Biochem. Physiol. 73C: 285
20. Stegeman J J, Pajor A M, Thomas P (1982) Biochem. Pharmacol. 31: 3978
21. Ahokas J T, Pelkonen O, Kärki N T (1977) Xenobiotica 7: 104
22. Schwen R J, Mannerung G J (1982) Comp. Biochem. Physiol. 71B: 445
23. Hofe E v, Puffer H W (1986) Arch. Environ. Contam. Toxicol. 15: 251
24. Ankley G T, Blazer V S, Reinert R E, Agosin M (1986) Aquatic Toxicol. 9: 91
25. Tate L G (1988) Arch. Environ. Contam. Toxicol. 17: 325
26. Goksøyr A, Andersson T, Hansson T, Klungsøyr J, Zhang Y, Förlin, L. (1987) Toxicol. Appl. Pharmacol 89: 347
27. Lindström-Seppa P, Koivuşaari U, Hänninen O (1983) Aquatic Toxicol. 3: 35
28. Verdina A, De Angelis G, Funari E, Testai E, Vittozzi L (1987) Comp. Biochem. Physiol. 88B: 619
29. Schell J D Jr, Cooper K O, Cooper K R (1987) Environ. Toxicol. Chem. 6: 717
30. Ahokas J T, Pelkonen O (1984) Mar. Environ. Res. 14: 59
31. Lindström-Seppa P, Koivusaari U, Hänninen O (1981) Comp. Biochem. Physiol. 69C: 259
32. Förlin L, Haux C, Karlsson-Norrgren L, Runn P, Larsson A (1986) Aquatic Toxicol. 8: 51
33. Eisele T A, Coulombe R A, Pawlowski N E, Nixon J E (1984) Aquatic Toxicol 5: 211
34. Elcombe C R, Fanklin R B, Lech J J (1979) ACS Symp. Series 99 (Pesticides and Xenobiotic Metabolism in Aquatic Organisms), 319
35. Kleinow K H, Haasch M L, Lech J J (1986) Aquatic Toxicol. 8: 231
36. Gregus Z, Watkins J R, Thompson T N, Harvey M E, Rozman K, Klaassen C D (1983) Toxicol. Appl. Pharmacol. 67: 430
37. Lindström-Seppa P, Koivusaari U, Hänninen O (1981) Comp. Biochem. Physiol. 68C: 121
38. Förlin L, Hansson T (1982) J. Endocrinol. 94: 245
39. Elcombe C R, Lech J J (1979) Toxicol. Appl. Pharmacol. 49: 437
40. Pesonen M, Celander M, Förlin L, Andersson T (1987) Toxicol. Appl. Pharmacol. 91: 75
41. French J S (1984) Mar. Environ. Res. 14: 407
42. James M O, Little P J (1981) Chem.-Biol. Interactions 36: 229
43. Foureman G L, White N B Jr, Bend J R (1983) Can. J Fish. Aquat. Sci. 40: 854
44. Souhaili-el Amri H, Batt A M, Siest G (1986) Xénobiotica 16: 351
45. Braune P U, Kremers P G, Kaninsky L S, de Groeve J, Albert A, Guengerich F. P (1986) Drug

Metab. Disp. 14: 437
46. Siddiqui M K J, Anjum F, Mahboob M, Quadri S S H (1986) J. Environ. Sci. Health B21: 115
47. Ade P, Chiesari E, Funari E, Ramundo Orlando A, Vittozzi L, Marabini L (1983) Arch. Toxicol. suppl. 6: 335
48. Davies J M, Bell J S, Houghton C (1984) Mar. Environ. Res. 14: 23
49. Bend J R, James M O, Dansette P M (1977) Ann. N. Y. Acad. Sci. 298: 505
50. Binder R L, Melancon M J, Lech J J (1984) Drug Metab. Rev. 15: 697
51. James M O, Khan M A Q, Bend J R (1979) Comp. Biochem Physiol. 62C: 155
52. Fair P H (1986) Arch. Environ. Contam. Toxicol. 15: 257
53. DeWaide J H (1971) Ph. D. Thesis, Nijmegen
54. Melancon M J, Lech J J (1983) Aquatic Toxicol. 4: 51
55. Gruger E H Jr, Hruby T, Karrick N L (1976) Environ. Sci. Technol. 10: 1033
56. Stegeman J J, Kloepper-Sams P J, Farrington J W (1986) Science 231: 1287
57. James M O, Heard C S, Hawkins W E (1988) Aquatic Toxicol. 12: 1
58. Laitinen M, Nieminen M, Hietanen E (1982) Acta Pharmacol. Toxicol 51: 24
59. Förlin L (1980) Toxicol. Appl. Pharmacol. 54: 420
60. Andersson T, Pesonen M, Johansson C (1985) Biochem. Pharmacol. 34: 3309
61. Gerhart E, Carlson R M (1978) Environ. Res. 17: 284
62. Koivusaari U, Harri M, Hänninen O (1981) Comp. Biochem. Physiol. 70C: 149
63. Bend J R, Foureman G L, Ben-Zvi Z, Albro P W (1984) Natl. Canc. Inst. Monogr. 65: 359
64. Cantrell E, Abreu M, Busbee D (1976) Biochem. Biophys. Res. Comm. 70: 474
65. Dehnen W, Tomingas R, Roos J (1973) Anal. Biochem. 53: 373
66. Nebert D W, Gelboin H V (1968) J. Biol. Chem. 243: 6242
67. Van Cantfort J, de Graeve J, Gielen J E (1977) Biochem. Biophys. Res. Comm. 79: 505
68. DePierre J W, Moron M S, Johannesen K A M, Fenster L (1975) Anal. Biochem. 63: 470
69. Nesnow S, Fahl W E, Jefcoate C R (1977) Anal. Biochem. 80: 258
70. Burke M D, Mayer R T (1974) Drug Metab. Disp. 2: 583
71. Ullrich V, Weber P (1972) Hoppe-Seyler's Z. Physiol. Chem. Bd. 353: 1171
72. Bakke J E, Larsen G L, Aschbacher P W, Rafter J J, Gustafsson J-Å, Gustafsson B E (1981) In: Sulfur and pesticide action and metabolism, p 165
73. Layiwola P J, Linnecar D F C, Knights B (1983) Xenobiotica 13: 27
74. Netter K J (1980) Pharmac. Ther. 10: 515
75. Lidman U, Förlin L, Molander O, Axelson G (1976) Acta Pharmacol. et Toxicol. 39: 262
76. Schoor W P, Srivastava M (1984) Mar. Environ. Res. 14: 448
77. Lech J J, Bend J R (1980) Environ. Health Persp. 34: 115
78. Williams D E, Bender R C, Morrissey M T, Selivonchick D P, Buhler D R (1984) Mar. Environ. Res. 14: 13
79. Williams D E, Okita R T, Buhler D R, Siler Masters B S (1984) Arch. Biochem. Biophys. 231: 503
80. Klotz A V, Stegeman J J, Walsh C (1984) Mar. Environ. Res. 14: 402
81. Bend J R, Foureman C L (1984) Mar. Environ. Res. 14: 405
82. Vodicnik M J, Elcombe C R, Lech J J (1981) Toxicol. Appl. Pharmacol. 59: 364
83. Hansson T, Rafter J, Gustafsson J-Å (1980) Biochem. Pharmacol. 29: 583
84. Fingerman S W, Brown L A, Lynn M, Short E C Jr (1983) Arch. Environ. Contam. Toxicol. 12: 195
85. Payne J F, Fancey L L (1982) Chemosphere 11: 207
86. Binder R L, Stegeman J J (1980) Biochem. Pharmacol. 29: 949
87. Förlin L, Andersson T, Koivusaari U, Hansson T (1984) Mar. Environ. Res. 14: 47
88. Law F C P, Addison R F (1981) Bull. Environ. Contam. Toxicol. 27: 605
89. Walton D G, Fancey L L, Green J M, Kiceniuk J W, Penrose W R (1983) Comp. Biochem. Physiol. 76C: 247
90. Kurelec B, Britvic S, Rijavec M, Müller W E G, Zahn R K (1977) Mar. Biol. 44: 211
91. Spies R B, Fetton J S, Dillard L (1984) Mar. Environ. Res. 14: 412
92. Melancon M J, Yeo S E, Lech J J (1987) Environ. Toxicol. Chem. 6: 127
93. Brodie B B, Gillette J R, LaDu B N (1958) Ann. Rev. Biochem. 27: 427
94. Adamson R H, Dixon R L, Francis F L, Rall D P (1965) Proc. Natl. Acad. Sci. USA 54: 1386
95. Hitchcock M, Murphy S D (1967) Biochem. Pharmacol. 16: 1801

96. James M O (1987) Environ. Health Persp. 71: 97
97. Sakai T, Gotoh O, Kawatsu H (1983) Bull. Jap. Soc. Sci. Fish. 49: 1835
98. Sakai T, Kawatsu H, Gotoh O (1983) Bull. Jap. Soc. Sci. Fish. 49: 679
99. Andersson T, Koivusaari U (1986) Aquatic Toxicol. 8: 85
100. Habig W H, Pabst M J, Jakoby W B (1974) J. Biol. Chem. 249: 7130
101. Sinnhuber R O, Hendricks J D, Wales J H, Putnam G B (1977) Ann. N. Y. Acad. Sci. 298: 398
102. Jimenez B D, Cirmo C P, McCarthy J F (1987) Aquatic Toxicol. 10: 41
103. Takimoto Y, Ohshima M, Miyamoto J (1987) Ecotoxicol. Environ. Safety 13: 104
104. Walker C H (1980) In: Bridges J W, Chasseaud L F (eds) Progress in drug metabolism vol. 5, John
     Wiley, New York, 113
105. Hansson T, Förlin L, Rafter J, Gustafsson J-Å (1982) In: Hietanen E, Laitinen M, Hänninen O (eds)
     Cytochrome P-450, biochemistry, biophysics and environmental implications. Elsevier, 217
106. Sivarajah K, Franklin C S, Williams W P (1978) J. Fish Biol. 13: 401
107. Pedersen M G, Hershberger W K, Zachariah P K, Juchau M R (1976) J. Fish. Res. Board Can. 33:
     666
108. Walker C H (1978) Drug Metab. Rev. 7: 295
109. Addison R F, Zinck M E, Leahy J R (1976) J. Fish. Res. Board Can. 33: 2073
110. Ludke J L, Gibson J R, Lusk C I (1972) Toxicol. Appl. Pharmacol. 21: 89
111. Addison R F, Zinck M E, Willis D E (1977) Comp. Biochem. Physiol. 57C: 39
112. Krieger R I, Lee P W (1973) Arch. Environ. Contam. Toxicol. 1: 112
113. Metcalf R L, Lu P-Y, Kapoor I P (1973) Water Resources Center WRC Res. Rep. 69. Illinois, USA
114. Reinbold K A, Metcalf R L (1976) Pestic. Biochem. Physiol. 6: 401
115. Wells D E. (1979) Anal. Chim. Acta 104: 253
116. Lu P-Y, Metcalf R L (1975) Environ. Health Persp. 10: 269
117. Chan T M, Gillett J W, Terriere L C (1967) Comp. Biochem. Physiol. 20, 731
118. Thornton S C, Diamond L, Baird W M (1982) J. Toxicol. Environ. Health 10: 157
119. Kurelec B, Matijasevic Z, Rijavec M, Alacevic M, Britvic S, Müller W E G, Zahn R K (1979) Bull.
     Environ. Contam. Toxicol. 21: 799
120. Varanasi U, Gmur D J, Krahn M M (1980) In: Polynuclear aromatic hydrocarbons; fourth
     international symposium on analysis, chemistry and biology, Batelle, Columbus p 455
121. Lee, R F, Sauerheber R, Dobbs G M (1972) Mar. Biol. 17: 201
122. Hellou J, King A (1987) Bull. Environ. Contam. Toxicol. 39: 182
123. Chin B H, Sullivan L J, Eldridge J E (1979) J. Agric. Food Chem. 27: 1395
124. Statham C N, Pepple S K, Lech J J (1975) Drug Metab. Disp. 3: 400
125. Gill S S (1980) Bull. Environ. Contam. Toxicol. 25: 697
126. Sudershan P, Khan M A Q (1980) J. Agric. Food Chem. 28: 291
127. Feroz M, Khan, M A Q (1979) Bull. Environ. Contam. Toxicol. 23: 64
128. van Veld P A, Bender M E, Roberts M H Jr, (1984) Aquatic Toxicol. 5, 33
129. Herbst E, Scheunert I, Idein W, Korte F (1978) Chemosphere 3: 221
130. Opperhuizen A, Voors P I (1987) Chemosphere 16: 953
131. Chui Y C, Addison R F, Law F C P (1986) Aquatic Toxicol. 8: 41
132. Tulp M Th M, Tillmanns G M, Hutzinger O (1977) Chemosphere 5: 223
133. Layiwola P J, Linnecar D F C, Knights B (1983) Xenobiotica 13: 107
134. Addison R F, Willis D E (1978) Toxicol. Appl. Pharmacol 43: 303
135. Rodgers C, Stalling D L (1972) Weed Sci. 20: 101
136. Statham C N, Lech J J (1976) Toxicol. Appl. Pharmacol. 36: 281
137. Metcalf R L, Sangha G K, Kapoor I P (1971) Environ. Sci. Technol. 5: 709
138. Greer G L, Paim U (1968) J. Fish. Res. Board Can. 25: 2321
139. Atchinson G J, Johnson M E (1975) Trans. Amer. Fish. Soc. 104: 782
140. Addison R F, Zinck M E (1977) J. Fish. Res. Board Can. 34: 119
141. Adamson R H, Guarino A M (1972) Comp. Biochem. Physiol. 42A: 171
142. Kapoor I P, Metcalf R L, Hirwe A S, Coats J R, Khalsa M S (1973) J. Agric. Food Chem. 21: 310
143. Kapoor I P, Metcalf R L, Nystrom R F, Sangha G K (1970) J. Agric. Food Chem. 18: 1145
144. Ernst W, Goerke H (1974) Mar. Biol. 24: 287
145. Pritchard J B, Guarino A M, Kinter W B (1973) Environ. Health Persp. 4: 45

146. Stalling D L, Hogan J W, Johnson J L (1973) Environ. Health Persp. 3, 159
147. Metcalf R L, Booth G M, Schuth C K, Hansen D J and Lu P -Y (1973) Environ. Health Persp. 4, 27
148. Melancon M J Jr, Saybolt J, Lech J J (1977) Xenobiotica 7: 633
149. Melancon M J Jr, Lech J J (1976) Drug Metab. Disp. 4: 112
150. Chambers H, Dziuk L J, Watkins J (1977) Pest Biochem. Physiol. 7: 297
151. Bruggeman W A (1983) Ph. D. Thesis, University of Amsterdam
152. Herbst E, Weisgerber I, Klein W, Korte F (1976) Chemosphere 5: 127
153. Sijm D T H M, Opperhuizen A (1988) Chemosphere 17: 83
154. Sudershan P, Khan M A Q (1981) Pest. Biochem. Physiol. 15: 192
155. Sikka H C, Appleton H T, Gangstad E O (1977) J. Agric. Food Chem. 25: 1030
156. Sikka H C, Ford D, Lynch R S (1975) J. Agric. Food Chem. 23: 849
157. Olson L E, Allen J L, Hogan J M (1977) J. Agric. Food. Chem. 25: 554
158. Sanborn J R, Metcalf R L, Yu C -C, Lu P -Y (1975) Arch. Environ. Contam. Toxicol. 3, 244
159. Penrose W R, Murphy R G, Dave L L, White M D, Gulliver W P, Walton D G (1979) Chemosphere 8: 509
160. Cravedi J -P, Tulliez J (1986) Arch. Environ. Contam. Toxicol. 15: 207
161. Cravedi J -P, Tulliez J (1987) Xenobiotica 17: 1103
162. Rao D M R, Murty A S (1982) Environ. Poll. (Series A) 27: 223
163. Rao D M R, Deri A P, Murty A S (1981) Pest. Biochem. Physiol. 15: 282
164. Sudershan P, Khan M A Q (1980) Pest. Biochem. Physiol. 14: 5
165. Kapoor I P, Metcalf R L, Hirwe A S, Lu P -Y, Coats J R, Nystrom R F (1972) J. Agric. Food Chem 20: 1
166. Feroz M, Khan M A Q (1979) Arch. Environ. Contam. Toxicol. 8: 519
167. Sanborn J R, Childers W F, Hansen L G (1977) J. Agric. Food Chem. 25: 551
168. Melancon M J, Lech J J (1984) Comp. Biochem. Physiol. 79C: 331
169. Foster G D, Crosby D G (1986) Environ. Toxicol. Chem. 5: 1059
170. Roubal W T, Collier T K, Malins D C (1977) Arch. Environ. Contam. Toxicol 5: 513
171. Reichert W L, Varanasi U (1982) Environ. Res. 27: 316
172. Krahn M M, Brown D W, Collier T K, Friedman A J, Jenkins R G, Malins D C (1980) J. Biochem. Biophys. Methods 2: 233
173. Varanasi U, Gmur D J, Reichert W L (1981) Arch. Environ. Contam. Toxicol. 10: 203
174. Potter J L, O'Brien R D (1964) Science 144: 55
175. Glickman A H, Shono T, Casida J E, Lech J J (1979) J. Agric. Food Chem. 27: 1038
176. Hirwe A S, Metcalf R L, Lu P-Y, Chio L C (1975) Pest. Biochem. Physiol. 5: 65
177. Muir D C G, Yarechewski A L, Knoll A, Webster G R B (1986) Environ. Toxicol. Chem. 5: 261
178. Melancon M J, Lech J J (1976) Bull. Environ. Contam. Toxicol. 15: 181
179. Cole R B, Metcalf R L (1987) Bull. Environ. Contam. Toxicol. 38: 96
180. Knowles C O, Benezet H J, Mayer F L (1980) J. Environ. Sci. Health B15: 351
181. Opperhuizen A, Baalhuis G H W, Gobas F A P C (in preparation)
182. Hinz R, Matsumura F (1977) Bull. Environ. Contam. Toxicol. 18: 631
183. Kanazawa J, Tomizawa C (1978) Arch. Environ. Contam. Toxicol. 7: 397
184. Ahokas J T, Pelkonen O, Kärki N T (1975) Biochem. Biophys. Res. Comm. 63: 635
185. Pohl R J, Bend J R, Guarino A M, Fouts J R (1974) Drug Metab. Disp. 2: 545
186. Pedersen M G, Hershberger W K, Juchau M R (1974) Bull. Environ. Contam. Toxicol. 12: 481
187. Elcombe C R, Lech J J (1978) Environ. Health Persp. 23: 309
188. Payne J F, Penrose W R (1975) Bull. Environ. Contam. Toxicol. 14: 112
189. Gruger E H Jr, Wekell M M, Numoto P T, Craddock D R (1977) Bull. Environ. Contam. Toxicol. 17: 512
190. Ahokas J T, Pelkonen O, Kärki N T (1976) Acta Pharmacol. Toxicol. 38: 440
191. Ahokas J T (1976) Res. Comm. Chem. Path. Pharmacol. 13: 439
192. Ahokas J T, Kärki N T, Oikari A, Soivio A (1976) Bull. Environ. Contam. Toxicol. 16: 270
193. Hitchcock M, Murphy S D (1966) Fed. Proc. 25: 687
194. Call D J, Brooke L T, Lu P-Y (1980) Arch. Environ. Contam. Toxicol. 9: 699
195. Bahig M E, Kraus A, Klein W, Korte F (1981) Chemosphere 10: 319
196. Oikari A, Anäs E, Kruzynski G, Holmbom B (1984) Bull. Environ. Contam. Toxicol. 33: 233

197. Statham C N, Lech J J (1975) J. Fish. Res. Board Can. 32: 515
198. Hellou J, Banoub J H, Payne J F (1986) Chemosphere 15: 787
199. Oikari A, Anäs E (1985) Bull. Environ. Contam. Toxicol. 35: 802
200. Layiwola P J, Linnecar D F C (1981) Xenobiotica 11: 167
201. Nagel R (1983) Xenobiotica 13: 101
202. Nagel R, Urich K (1983) Xenobiotica 13: 97
203. Kobayashi K, Kimura S, Shimizu E (1977) Bull. Jap. Soc. Sci. Fish. 43: 601
204. Pritchard J B, Andersson J B, Rall D P, Guarino A M (1980) Comp. Biochem. Physiol. 65C: 99
205. Schultz D P, Harman P D, Luhning C W (1979) J. Agric. Food Chem. 27: 328
206. Kleeman J M, Olson J R, Chen S M, Peterson R E (1986) Toxicol. Appl. Pharmacol. 83: 402
207. Lech J J (1973) Toxicol. Appl. Pharmacol. 24: 114
208. Lech J J, Statham C N (1975) Toxicol. Appl. Pharmacol. 31: 150
209. Kobayashi K, Kimura S, Akitake H (1976) Bull. Jap. Soc. Sci. Fish. 42: 171
210. Akitake H, Kobayashi K (1975) Bull. Jap. Soc. Sci. Fish. 41: 321
211. Kobayashi K, Akitake H (1975) Bull. Jap. Soc. Sci. Fish. 41: 93
212. Kobayashi K, Akitake H (1975) Bull. Jap. Soc. Sci. Fish. 41: 87
213. Kobayashi K, Akitake H (1975) Bull. Jap. Soc. Sci. Fish. 41: 1271
214. Huang K C, Collins S F (1962) J. Cell Comp. Physiol 60: 49
215. Stalling D L, Huckins J N (1978) J. Agric. Food Chem 26: 447
216. James M O, Bend J R (1976) Xenobiotica 6: 393
217. Hunn J B, Schoettger R A, Willford W A (1968) J. Fish. Res. Board Can. 25: 25
218. Southworth G R, Keffer C C, Beauchamp J J (1981) Arch. Environ. Contam. Toxicol. 10: 561
219. Sanborn J R, Childers W F, Metcalf R L (1975) Bull. Environ. Contam. Toxicol. 13: 209
220. Metcalf R L, Sanborn J R, Lu P-Y, Nye D (1975) Arch. Environ. Contam. Toxicol. 3, 151
221. Metcalf R L (1977) Fate of pollutants in the air and water environment. In: Suffet I H (ed) Advances in environmental science and technology. vol 8, Wiley, New York
222. McLean S, Sameshima M, Katayama T, Iwata I, Olney C E, Simpson K L (1984) Bull. Jap. Soc. Sci. Fish. 50: 1419
223. Coats J R, Metcalf R L, Kapoor I P (1974) Pest. Biochem. Physiol. 4: 201
224. Melancon M J Jr, Lech J J (1978) Arch. Environ. Contam. Toxicol. 7: 207
225. Varanasi U, Uhler M, Stranahan S I (1978) Toxicol. Appl. Pharmacol. 44: 277
226. Darrow D C, Addison R F (1979) Bull. Environ. Contam. Toxicol. 22: 265
227. Krahn M M, Myers M S, Burrows D G, Malins D C (1984) Xenobiotica 14: 633
228. LeBon A-M, Cravedi J-P, Tulliez J (1987) Ecotoxicol. Environ. Safety 13: 274
229. Guiney P D, Peterson R E, Melancon M J Jr, Lech J J (1977) Toxicol. Appl. Pharmacol. 39: 329
230. Melancon M J, Lech J J (1980) J. Toxicol. Environ. Health 6: 645
231. Oikari A O J (1986) Bull. Environ. Contam. Toxicol. 36: 429
232. Oikari A, Kunnamo-Ojala T (1987) Aquatic Toxicol. 9: 327

Abbreviations and Common Names

AH	anilinehydroxylase
AHH	arylhydrocarbonhydroxylase
ANF	α-naphtoflavone (7,8-benzoflavone)
APND	aminopyrine-N-demethylase
AU	arbitrarily unit
BEND	benzphetamine-N-demethylase
BNF	β-naphtoflavone (5,6-benzoflavone)
BP	benzo(a)pyrene
BP-6	flame retardant, containing hexabromobiphenyl
carbaryl	1-naphthyl-N-methylcarbamate
chlorpyrifos	O-(3,5,6-trichloro-2-pyridyl)-O,O-diethylthiophosphate
CoA	coenzyme A

2,4-D	2,4-dichlorophenoxyacetic acid
DBA	1,2,3,4-dibenzanthracene
DBH	p,p'-dichlorobenzhydrol
DBP	p,p'-dichlorobenzophenone
DDA	2,2-bis(p-chlorophenyl)acetic acid
DDD	2,2-bis(p-chlorophenyl)-1,1-dichloroethane
DDE	2,2-bis(p-chlorophenyl)-1,1-dichloroethene
DDMS	2,2-bis(p-chlorophenyl)-1-chloroethane
DDMU	2,2-bis(p-chlorophenyl)-1-chloroethene
DDNU	1,1-bis(p-chlorophenyl)-ethene
DDOH	2,2-bis(p-chlorophenyl)-ethanol
DDT	2,2-bis(p-chlorophenyl)-1,1,1-trichloroethane
DEHP	di-ethylhexylphthalate
dicofol	2-hydroxy-2,2-bis(p-chlorophenyl)-1,1,1-trichloroethane
ECOD	7-ethoxy-coumarin-O-dealkylase
EMND	ethylmorphine-N-demethylase
EPN	O-(p-nitrophenyl)-O-ethylphosphothioate
EROD	7-ethoxyresorufin-O-dealkylase
FU	fluorescence unit
GSH-t	glutathione transferase
ia	intra-arterial
im	intramuscular
ip	intraperitoneal
iv	intravenous
3-MC	3-methylcholanthrene
ME	model ecosystem
MEHP	mono-ethylhexylphthalate
methiochlor	2,2-bis(p-methylthiophenyl)-1,1,1-trichloroethane
MFO	mixed-function oxidase
MO	monooxygenase
NADPH	nicotinamide dinucleotide phosphate
PAPS	3-α-phospho-adenylsulphate
parathion	O,O-diethyl-O-phenylthiophosphate
PB	phenobarbital
PCB	polychlorinated bipheny
PCN	pregnenolone-16α-carbonitrile
PNAOD	p-nitroanisole-O-demethylase
PNPOD	p-nitrophenetole-O-deethylase
pristane	2,6,10,14-tetramethylpentadecane
prolan	1,1-bis(p-chlorophenyl)-2-nitropropane
propoxur	o-isopropyl-N-methylcarbamate
SDS-PAGE	sodium dodecyl sulphate polyacrylamide gel electrophorese
TCDD	tetrachlorodibenzo-p-dioxin
TFM	trifluoromethyl-4-nitrophenol
trifluralin	2,6-dinitro-4-(a,a,a-trifluormethyl)-N,N-di-isopropylaniline
UDPGA	uridinediphosphoglucuronic acid
UDPGA-t	uridinediphosphoglucuronic acid transferase

Appendix. Fishes and Their Scientific Names

Name	Scientific name	Name	Scientific name
Alewife	*Alosa pseudoharengus*	Gudgeon	*Gobio gobio*
American eel	*Anguilla rostrata*	Guppy	*Poecilia reticulata*
American shad	*Alosa sapidissima*	Gurnard	*Eutrigla gurnardus*
Argentine	*Argentinus sp.*	Haddock	*Melanogrammus*
Atlantic cod	*Gadus morhua*		*aeglefinnus*
Atlantic salmon	*Salmo salar*	Hagfish	*Myxine glutinosa*
Atlantic stingray	*Dasyatis sabina*	Hake	*Merluccius merluccius*
Barracuda	*Sphyraena barracuda*	Horse mackerel	*Trachurus trachurus*
Bitterling	*Rodeus sericeus amarus*	Houting	*Coregonus oxyrhynchus*
Black drum	*Pogonias cromis*	Ide	*Leuciscus idus*
Black Sea bass	*Centropristis striatus*	Jack	*Caranx hippos*
Bluegill sunfish	*Lepomis macrochirus*	Japanese medaka	*Oryzias latipes*
Blue whiting	*Micromesistius*	Jack crevalle	*Caraux hippos*
	poutassou	Killifish	*Fundulus heteroclitus*
Bluntnose stingray	*Dasyatis sayi*	Killifish	*Oryzias latipes*
Bullhead	*Cottus gobio*	King cobra guppy	*Poecilia reticulata*
Bream	*Abramis brama*	King of Norway	*Hemitrypterus americanus*
Brook trout	*Salvelinus fontinalis*	Kissing gourami	*Helostama temmincki*
Brown trout	*Salmo trutta*	Lake trout	*Salmo trutta lacustris*
Bullhead	*Ictalurus melas*	Largemouth bass	*Micropterus salmoides*
California killifish	*Fundulus parvipinnis*	Largescale sucker	*Catostomus macrocheilus*
Capelin	*Mallatus villosus*	Large skate	*Raja occellata*
Carp	*Cyprinus carpio*	Lavaret	*Coregonus lavaretus*
Catfish	*Heteropneustes fossilis*	Lemon shark	*Negaprion brevirostris*
Catfish	*Mystus carasius*	Lemon sole	*Microstomus kitt*
Catfish	*M. vittatus*	Ling	*Molva molva*
Channa	*Ophiocephalus punctatus*	Little skate	*Raja erinacea*
Channel catfish	*Ictalurus punctatus*	Loach	*Nemachilis barbatula*
Chinook salmon	*Oncorhynchus*	Long rough dab	*Hippoglosoides*
	tschawytscha		*plattesoides*
Cod	*Gadus morrhua*	Mackerel	*Scomber scombrus*
Coho salmon	*Oncorhynchus kisutch*	Mangrove snapper	*Lutjanus griseus*
Common dab	*Limanda limanda*	Megrim	*Lepidorhombus*
Crayfish	*Astacus astacus L.*		*whiffiagonis*
Crayfish	*Procamburus clarkii*	Minnow	*Phoximus phoximus*
Crucian carp	*Carassius carassius*	Mosquitofish	*Gambusia affinis*
Cunner	*Tantogolabrus adspersus*	Mudsucker	*Gillichthys mirabilis*
Dace	*Leuciscus leuciscus*	Mullet	*Mugil cephalus*
Dogfish	*Scylorhinus canicula*	Mummichog	*Fundulus heteroclitus*
Dogfish shark	*Squalus acanthias*	Northern pike	*Esox lucius*
Eel	*Anguilla rostrata*	Nurse shark	*Ginglymostoma cirratum*
English sole	*Parophrys vetulus*	Pacific lamprey	*Lampreta tridentata*
Gulf killifish	*Fundulus grandis*	Perch	*Perca fluviatilis L.*
Fathead minnow	*Pimephales promelas*	Pigfish	*Orthopristis chrysopterus*
Flathead fatfish	*Pylodictus olivaris*	Pike	*Esox lucius*
Golden orfe ide	*Leuciscus idus*	Poor cod	*Trisopterus minutus*
Golden shiner	*Notemigonus*	Rainbow trout	*Salmo gairdneri*
	chrysoleucas	Red fish	*Sebastes marinus*
Goldfish	*Carassius auratus*	Red killifish	*Oryzias latipes*
Green sunfish	*Lepomis cyanellus*	Roach	*Rutilis rutilis L.*

Name	Scientific name	Name	Scientific name
Rock sole	*Lepidopsetta bilineata*	Sucker	*Catostomus commersoni*
Rudd	*Scardinius erythrphtalmus*	Sunfish	*Lepomis gibbosus*
Sailfin mollie	*Poecilia latipinna*	Tench	*Tinca tinca*
Saithe	*Polachius virens*	Thornback ray	*Raja clavata*
Salmon	*Salmo salar*	Thorny skate	*Raja radiata*
Sand dab	*Citharichthys stignaeus*	Three-spined	
Sand goby	*Gillichthys mirabilis*	stickleback	*Gasterosteus aculeatus*
Sand goby	*Gobio Kessleri*	Topmouth gudgeon	*Pseudorasbora parva*
Sculpin	*Callionymus lyra*	Torsk	*Brosme brosme*
Sculpin	*Myoxocephalus scorpius*	Twait	*Alosa finta*
Sculpin	*Oligocottus maculosus*	Vendace	*Coregonus albula*
Scup	*Stenotomus chrysops*	Walbaum	*Tautogolabrus adspersus*
Scup	*Stenotomus versicolor*	Whitch	*Glyptocephalus*
Sea bass	*Micropterus salmoides*		*cynoglossus*
Sea lampry	*Petromyzon marinus*	White bream	*Blicca bjoerkna*
Sepat siam	*Trichogaster pectoralis*	White crappie	*Pomoxis annularis*
Sheepshead	*Archosargus*	White sturgeon	*Acipenser transmontanus*
	probatocephalus	Whiting	*Merlangius merlangus*
Sheepshead minnow	*Cyprinodon variegatus*	Winter flounder	*Pseudopleuronectes*
Skate	*Raja erinacea*		*americanus*
Smallmouth bass	*Micropterus dolomieu*	Winter skate	*Raja ocellata*
Sockeye salmon	*Oncorhynchus nerka*	Winter skate	*Sarotherodon niloticus*
Sole	*Solea solea*	Winter skate	*Seriolo quiquerdita*
Southern flounder	*Paralichthys lethostigma*	Yellow bullhead	
Speckled sanddab	*Citharichthys stignaeus*	catfish	*Ictalurus natalis*
Splake	*Salvelinus fontinalis* x	Yellow perch	*Perca flavescens*
Splake	*Salvelinus namaycush*	Yellowtail snapper	*Ocyurus chrysurus*
Spot	*Leiostomus xanthurus*	Zebrafish	*Brachydanio rerio*
Spur dog	*Squalus acanthias*	(deep-sea fish)	*Coryphaenoides armatus*
Starry flounder	*Platichtlys stellatus*	u	*Lophius americanus*
Steelhead trout	*Salmo gairdneri gairdneri*	u	*Macrognathus aculeatum*
Stingray	*Dasyatis americana*	u	*Tilapia mossambica*
Stingray	*Dasyatis sabina*		

u = unknown

Subject Index

The Handbook of
Environmental Chemistry
Edited by O. Hutzinger

Volume 2

Reactions and Processes

Part D

1988. XI, 210 pp. 47 figs. 55 tabs.
ISBN 3-540-15547-3

Contents: *R. Herrmann,* Bayreuth, FRG:
Hydrology. – *N. O. Crossland,* Sittingbourne,
UK; *C. J. M. Wolff,* Amsterdam, The Nether-
lands: Outdoor Ponds: Their Construction,
Management, and Use in Experimental Eco-
toxicology. – *T. Mill,* Menlo Park, USA;
W. Mabey, San Francisco, USA: Hydrolysis of
Organic Chemicals. – *M. Waldichuk,*
Vancouver, Canada: Exchange of Pollutants
and Other Substances Between the Atmo-
sphere and the Oceans. – *P. B. Tinker,*
Swindon, UK; *P. B. Barraclough,* Harpenden,
UK: Root-Soil Interactions. – *C. M. Menzie,*
Washington D.C., USA: Reaction Types in the
Environment

Springer-Verlag Berlin
Heidelberg New York London
Paris Tokyo Hong Kong

Springer

The Handbook of
Environmental Chemistry
Edited by O. Hutzinger

Volume 4

Air Pollution

Part B

1989. XI, 262 pp. 93 figs.
ISBN 3-540-50915-1

Contents: *H. Brauer,* Berlin: Air Pollution Control Equipment. – *J. S. Gaffney, N. A. Marley,* Los Alamos, NM; *E. W. Prestbo,* Seattle, WA/USA: Peroxyacyl Nitrates (Pans): Their Physical and Chemical Properties. – *R. Harkov,* Somerset, NJ/USA: Semivolatile Organic Compounds in the Atmosphere. – *F. W. Lipfert,* Northport, NY/USA: Air Pollution and Materials Damage. – *G. E. Shaw,* Fairbanks, AK; *M. A. K. Khalil,* Beaverton, OR/USA: Arctic Haze

Springer-Verlag Berlin
Heidelberg New York London
Paris Tokyo Hong Kong

Springer